SCIr

NOV 0 8 2006

Environmental Technologies Handbook

Edited by Nicholas P. Cheremisinoff

Government Institutes
An imprint of
The Scarecrow Press, Inc.
Lanham, Maryland • Toronto • Oxford
2005

Government Institutes

Published in the United States of America
by Government Institutes, an imprint of The Scarecrow Press, Inc.
A wholly owned subsidiary of
The Rowman & Littlefield Publishing Group, Inc.
4501 Forbes Boulevard, Suite 200
Lanham, Maryland 20706
govinst.scarecrowpress.com

PO Box 317
Oxford
OX2 9RU, UK

British Library Cataloguing in Publication Information Available

Library of Congress Cataloging-in-Publication Data

Environmental technologies handbook / edited by Nicholas P. Cheremisinoff.
    p. cm.
  Includes bibliographical references and index.
  ISBN 0-86587-980-X (cloth : alk. paper)
  1. Factory and trade waste—Handbooks, manuals, etc. 2. Environmental
  engineering—Handbooks, manuals, etc. 3. Pollution
prevention—Handbooks, manuals, etc. I. Cheremisinoff, Nicholas P.
TD897.5.E584 2005
628—dc22

              2004018368

Contents

Illustrations

Tables

Preface

As we enter into the twenty-first century, the level of sophistication needed to address pollution management by industry is increasing. Challenges to develop new and cost-effective technologies are driven by an increasingly higher bar of achievement brought on by ever-tightening regulations throughout the world and the constant need to keep control costs down.

Traditionally the costs for environmental management have been universally accepted as a part of the cost of doing business. But this view began changing during the 1990s. The change in philosophy, which has caused a shift in corporate mentality away from more traditional technologies and environmental management practices, can be related entirely to economic factors, namely:

- Sustainability
- Social responsibility
- Regulation

Industry is faced with the fact that resources are limited. Whether those resources are minerals, petroleum, water, or other natural resources consumed in manufacturing, they are finite. While there are technologies on the horizon that will address certain limited resources, such as the transition to those that are based upon a hydrogen economy, the R&D efforts, infrastructure investments needed to support the technologies, and time required for commercialization are realistically almost a half a generation or more away. That means that present resources must be stretched as far as possible. And as regions of the world, which we formerly considered Third World or "transitioning economies" are being developed at accelerated rates that are unheard of historically, the demands on resources potentially could outstrip them.

As an example, Jordan is severely limited in water resources. Projections show that by the year 2020, groundwater reserves will be completely depleted. Industry consumes about 5 percent of the country's water for manufacturing purposes, with plans to grow various industry sectors at a rate where consumption could be as high as 10 percent or more within a decade. If we look toward Eastern Europe, less than a decade ago that region was struggling with less than 10 percent industry productivity due to the collapse of Communism. Now, as some of these countries are entering the European Union, industry activities have risen above 80 percent productivity in certain heavy-industry sectors such as steel, machine building, and pharmaceutical and chemical manufacturing. These few examples point toward the ever-growing need to manage environmental affairs with technologies that will ensure that business and societies can be maintained and the baton passed over to the next generation with a base to work with.

While most of us tend to think it is the social responsibility of corporations to act in a responsible manner, the fact remains that this is a financially driven issue as well. Corporations like to think of themselves as being caretakers of the environment and society. This comes across clearly when one considers the environmental policy statements presented by many multinational corporations. Such companies promote principles of ISO 14001, "Responsible Care," and other popularized environmental management systems. While this is all well and good, let's recognize it for what it is. Social responsibility or being a good corporate environmental citizen is a matter of good business sense and even sustainability. Companies with poor track records with regard to managing their environmental affairs can suffer from public boycotts of their products, are the targets of bad press and even focused legislation aimed at restricting their manufacturing or imposing environmental standards that are too excessive to meet, may face a loss in investor confidence, are denied favorable insurance premium rates, and can have difficulty qualifying for favorable terms on loans and lines of credit. And they can lose out in market share to competitors that the public perceives as being more "green." Look to the food-processing industry in the United Sates today: try to find an article that does not refer to this sector as "Big Food" with big environmental problems.

Little needs to be said about environmental regulations. There is not a technologically developed or developing country today where industry is not wrestling with complex environmental legislation with increasing pressures from regulators. Corporations are forced to maintain full-time staffs to address regulations, to deal with regulators on a daily basis, and to maintain the records of their operations—and what is not regulated today could and likely will become an environmental liability in the not-too-distant future.

The answer that has been promoted over the last decade is pollution prevention. This Holy Grail, so to speak, has been discovered as a means by which corporations can save money and reduce their costs by reducing or eliminating pollution at the source. And clearly this has a lot of merit—but pollution prevention alone is not enough. There are countless situations where pollution control technologies are the only approaches that make sense until major technological changes are made to certain manufacturing operations. And here lies a problem for industry. If we look at the mainstay technologies relied upon for some of the traditional environmental problems industry continues to face, many of these are inadequate. In fact, it has been nearly five decades since major technological innovations have been made to pollution abatement equipment.

This book is intended as a general reference for industry, with a focus on some of the newer technologies and emerging concepts. The book contains chapters that are individual discussions on technologies and concepts that will aid pollution control and prevention experts. Each chapter has been written by an expert in his or her respective field. While some of the conventional concepts and technologies are discussed, the user will find a greater slant toward emerging concepts and technologies—some cutting-edge, others commercial but still relatively new.

There are seven chapters to the volume. Chapter 1 focuses on human health risk assessment. This chapter provides an extensive coverage of the concepts of risk assessment to environmental media and relates advanced concepts of dose-response exposures to risk reduction practices. Chapter 2 covers leading-edge technologies for air quality monitoring and new instrumentation. Chapter 3 provides selection criteria for air pollution control technologies. Cost and operational data for technologies are included in this chapter. Chapter 4 covers thermal methods of waste treatment. This chapter provides extensive coverage of cutting-edge technologies for solid- and liquid-waste processing. Chapter 5 addresses advanced wastewater treatment technologies. These are new technologies now becoming available for industry. Chapter 6 covers conventional technologies for wastewater treatment, with a focus given to suspended matter removal. Finally, chapter 7 addresses cost accounting tools. All technologies and new projects need to be carefully assessed in terms of their cost competitiveness. This chapter explains the methodology and computation methods available for life-cycle costing analyses and proper environmental cost accounting.

Each chapter has been prepared individually by experts in their respective fields, and as such, authors are responsible for their own statements. All chapters have been prepared in good faith, exercising due care and attention. However, no representation or warranty, expressed or implied, is made as to the relevance, accuracy, completeness, or fitness for purpose of this volume in respect of any particular user's circumstances. Users of this volume should satisfy themselves concerning its application to, and where necessary seek expert advice about, their situation. The authors, organizations for which authors work or report to, and the publisher shall not be liable to any person or entity with respect to any liability, loss, or damage caused or alleged to have been caused directly or indirectly by this publication.

Chapter 1
Human Health Risk Assessment

Pavel Muller and Angela Li-Muller

1 Background

Human health risk assessment has advanced from its modest origin in the 1980s into a large and active field due to growing concerns about toxic chemicals, particularly carcinogens, present in the environment. Anybody writing on this subject is bound to barely scratch the surface. Facing various tradeoffs, we have decided to focus our discussion on the risk assessment of nonradiological chemicals and to target this review to environmental practitioners who may need to decide whether the use of risk assessment is appropriate or who need to understand the outcome of the assessment but are not risk assessment specialists per se. We attempt to highlight those aspects of the assessment that impact the outcome the most and to provide references for further reading. There are other risk assessment primers available and the readers may wish to consult alternative sources such as American Chemical Society 1998; Health Canada and Ontario Ministry of Health 1997; Kamrin, Katz, and Walter 2000; McColl et al. 2000; and USEPA 1991a.

Like most technical subjects, risk assessment is rich with jargon. The situation is complicated by subtle differences in terminology and protocol used by different jurisdictions, especially outside of the United States. In this chapter, we try to avoid jargon whenever possible and focus on basic concepts and the underlying toxicology. However, many terms are used frequently in risk assessment, and we will clarify these terms so that readers can understand existing assessments. The principles and basic risk assessment processes are the same around the world and can usually be traced back to the famous "Red Book" (National Research Council 1983). For clarity's sake, we will adopt the terminology of the U.S. Environmental Protection Agency (USEPA) in our discussion. Readers operating in Canada, Europe, and Australia may have to learn slightly different terms, which are roughly, but not exactly, comparable to the USEPA terms.

In this chapter, we rely mostly on USEPA sources; thus many processes, which are intended to be generic, may be colored by USEPA's regulatory processes. U.S. sources are used because the United States is a pioneer in risk assessment and an early adopter in applying the process for regulatory purposes. USEPA has become an innovating and trendsetting organization in environmental risk assessment. In addition, USEPA has always been generous in distributing its publications (initially as hard copies and recently also in electronic forms). Until recently, organizations in other countries either did not widely publicize their reports or made them available only at prices that few were able to afford. The situation has changed and excellent English-language reports are now freely available from the World Health Organization (WHO) and its various programs (e.g., International Program on Chemical Safety [IPCS] and International Agency for Research on Cancer [IARC]), the European Union (EU), Canada, Australia, and other countries. Furthermore, WHO and the EU, given their multinational mandates, provide overviews of risk assessment–related processes and approaches in different countries and appear to draw on international expertise more than USEPA does. We attempt to cite some of the key documents produced outside of the United States in order to provide a more international perspective.

1.1   What Is Risk and What Is Hazard

Outside of the field of risk assessment, the difference between *hazard* and *risk* is often blurred. In risk assessment, however, there is a clear distinction. USEPA defines hazard as *a potential source of harm* (USEPA 1999a). In contrast, it defines risk as *the probability of injury, disease, or death from exposure to a chemical agent or a mixture of chemicals*. Thus, hazard is a descriptive, qualitative term describing the nature of a potential adverse effect, while risk is a quantitative term describing the *probability* of this adverse effect. Risk in effect is a function of hazard and exposure, while risk assessment attempts to determine the value of the risk for a given situation.

Environmental and even occupational risks can be very small and hard to imagine. Table 1.1 illustrates very small probabilities by showing the chance of drawing a cer-

Table 1.1 Probability of being dealt spades in a hand of 13 cards (from a 52-card deck)

Spades/hand	Probability		Likelihood in words
	Scientific notation	Decimal	
5 spades	1.3 E-1	0.13	about 1 in 10
7 spades	8.8 E-3	0.0088	less than 1 in 100
8 spades	1.2 E-3	0.0012	about 1 in 1,000
9 spades	9.3 E-5	.000093	less than 1 in 10,000
10 spades	6.1 E-6	.0000061	better than 1 in a million
13 spades	1.6 E-12	.0000000000016	about 1 in a quadrillion

tain number of spades in a 13-card hand drawn from a deck of 52 cards. This example is not intended to trivialize potential risk from environmental exposure. Rather, it is meant to give readers a yardstick that could be useful in visualizing very low probabilities.

1.2   What Human Health Risk Assessment Is and How It Is Used

In general terms, the purpose of a human health risk assessment is to define the nature of the adverse effect (hazard) and determine the likelihood and magnitude of the adverse effect (risk) associated with a defined exposure to hazardous agents such as chemicals or radiation. USEPA defines human health risk assessment as "the determination of potential adverse health effects from exposure to chemicals, including both quantitative and qualitative expressions of risk. The process of risk assessment involves four major steps: hazard identification, dose-response assessment, exposure assessment, and risk characterization" (USEPA 1999a).

Although it is usually possible to arrive at some numerical estimate of human health risk, estimates in absolute terms (e.g., the number of people expected to die of cancer as a result of exposure to benzene in town X) are usually not justified. Risk assessment is based on a combination of actual data and what is termed "expert judgment" when data are not available. For example, genotoxic carcinogens are assumed to increase cancer risk linearly with increasing dose (a convention that will be discussed later in the section on dose-response assessment). Direct measurements to support this assumption are not possible because the effects at low doses are too small to be measured, but our understanding of the mechanism of action of genotoxic carcinogens supports this convention. The assumption is considered to be conservative and is generally adopted. As risk assessment typically incorporates *default* assumptions, which either cannot be or were not tested, risk assessment is better suited for comparing risks than for determining the level of absolute risk:

> It should be emphasized that because of the uncertainties in estimating dose-response relationships for single compounds, and the additional uncertainties in combining the individual estimate to assess response from exposure to mixtures, response rates and hazard indices may have merit in comparing risks but should not be regarded as measures of absolute risk. (USEPA 1986b)

Risk assessment is not a single standard spreadsheet-based process that accepts certain inputs and provides fixed outputs. Rather, it is a tool that needs to be adjusted to reflect the purpose for which the assessment is undertaken; in particular, the level of conservatism and the level of complexity must meet the purpose of the assessment. For example, if the goal is to exclude any safety concern, the risk assessment can be simple and conservative. If the purpose is to estimate realistically the risk facing a population, the assessment has to be detailed and must minimize consistent bias toward either permissiveness or conservatism.

Risk assessment is often conducted according to prescribed regulatory guidelines to determine if the risk meets some preset regulatory criteria. It can be used as an environmental management tool, as a risk communication tool, and generally as a guide for environmental decision making. How risk assessment is being applied and what factors need to be considered for each application will be discussed in greater detail in sections 4 and 5.

1.3   Human Health Risk Assessment as a Tool in Environmental Management

Human health risk assessment is a useful tool in directing environmental management strategies in the macro regulatory environments as well as in a site-specific or case-specific situation. Many government regulations are risk based. Risk assessment often serves as the scientific basis for setting environmental standards, other government policies, and regulations for the protection of public health. Upon determining through risk assessment that a population is at risk, steps can be taken to mitigate or reduce that risk. Risk assessment can also be used to help select the management option that will be most proficient in risk reduction. Where risk assessment fits into the overall risk management process is described below.

1.4   Typical Steps in Risk Management

1.4.1   Scoping, Planning, and Data Acquisition

Most risk management processes involve data acquisition, usually including environmental monitoring and modeling the levels of environmental contaminants. If data acquisition is conducted for regulatory purposes, there usually are applicable guidelines for sample collection and analysis. If the environmental levels fall below environmental guidelines or standards, no regulatory action is required and risk assessment is often not conducted.

The discussion of monitoring and modeling is outside the scope of this review. However, if the risk manager anticipates a need for a risk assessment, it is highly recommended to consult with the risk assessor to ensure that the data needed for the eventual risk assessment are indeed collected. For instance, soil sampling is usually driven by the need of hydrogeologists to understand groundwater flow and subsurface structure, and key data needed for a risk assessment may not be collected. Another example relates to conducting soil monitoring at depth, which may not be relevant because people are most likely exposed to surface or near-surface soils. Sampling for volatile and semivolatile contaminants needs to be adequate in areas near existing or planned buildings so that risk assessors can estimate infiltration of volatile substances into the indoor air. Risk assessors also need to ensure that the testing is conducted for those contaminants, which will need to be studied in the risk assessment.

1.4.2   Risk Assessment

Upon review of the collected data, the risk managers may decide to go ahead with ecological risk assessment, human health risk assessment, or both. It is a frequent practice—and often a regulatory requirement—that risk management solutions be sufficiently protective of both human and nonhuman biota (Ontario Ministry of Environment and Energy 1997a; CCME 1997; USEPA 1980).

Risk assessment requires multidisciplinary effort. While expertise in chemistry, environmental engineering, geology/hydrogeology, soil sciences, atmospheric sciences, biostatistics, and communications may be needed for both human health and ecological assessments, the two types of risk assessment do require different types of core expertise in toxicology/biology, which rarely reside in the same individual. Therefore, different experts usually prepare the two types of assessments, even if they are conducted by the same organization.

The conclusion of a human health risk assessment may include a list of recommendations (objectives) the risk manager needs to meet in order to achieve a risk level low enough to be acceptable for the protection of presently and potentially exposed individuals. For example, the recommendation may stipulate reduction of volatile contaminant levels in the indoor air to a specified level or creation of a physical barrier to block an exposure pathway.

1.4.3   Risk Management and Risk Communication

The next step after conducting a risk assessment is to develop a risk management plan to implement the assessment's recommendations. Social, technical, political, economic, and other factors are also often taken into consideration in developing the risk management plan. The plan may involve regulation or policy at the regulatory level or engineering solutions on a site-specific basis.

Risk assessors have an important role in risk communication. The purpose of conducting risk assessment, the target audience, and the regulatory environment all influence the nature of risk communication. For example, when a risk assessment is prepared for a public or private organization for internal use only, risk communication may involve a written and verbal presentation of the process, the assessment findings, and an interpretation of the findings in a risk characterization section. For other risk assessments that involve multiple stakeholders, including the public, it is generally recommended that the risk communication process be more extensive and involve the stakeholders early on, starting at the planning stage. If risk assessment is conducted to meet regulatory requirements, the extent of risk communication is often specified by the regulatory agency and contained in the agency's risk assessment guideline.

The attitude of the public toward communication by professionals on health and safety has evolved over the years. Several decades ago, the public was expected mostly to accept by faith the opinion of professionals on environmental and occupational safety matters. With better education and better accessibility to information, the public's expectations have changed in the last two decades. People are more engaged and demand more involvement. At the same time, how they perceive risk is also better understood. As a result, most regulatory agencies are now trying to make the health risk assessment process as consultative and transparent as possible.

There is good guidance on risk communication (see the section below for resources); however, risk communication is outside the scope of this review.

1.5   Key Sources of Information

Each jurisdiction typically develops its own guidance for risk assessment and risk management, often reflecting the regulatory environment in which the guidance was developed and the specific applications the guidance was intended to address. Nevertheless, many of these guidance documents have wider applicability. Guidance developed for assessing risks from contaminated soils, for example, may point to a similar approach that would be acceptable for contamination present in other environmental media. The best of the guidance documents usually contain background information on the data and judgment used in the development of the guidance. This information is valuable for fine-tuning risk assessment procedures for applications in jurisdictions other than those originally intended.

USEPA's Superfund website, http://www.epa.gov/superfund/, is probably the best site for initial guidance on risk assessment. The *Risk Assessment Guidance for Superfund (RAGS)* is available online and is an excellent place to start, particularly volume 1, *Human Health Evaluation Manual*, parts A (USEPA 1989) and D (*Standardized Planning, Reporting, and Review of Superfund Risk Assessments*, USEPA 2001a). USEPA's National Center for Environmental Assessment (NCEA) provides guidance on various topics in risk assessment (USEPA 2003g). For those who wish to start with risk assessment guidelines for air, the California Environmental Protection Agency (CalEPA) guideline for the "Hot Spots" program (CalEPA 2002) is a good first stop.

Guidelines are published as part of the *Environmental Health Criteria* (EHC) series by IPCS, a program sponsored jointly by WHO, the United Nations Environment Program (UNEP), and the International Labor Organization. The EHC series contains a number of guidance documents for various aspects of risk assessment, but the key report is *Environmental Health Criteria 210: Principles for the Assessment of Risks to Human Health from Exposure to Chemicals* (IPCS 1999).

In Europe, there are guidelines published by the European Commission (1996), and more recent guidelines released by the commission's European Chemicals Bureau (2003a, 2003b, 2003c). Dutch soil guidelines are contained in a report by Janssen and Speijers (1997).

In Australia, the national approach to environmental health risk assessment is provided in *Environmental Health Risk Assessment: Guidelines for Assessing Human Health Risks from Environmental Hazards* (enHealth Council 2002). Soil guidelines are contained in the report *Health-Based Soil Investigation Levels* (enHealth Council 2001b). Other reports useful to risk assessors may be found at the enHealth Council website, http://enhealth.nphp.gov.au/.

The *Canadian Environmental Quality Guidelines* by the Canadian Council of Ministers of the Environment (CCME 2002) can be purchased directly from CCME. The package contains guidelines for all major environmental media, including soil guidelines. Soil guidelines can also be downloaded directly from the CCME website (CCME 1996). Most of the other environmental guidelines in the CCME package have been designed to be protective of ecology rather than humans. Also available is a guidance report from Health Canada for conducting human health risk assessments on existing substances in the Priority Substances List under the Canadian Environmental Protection Act (CEPA) program (Health Canada 1994).

There are several excellent sources useful for those wishing to learn more about risk communication. The excellent *Primer on Health Risk Communication Principles and Practices* was published by the Agency for Toxic Substances and Disease Registry (2001). A more detailed source is available from USEPA (2003o); this report contains an extensive bibliography of risk communication resources produced in 1995. Additional references include:

- *Communication in Risk Situations: Responding to the Communication Challenges Posed by Bioterrorism and Emerging Infectious Diseases* by the Association of State and Territorial Health Officials (2002)

- *Seven Cardinal Rules of Risk Communication* (Covello and Allen 1988)

- *Risk Management: Guideline for Decision Makers* by the Canadian Standards Association (2002)

- *Explaining Environmental Risk* (Sandman 1986)

- *Media Relations Training Seminar* (Spurway 2001)

- *Communicating in a Crisis: Risk Communication Guidelines for Public Officials* by the U.S. Department of Health and Human Services (2002)

- *Community Involvement in Superfund Risk Assessments,* supplement to part A of volume 1, *Human Health Evaluation Manual,* of *Risk Assessment Guidance for Superfund* by USEPA (1999c).

2   History of Environmental Risk Assessment

A sixteenth-century German physician, Paracelsus, is credited with the discovery of the dose-response relationship and is sometimes considered the father of modern toxicology. He is most often quoted as saying: "All things are poison and not without poison; only the dose makes a thing not a poison" (for review, see Krieger 2001). His conclusion was important, because it rejected a simplistic separation of chemicals into toxic and nontoxic. It was recognized that the magnitude and the nature of effects were dependent on the magnitude of exposure. These concepts remain the cornerstones of modern pharmacology and toxicology today (Goodman et al. 2001; Kalant, Roschlau, and Roschlau 1997). In occupational toxicology, this concept has been used to develop threshold limit values (TLVs). TLVs were first published by the American Conference of Governmental Industrial Hygienists (ACGIH) in the 1950s and have been updated ever since.

The scientific evidence for dose-response relationship was further developed in the nineteenth and early twentieth centuries (for review, see Calabrese and Baldwin 2000a, 2000b), mostly from experiments using plants and fungi. During the latter part of the nineteenth century, two German scientists—Hugo Arndt, a physician, and Rudolph Schulz, a biologist—proposed the Arndt-Schulz Law, causing much controversy. The Arndt-Schulz Law states that for every substance, small doses stimulate, moderate doses inhibit, and large doses kill (for review, see Calabrese and Baldwin 2000a, 2000b). Schulz generalized his conclusions based on experimentation with the effects of chemicals on yeast fermentation to include medical applications. Arndt, on the other hand, came to similar conclusions from his clinical observations.

The Arndt-Schulz Law became the basis of homeopathic medicine. Homeopathic practitioners believed that ingesting minute amounts of a substance that in larger amounts might cause a specific problem in a healthy person, could cure that same health problem in an afflicted person. As this medical approach fell out of favor in the mainstream medical and scientific community, the Arndt-Schulz Law also became marginalized. Recently, however, the concept of stimulatory effect by chemicals (chemical hormesis) or radiation (radiation hormesis) in low doses is again being revived (Calabrese and Baldwin 2000a, 2000b; Welshons et al. 2003). Readers who are interested in learning more about the subject can look up other publications authored by Edward Calabrese or search the literature under "hormesis."

From early on, the approaches to the assessment of carcinogens and other substances differed. It was assumed, but not proven, that at least some carcinogens induced a risk of cancer, albeit small, at any dose greater than zero, while most chemicals elicit *threshold effects.*[1] The decision as to whether to treat an effect as *threshold* or *non-threshold*[2] often has great impact on the potency estimate for the chemical and can shape the risk management strategy. Non-threshold estimates tend to be far more conservative in many circumstances, thus driving the risk assessment.

2.1 Threshold Chemicals

Initial concepts of the dose-response curve and of Lethal Dose$_{50}$ (LD$_{50}$) as the dose that produces lethality in 50 percent of test animals were developed in the 1920s (Trevan 1927). The U.S. Food and Drug Administration (FDA) was an early leader in regulatory toxicology partly because of its responsibility under the federal Food, Drug, and Cosmetic Act of 1938 to ensure that new drugs are safe before marketing.[3] The FDA developed new animal tests to study the safety of chemicals and cosmetics, such as the well-known Draize Test for identifying eye irritants (Draize, Woodward, and Calvery 1944). In contrast to TLVs, which were biased toward short-term occupational effects, the FDA emphasized from the beginning the use of laboratory experimental data.

In 1949, the FDA released its landmark guidance to the industry, titled *Procedures for the Appraisal of the Toxicity of Chemicals in Food* (U.S. Food and Drug Administration 1949). This publication was the first comprehensive guide to testing chemicals on animals. The procedures included measures of blood pressure, heart rate, and respiration; tests of acute, subacute, and chronic toxicity; and studies of reproductive effects, hematology, absorption, excretion, and distribution. This standardized system of tests was generally well accepted and this type of information is used to this day in the hazard identification and dose-response assessment components of risk assessment.

The next major development was in the interpretation of these data for regulatory purposes. Lehman and Fitzhugh (1954) described a process for estimating the threshold level of human daily intake of a chemical, below which there would be no significant risk of adverse effects. The authors started with a dose that produced no adverse effect, and then applied a "safety factor" (SF) of 100 to estimate a dose that is unlikely to cause an adverse effect in humans. Dourson and Stara (1983) reviewed this methodology and concluded that this factor was appropriate for many chemicals, based on some experimental data and theoretical considerations (for more details, see USEPA 1993a).

The basic concepts of the approach proposed by Lehman and Fitzhugh were utilized by USEPA and other agencies to establish "safe doses" for a wide range of chemicals. Initially, USEPA established *acceptable daily intake* (ADI) as the dose below which no adverse effects are expected. More recently, it gave a definition of ADI as "the amount of a chemical a person can be exposed to on a daily basis over an extended period of time (usually a lifetime) without suffering deleterious effects" (USEPA 2003e). ADI is calculated as shown:

$$ADI \text{ (human dose)} = NOAEL \text{ (experimental dose)}/SF \text{ (safety factor)},$$

where NOAEL (no-observed-adverse-effect level) is "the highest exposure level at which there are no statistically or biologically significant increases in the frequency or severity of adverse effect between the exposed population and its appropriate control; some effects may be produced at this level, but they are not considered adverse, nor precursors to adverse effects" (USEPA 2003e).

SFs often consist of powers of 10, each factor representing a specific area of uncertainty inherent in the available data. The number of safety factors and their individual values are selected based on considerations such as the adequacy of and confidence in the data. For example, a factor of 10 may be introduced to account for the possible difference in responsiveness between humans and animals in prolonged exposure studies. A second factor of 10 may be used to account for variation in susceptibility among individuals in the human population. The resultant SF of 100 (10 × 10) has been judged to be appropriate for many chemicals. For chemicals with databases that are less complete (e.g., only results of subchronic studies are available), an additional factor of 10 (leading to an SF of 1,000) might be judged to be more appropriate. For chemicals that have well-characterized dose-response data in sensitive humans (e.g., effect of fluoride on human teeth), an SF as small as one might be selected (USEPA 1993a).

2.2 Non-Threshold Chemicals

From the beginning, many scientists believed that carcinogens, particularly mutagenic carcinogens, needed to be treated differently from other contaminants in that no threshold to their toxicity should be implied. Other scientists maintained that the threshold model was applicable to all contaminants, including mutagenic carcinogens. The following quote from the National Academy of Sciences illustrates the argument.

> Those who argue that safety factors are inadequate and that almost no thresholds can be determined, or theoretically developed, suggest that even one or a very few molecular events have a finite probability of initiating a successful malignant or neoplastic transformation in a cell, and that this can lead to a lethal cancer. Although one malignant cell can lead to death by cancer, many liver or kidney cells can be killed or damaged (but not malignantly transformed), without causing any detectable disease. Furthermore, man is never exposed to one carcinogen at a time, but is exposed to low concentrations of many at the same time.
>
> Accordingly, we have adopted a "nonthreshold" approach for estimating risks from pollutants that have been shown to be carcinogenic in laboratory animals. (National Academy of Sciences 1977)

The non-threshold model is based on the assumption that a mutagenic mechanism produces a lesion in the DNA of a single cell that eventually leads to cancer. Besides the mechanistic arguments, some observations were consistent with such a non-threshold model. For example, in the 1960s and 1970s, asbestos, vinyl chloride, and other substances were found to cause cancer in humans at doses below those that were previously considered to be safe (Crump 1996).

In the late 1950s and mid-1970s, regulatory agencies in the United States generally accepted the non-threshold model (Wilson 1996). The well-known "Delaney Clause," enacted as part of the 1958 amendments to the Food, Drug, and Cosmetic Act, prohibited approval and the use of intentional ("direct") food additives that "induce cancer." The clause was subsequently rescinded when the new Food Quality Protection Act was signed into law in 1996 (USEPA 1996a).

2.3 Recent Developments

In recent years, there has been a continuing shift in how carcinogens and noncarcinogens are considered in risk assessment. While some jurisdictions (WHO 1995; Health

Canada 1994) recognized early on that some cancer effects could have a threshold based on the mechanism of action, USEPA traditionally considered all carcinogenic effects as having no threshold, with linear dose-response relationships (Anderson and U.S. Environmental Protection Agency Carcinogen Assessment Group 1983; USEPA 1986a). It is only recently that USEPA has changed its view. In its latest draft carcinogenic guidelines (2003c), USEPA recommended a non-threshold approach only for mutagenic carcinogens and as a conservative default. This approach signals a welcome departure from its traditional policy.

Likewise, there is also a shift in the discussion regarding the conduct of risk assessment for noncarcinogenic effects. For example, the absence of a discernable threshold for lead neurotoxicity in young children and cardiovascular toxicity in adults led some investigators to question whether some noncarcinogenic effects might not have a threshold (Ontario Ministry of the Environment and Energy 1994; CalEPA 1997).

3   Components of Human Health Risk Assessment

The human health risk assessment process consists of four major steps: hazard identification/problem formulation, dose-response assessment, exposure assessment, and risk characterization. In the first step, one identifies (qualitatively) the problem, including the contaminants of concern and their potential adverse health effects, the receptors of concern (critical or indicator receptors), and the significant (critical) exposure pathways. In the second stage, the dose-response relationships for the critical adverse effects of the hazardous agents in the receptor are evaluated. The next step is to evaluate the exposure of the critical receptors to the hazardous agents under certain exposure scenarios (including the risk factor). Finally, during risk characterization, one evaluates health risk under various exposure scenarios by integrating the results from exposure and dose-response assessment. Each of these steps will be discussed in greater detail in the following subsections.

3.1   Hazard Identification and Problem Formulation

3.1.1   Description

USEPA defines *hazard assessment* somewhat narrowly as the process of determining whether exposure to an agent can cause an increase in the incidence of a particular adverse health effect (e.g., cancer, birth defect) and whether the adverse health effect is likely to occur in humans (USEPA 1999a). Hazard assessment may also include other qualitative descriptive information that is relevant, including discussions on the sources and fates of contaminants, their toxicokinetics, and mechanisms of action.

The prestigious Presidential/Congressional Commission on Risk Assessment and Risk Management describes *hazard identification* as the determination of the identities and quantities of chemicals present in the environment or manufactured for various uses and the types of hazards they may pose to human health (Omenn 1997). Hazard identification can also be thought of as the planning stage for the entire risk assessment. It consists of three steps: problem formulation, conceptual model, and analysis plan. These steps were not defined in the original National Research Council report

(1983); however, they were built into later ecological risk assessments (USEPA 1998a), integrated decision-making processes (USEPA 2000e), and cumulative risk assessments (USEPA 2003d).

USEPA describes *problem formulation* as

> a systematic planning step that identifies the major factors to be considered in a particular assessment. It is linked to the regulatory and policy context of the assessment. Problem formulation is an iterative process within which the risk assessor develops preliminary hypotheses about why adverse effects might occur or have occurred. It provides the foundation for the technical approach of the assessment. The outcome of the problem formulation process is a conceptual model. (USEPA 2003d)

A *conceptual model* is described as "a 'model' of a site developed at scoping using readily available information. Used to identify all potential or suspected sources of contamination, types and concentrations of contaminants detected at the site, potentially contaminated media, and potential exposure pathways, including receptors" (USEPA 2003k).

The conceptual model generally serves the following functions:

- Identifies hazardous agents to be considered in the assessment, their levels in various environmental *media* (e.g., soil, air, surface water) and in different *microenvironments* (e.g., outdoor air, indoor home, indoor work) where people (*receptors*) may be exposed to the selected contaminants

- Identifies significant contaminant sources and describes how the levels of contaminants may change over time as a result of release from on-site or off-site sources (e.g., volatilization, fugitive dust emission, surface runoff/overland flow, leaching to groundwater), migration toward environmental sinks where the contaminants may accumulate (e.g., sediment may be a sink for stable hydrophobic organics or metals), degradation in the environment, or migration outside of the area considered in the assessment

- Identifies the source-to-receptor exposure pathways, taking into consideration environmental fate and transport of the hazardous agents

- Characterizes the populations (receptors), including the subpopulations, which may be considerably exposed or vulnerable to the selected contaminants

The conceptual model is often presented as a collection of tables and figures, which are usually not self-explanatory. On the other hand, a pictorial representation of the conceptual model can be quite effective. It is also important to provide an explanation of the conceptual model and the implication for carrying through the risk assessment. The analysis plan should provide an overview of the intent and a high-level plan of the assessment. It should state how results would be interpreted (e.g., how cancer risk slightly exceeding one in a million would be interpreted).

The terms *hazard identification* and *problem formulation* are often used interchangeably. The key to this component of the assessment is:

- to provide complete information about the past, current, and antici-pated future conditions of the assessed site or situation
- to assess and interpret these facts
- to summarize the conclusions and to present an assessment plan

Originally intended only for assessment of contaminated soils under the Super-fund program, the USEPA document *Standardized Planning, Reporting, and Review of Superfund Risk Assessments* (USEPA 2001a) provides detailed guidance on hazard identification. This document is recommended as the first source of information for planning a risk assessment.

One of the outputs of hazard identification is a summary of the toxic effects associ-ated with the substances or mixtures of concern, including a review of the weight of evidence for carcinogenicity. This type of information can be obtained from sources that are listed below in section 3.1.4.

Although hazard identification can have great impact on the outcome of the risk assessment, it is often not given the attention it deserves. A strong and well-developed hazard identification section usually signals good quality for the overall risk assess-ment.

3.1.2 Weight of Evidence for Carcinogenicity

Scientific publications vary in quality; some reports may even contradict each other. There may not be enough data to make a definitive conclusion about the ability of a substance to induce cancer in humans. Many agencies have therefore developed weight-of-evidence schemes to help evaluate the strength of the available evidence in a consistent manner to determine whether a substance (or a mixture of substances or a process) is carcinogenic to humans. The International Agency for Research on Can-cer (IARC), part of WHO, is the first organization that has developed such a plan, whereby a panel of international experts systematically evaluates the evidence for and against carcinogenicity of a given agent. The IARC panel then publishes the carcino-genic ranking for a given agent, summarizing the weight of evidence for its potential to induce cancer in humans.

IARC classifies substances, mixtures, and processes into five categories.

Group 1: Carcinogenic to humans

Group 2(a): Probably carcinogenic to humans

Group 2(b): Possibly carcinogenic to humans

Group 3: Not classifiable as to carcinogenicity

Group 4: Probably not carcinogenic to humans

IARC bases its classification solely on human epidemiological and animal experimen-tal studies. Potency and mechanism of action are not usually considered in the assess-ment. Readers who are interested in a more detailed description of the classification can find the information in the preambles to IARC monographs or their Internet-based summaries for individual substances.

Although the IARC ranking continues to be highly respected, other agencies have developed similar ranking schemes. Of these, the one published by USEPA (1986a) is probably the most influential. USEPA introduced the following letter designations in 1986:

A: Human carcinogen

B: Probable human carcinogen

B1: Limited evidence of carcinogenicity in humans

B2: Sufficient evidence of carcinogenicity in animals with inadequate or lack of evidence in humans

C: Possible human carcinogen

D: Not classifiable as to human carcinogenicity

E: Evidence of noncarcinogenicity for humans

In 1996, USEPA replaced its ranking system based on letters with a new descriptive scheme, which takes into account a wider range of data (USEPA 1996b). In the *Draft Final Guidelines for Carcinogen Risk Assessment* (USEPA 2003c), USEPA recommended that "standard descriptors" be utilized in addition to the descriptive narrative. The standard descriptors include "likely to be carcinogenic to humans," "suggestive evidence of carcinogenicity, but not sufficient to assess human carcinogenic potential," "data are inadequate for an assessment of human carcinogenic potential," and "not likely to be carcinogenic to humans."

The USEPA's 1986 letter designations are still widely used, in part because the evaluations based on this system continue to be reported in the Integrated Risk Information System (IRIS) database. The IARC and 1986 USEPA ranking schemes are quite similar. Although both organizations place greater emphasis on good human epidemiological data over animal data, USEPA has traditionally placed heavier emphasis on animal data than IARC. Even though the 1996 USEPA descriptive rankings have been in use for a few years, the number of agents so ranked is relatively small and thus it is not yet as widely used as the older scheme. Unlike the earlier IARC and USEPA rankings, the newer USEPA classification puts more emphasis on the mode of action.

In Canada, Health Canada has developed a carcinogen-ranking scheme under the Canadian Environmental Protection Act based on the IARC ranking scheme. Health Canada (1994) classified the substances on the first Priority Substances List into six main groups:

I: Carcinogenic to humans

II: Probably carcinogenic to humans

III: Possibly carcinogenic to humans

IV: Unlikely to be carcinogenic to humans

V: Probably not carcinogenic to humans

VI: Unclassifiable with respect to carcinogenicity in humans

Health Canada's scheme consisted of more categories and subcategories and was not very compatible with those of IARC and USEPA. Health Canada distinguished between genotoxic and nongenotoxic carcinogens and gave the latter group a lower ranking. Although this treatment may be sensible in many circumstances, it leads to some surprises. For example, Health Canada has not classified dioxins. A low rank of IIIC ("possibly carcinogenic to humans") would probably have to be assigned to this potent family of carcinogens using Health Canada's ranking scheme (for further details, see Meek et al. 1994).

The Health Canada categories have now been replaced with a more narrative approach to more clearly delineate the basis for the considered carcinogenic potential of the substances to humans (Liteplo 2004).

Some U.S. states, including California, have their own rankings. So do many European countries (see Moolenaar 1994). A comparison of the key ranking schemes is summarized in table 1.2.

3.1.3 Assessing Evidence for Mutagenicity

It is important to distinguish between carcinogens that react with DNA and are directly mutagenic and other carcinogens that are not mutagenic. Whether a carcinogen is mutagenic often has implication for how it is treated in risk assessment. USEPA assumes that mutagenic carcinogens have dose-response curves, which after subtracting background exposures and background effects, pass through the origin and are linear at low doses (USEPA 2003c). Carcinogens that are nonmutagenic, weakly mutagenic, or mutagenic only at high doses or that have weak weight of evidence for mutagenicity are now assumed to have nonlinear dose-response relationships (i.e., to be threshold agents). These decisions are made during dose-response assessments, but are discussed here because hazard identification involves assessment of mutagenicity.

3.1.4 Key Sources of Information

Table 1.3 lists the key sources of information for hazard identification of contaminants. Several sources are particularly valuable, because they cover all the compo-

Table 1.2 Comparison of three well-known weight-of-evidence classification schemes for carcinogens

Strength/Type of Evidence	Weight-of-Evidence Classification		
	USEPA[a]	IARC (WHO)[b]	Health Canada[c]
Strong human evidence	A	1	I
Some human and animal evidence	B1	2A	IIIB[d]?
Little or no human evidence, but strong animal evidence	B2	2B	II?
Weak evidence from human and animal data	C	—	III (except IIIB?)
Little evidence for or against carcinogenicity	D	3	VI
Good evidence for absence of carcinogenicity	E	4	V, (IV?)

? – Indicates imperfect fit.
[a] USEPA 1986
[b] International Agency for Research on Cancer 2003
[c] Health Canada 1994
[d] IIIB (a subcategory of III) is defined by some evidence of carcinogenicity in animals but with limited data, inadequate epidemiological studies to assess carcinogenicity in humans, limited evidence for genotoxicity, or mixed results (Health Canada 1994).

Table 1.3 Key sources of information for hazard identification and dose-response assessments

Sources	Links
ATSDR—*ToxFAQs*	http://www.atsdr.cdc.gov/toxfaq.html
ATSDR—*Toxicological Profiles*	http://www.atsdr.cdc.gov/toxpro2.html
Australian Government—*National Industrial Chemicals Notification and Assessment Scheme (NICNAS)*	http://www.nicnas.gov.au/publications/CAR/
Chemfinder.com	http://chemfinder.cambridgesoft.com/
ECB—*European Union Risk Assessment Reports*	http://ecb.jrc.it/existing-chemicals/
enHealth Council—*National Environmental Health Monographs*	http://enhealth.nphp.gov.au/council/pubs/ecpub.htm
Environmental Defence Fund (EDF)	http://www.environmentaldefense.org/home.cfm
Health Canada—*Priority Substances List Assessment Reports*	http://www.hc-sc.gc.ca/hecs-sesc/exsd/psap.htm *Journal of Environmental Science and Health,* part C, *Environmental Carcinogenesis and Ecotoxicology Reviews,* vol. C12, no. 2 (1994); vol. C19, no. 1 (2001)
Health Canada—*Proposed Regulatory Decision Documents (PRDDs)*	http://www.hc-sc.gc.ca/pmra-arla/english/pubs/ prdd-e.html
Health Canada—*Regulatory Notes (REG)*	http://www.hc-sc.gc.ca/pmra-arla/english/pubs/ reg-e.html
IPCS—*Concise International Chemical Assessment Documents (CICADs)*	http://www.inchem.org/pages/cicads.html
IPCS—*Environmental Health Criteria (EHC) Monographs*	http://www.inchem.org/pages/ehc.html
IPCS—*Joint Meeting on Pesticide Residues (JMPR)*	http://www.inchem.org/pages/jmpr.html
IPCS—*Pesticide Data Sheets (PDSs)*	http://www.inchem.org/pages/pds.html
Registry of Toxic Effects of Chemical Substances (RTECS)	http://ccinfoweb.ccohs.ca/rtecs/search.html
Risk Assessment Information System (RAIS)	http://risk.lsd.ornl.gov/rap_hp.shtml
Syracuse Research Corporation—*Toxic Substance Control Act Test Submission (TSCATS) Database*	http://esc.syrres.com/efdb/TSCATS.htm
TOXNET[a]	http://toxnet.nlm.nih.gov/
USEPA—*Chemical Summaries/Fact Sheets*	http://www.epa.gov/chemfact/
USEPA—*Integrated Risk Information System (IRIS)*	http://www.epa.gov/iris/
USEPA—National Center for Environmental Assessment (NCEA)	http://cfpub.epa.gov/ncea/
USEPA—*Reregistration Eligibility Decisions (REDs)/ Fact Sheets*	http://cfpub.epa.gov/oppref/rereg/status.cfm? show = rereg
USEPA—*Risk Assessment Guidance for Superfund (RAGS),* part A	http://www.epa.gov/superfund/programs/risk/ragsa/

[a] Databases such as HSDB, TOXLINE, Gene-Tox, and CCRIS can be accessed through TOXNET.

nents of risk assessment. The best single source in North America is the Agency for Toxic Substances and Disease Registry (ATSDR), which publishes *Toxicological Profiles* (2003a) and *ToxFAQs* (2003b). The toxicological profiles are very comprehensive and are intended for experts, although some sections are written for nonexperts. *ToxFAQs* is brief and written for nonexperts, but contains key information needed for hazard identification. *ToxFAQs* does not contain information on physicochemical properties. The level of expertise required for USEPA's *Chemical Summaries* (2003b) is between that for ATSDR's *Toxicological Profiles* and *ToxFAQs*. The *Chemical Fact Sheets* published by USEPA (2003b) are similar to *ToxFAQs*.

IPCS publishes the very comprehensive *Toxicological Profiles and Environmental Health Criteria Monographs* (2003b), as well as the brief *Concise International Chemical Assessment Documents (CICADs)* (2003a). The European Chemicals Bureau (ECB) issues the wide-ranging *European Union Risk Assessment Reports* (2003a), while Australia distributes *National Environmental Health Monographs* (enHealth Council 2003) and *Public Chemical Assessment Reports* from the National Industrial Chemicals Notification and Assessment Scheme (2003).

In Canada, in-depth health risk assessments have been conducted for substances on the first and second Priority Substances Lists under the Canadian Environmental Protection Act. The links to these reports may be found currently at http://www.hc-sc.gc.ca/hecs-sesc/exsd/psap.htm. In spring 2004, Health Canada was to commence publication of a new series of "screening" assessment reports on large numbers of existing substances in Canada. The objective of these reports is to take an initial assessment of whether the substances pose a risk to human health. Summaries for the hazard characterizations of substances on the Priority Substances Lists under CEPA may also be found in *Environmental Carcinogenesis and Ecotoxicology Reviews* (*Journal of Environmental Science and Health* C12, no. 2 [1994], and C19, no. 1 [2001], respectively). In addition, some of these assessments are contained in the Canadian Council of Ministers of the Environment's *Canadian Environmental Quality Guidelines* (CCME 2002).

The sources described in the previous paragraphs review in detail properties of many substances. The quality of these reviews generally ranges from good to very good. The topics covered include:

- physicochemical properties
- environmental sources, typical levels, and fate
- toxicity, toxicokinetics, and mechanism of action

Two other sources offer less-complete information, but their websites are extremely well designed and the information that they do provide is both reliable and conveniently presented. The Risk Assessment Information System (RAIS) website (2003) attempts to provide a convenient one-stop resource site for the conduct of risk assessment. It contains guidance for conducting risk assessment as well as parameters needed in the assessment, including hazard identification for many radioactive and nonradioactive substances. The Environmental Defence Fund (EDF) is more tailored to nonexperts, but its website provides well-organized and credible information regarding the toxicity of chemicals, the quantity released in the United States, and other relevant information (Environmental Defence Fund 2003a, 2003b). The EDF website uses respected sources of information, and its data sources are well referenced.

USEPA has also published reviews for selected contaminants. Some of these reviews can be found on the National Center for Environmental Assessment (NCEA) website (USEPA 2003g) or the Integrated Risk Information System (IRIS) website (USEPA 2003f). The scope of these reports differs, but generally includes toxicity, toxicokinetics, and mechanism of action. Information regarding physicochemical properties, sources, and fate is less likely to be found in these reports.

Integrated assessments of sufficient quality may not be available for some substances. One can, however, extract relevant information from well-researched databases. TOXNET is a collection of databases and is an excellent starting point. The most useful databases within TOXNET for hazard identification are the Hazardous Substances Data Bank (HSDB 2003), TOXLINE (2003), and PubMed (2003). HSDB has all the information needed to prepare hazard identification, including physicochemical properties, environmental sources and fate, toxicity, toxicokinetics, and mechanism of action. TOXLINE and PubMed provide references and abstracts in the biomedical field. Also accessible through TOXNET are Gene-Tox and the Chemical Carcinogenesis Research Information System (CCRIS), which contain useful information on carcinogenicity and mutagenicity. These two databases are specialized and can be used in addition to the other more general sources.

The Registry of Toxic Effects of Chemical Substances (RTECS) database (2003) contains summaries of toxicological findings related to individual toxic substances or mixtures with citations from international journals, textbooks, technical reports, scientific proceedings, and compendia. The Environmental Fate Data Base (EFDB) from Syracuse Research Corporation (SRC) provides references for environmental fate, absorption and adsorption factors, and other important factors (Syracuse Research Corporation 2003a). SRC offers two additional databases: the K_{ow} database (2003b) and Toxic Substances Control Act Test Submissions (TSCATS) database (2003c). TSCATS is similar to RTECS and offers references and abstracts on unpublished, nonconfidential studies covering chemical testing results and adverse effects of chemicals on health and ecological systems. The studies are submitted by U.S. industry to USEPA under several sections of the Toxic Substance Control Act. The K_{ow} database provides K_{ow} values, which are useful for assessing the environmental fate of chemicals, chemical absorption and distribution in the organism, and bioaccumulative potentials. The Chemfinder website provides links to useful data needed in hazard identification.

There are several good sources of information for pesticides. IPCS publishes *WHO/ FAO Data Sheets on Pesticides* (2003d), also known as *Pesticide Data Sheets*, and more concise assessments of pesticide toxicity in its *Joint Meeting on Pesticide Residues (JMPR)* (IPCS 2003c). *Pesticide Data Sheets* contains minimum information on physicochemical properties and a good overview of toxic effects, while *JMPR* provides summaries of original toxicity studies. USEPA publishes reregistration eligibility decision (RED) reports for pesticides (2003i). These reports contain a complete set of information for hazard identification. USEPA also publishes pesticide fact sheets (2003h); these cover a broader range of active ingredients in pesticides, but only some fact sheets contain (briefly) all of the elements required for hazard identification. Health Canada publishes documents similar to USEPA's reregistration eligibility documents in the *Proposed Regulatory Decision Documents (PRDD Series)* (2003c) and *Regulatory Notes (REG Series)* (2003d).

Part A of the *Risk Assessment Guidance for Superfund (RAGS)* (USEPA 1989) provides general discussion on the data collection and evaluation aspect of hazard identification.

3.1.5 Issues in Hazard Identification and Problem Formulation

This section highlights some of the issues that are often weak or missing in risk assessments.

Hazard identification represents the descriptive, nonquantitative component of a risk assessment. As a result, often less attention is paid to hazard identification than to the quantitative components. In reality, hazard identification is very important, because this stage is when most of the planning and scoping of the assessment takes place. The risk assessor will determine whether all the necessary information is available to carry through the risk assessment as intended. For these reasons, hazard identification requires significant involvement of experienced risk assessors.

With respect to risk assessments involving brownfields sites, comprehensive checklists for reviewers have been published (Ontario Ministry of the Environment 1999; USEPA 1989), which may assist in ensuring that the hazard identification component is complete. Depending on the historical use of a site, one can expect to find certain contaminants. It is important that the sampling program includes contaminants *potentially* present at the site based on a review of site information and history as well as contaminants *expected* to be present; otherwise, any conclusion made regarding the safety of the site could be erroneous. Volume 1, part D, of the *Risk Assessment Guidance for Superfund* (USEPA 2001a) includes in its appendices a *Guide to Contaminants Found at Typical Brownfields Sites*, which could be a very useful resource for planning a risk assessment

One must also consider the environmental fate of the chemicals over time. It is important not only to assess the current situation, but also to look for clues about past and especially future conditions. For example, several chlorinated solvents (e.g., tetrachloroethene or trichloroethene) gradually degrade into vinyl chloride in an anaerobic subsurface environment. Depending on the local soil and groundwater conditions, vinyl chloride may appear and increase over time. Vinyl chloride can become a major concern, as it is more toxic than its precursors and must be taken into consideration in a risk assessment.

3.2 Dose-Response Assessment

3.2.1 Description

Dose-response assessment estimates the relationship between the magnitude of exposure and magnitude/probability of occurrence of an effect. In its IRIS database, USEPA (2003e) defines dose-response assessment as the "determination of the relationship between the magnitude of an administered, applied, or internal dose and a specific biological response. Response can be expressed as measured or observed incidence or change in level of response, percent response in groups of subjects (or populations), or the probability of occurrence or change in level of response within a population." Development of dose-response relationships from epidemiological studies or animal experimental data is a specialized discipline, and many risk assessors have not been involved in such an endeavor. Most risk assessments are conducted using estimated potency factors[4] developed by expert teams working for various regulatory or quasi-regulatory organizations such as WHO.

3.2.2 Current Estimated Potency Factors: USEPA

Although the terms *acceptable daily intake* and *safety factor* continue to be used in many jurisdictions for threshold effects, USEPA has since replaced them with terms that are more neutral with respect to implied safety and has provided a more detailed and prescriptive guidance for the use of these factors. USEPA now names the daily oral dose that is not expected to pose significant risk to humans the *reference dose* (RfD). RfD is

> an estimate (with uncertainty spanning perhaps an order of magnitude) of a daily oral exposure to the human population (including sensitive subgroups) that is likely to be without an appreciable risk of deleterious effects during a lifetime. It can be derived from a NOAEL, LOAEL, or benchmark dose, with uncertainty factors generally applied to reflect limitations of the data used. Generally used in EPA's noncancer health assessments. (USEPA 2003e)

Reference concentration (RfC) applies to the inhalation route of exposure. RfC is defined as

> an estimate (with uncertainty spanning perhaps an order of magnitude) of a continuous inhalation exposure to the human population (including sensitive subgroups) that is likely to be without an appreciable risk of deleterious effects during a lifetime. It can be derived from a NOAEL, LOAEL, or benchmark concentration, with uncertainty factors generally applied to reflect limitations of the data used. Generally used in EPA's noncancer health assessments. (USEPA 2003e)

The initial shape of a non-threshold dose-response curve for carcinogens was traditionally assumed to be linear based on mechanistic grounds. A linear no-threshold dose response was first proposed in 1976 (Guess and Crump 1976), refined over time (reviewed by Crump 1996), and eventually used as the default by USEPA (Anderson and U.S. Environmental Protection Agency Carcinogen Assessment Group 1983; USEPA 1986a). Most of the existing and currently available assessments used the linearized multistage model developed by Crump and his colleagues (see Crump 1996), even though other non-threshold models that can be fitted to the dose-response data and are linear at low dose have been proposed over the years (see, for example, Krewski et al. 1993; Moolgavkar and Knudson 1981).

Most of the models create a dose–risk function, which approaches zero at extremely low doses and increases almost linearly at low doses. The linear component of the dose-effect relationship often lies near and below the lowest dose at which the effect was observed in epidemiological or experimental studies, becoming nonlinear above this dose.

The slope of the linear component of a dose-response model is the *slope factor* and is expressed as

$$\text{risk} / ((\text{mg of chemical} / \text{body weight in kg}) \times \text{number of days})$$

for exposures by the oral and dermal routes.[5] For drinking water, the slope is referred to as the *unit risk* and is expressed as

$$(\mu\text{g of chemical} / \text{consumed liquids in liters})^{-1}$$

For the inhalation route of exposure, the initial linear slope is also called *unit risk* and is expressed as

$$(\mu g \text{ of contaminant / inhaled volume in } m^3)^{-1}$$

IRIS, which is the official USEPA database for estimated potency factors, lists RfDs, RfCs, slope factors, and unit risks. All of these values apply to lifetime (chronic) exposure to contaminants. The basic underlying assumption in constructing these factors is that the exposure is constant over a lifetime and even if the exposure were intermittent or shorter than a lifetime, it is acceptable to amortize (average) the total exposure over a lifetime. Although the assumption for averaging exposure is routinely used, its limitations are recognized. For example, if the total exposure a person had to a carcinogen occurred a week before death, it is unlikely that the person's death was as a result of this exposure.

RfDs and RfCs

A good overview of USEPA's traditional approach to developing RfDs and RfCs can be found in the report *Reference Dose (RfD): Description and Use in Health Risk Assessments* (USEPA 1993a). USEPA developed RfDs and RfCs for contaminants that do not cause cancer. Until recently, USEPA treated all carcinogens as non-threshold, assuming that even an extremely small dose is capable of increasing (slightly) cancer risk. However, even carcinogens may have other (non-cancer) effects (called *endpoints*) for which RfDs and RfCs can be developed.

RfDs have been developed for oral exposure and oral RfDs are often used as a starting point to estimate dermal RfDs. RfDs are expressed in standard units (milligrams of contaminant ingested or applied to the skin per kilogram of body weight per day). RfCs have been developed in a similar manner, the main difference between RfDs and RfCs being the route of exposure for which they are developed and the units in which they are expressed. RfCs are expressed in terms of mg of contaminant per m^3 of air (see table 1.5 later in this chapter).

The traditional starting point for deriving RfDs or RfCs was the NOAEL (no-observed-adverse-effect level). USEPA (1993a) describes NOAEL as "the highest experimentally determined dose without a statistically or biologically significant adverse effect." When NOAEL was unavailable, the lowest-observed-adverse-effect (LOAEL) level was used instead. LOAEL, also referred to as the lowest-effect level (LEL), is defined as "the lowest exposure level at which there are statistically or biologically significant increases in frequency or severity of adverse effects between the exposed population and its appropriate control group" (USEPA 2003e). Note that identification of a NOAEL and a LOAEL does not require statistical criteria and that *biologically significant* criteria are acceptable.

RfDs and RfCs are developed from NOAELs or LOAELs by applying an uncertainty factor (UF) and a modifying factor (MF), as shown:

$$\text{RfD or RfC} = \text{NOAEL} / (\text{UF} \times \text{MF})$$

The uncertainty factor is

> one of several, generally 10-fold factors, used in operationally deriving the RfD and RfC from experimental data. UFs are intended to account for (1) the variation in sensitivity among the members of the human population, i.e., interhuman or intraspecies variability; (2) the uncer-

tainty in extrapolating animal data to humans, i.e., interspecies variability; (3) the uncertainty in extrapolating from data obtained in a study with less-than-lifetime exposure to lifetime exposure, i.e., extrapolating from subchronic to chronic exposure; (4) the uncertainty in extrapolating from a LOAEL rather than from a NOAEL; and (5) the uncertainty associated with extrapolation from animal data when the data base is incomplete. (USEPA 2003e)

The modifying factor is "a factor used in the derivation of a reference dose or reference concentration. The magnitude of the MF reflects the scientific uncertainties of the study and database not explicitly treated with standard uncertainty factors (e.g., the completeness of the overall database). A MF is greater than zero and less than or equal to 10, and the default value for the MF is 1" (USEPA 2003e).

The uses of UF and MF are defined as presented in table 1.4.

In 2002, the USEPA Risk Assessment Forum reviewed the RfD and RfC development process (USEPA 2002a). The Technical Panel considered the purpose of the MF to be sufficiently subsumed in the general database UF and recommended that the use of MF be discontinued. At the present time, UF is also applied to account for inadequacy of reproductive studies.

The Office of Pesticide Programs of USEPA also applies an appropriate Food Quality Protection Act (FQPA) safety factor when developing aggregate risk assessments[6] and regulatory decisions for single ingredients of pesticide products (USEPA 2002b). It should be noted that these factors are not adopted for other USEPA programs and activities. The factor is intended to protect infants and children from pre- and postnatal toxicity and to account for completeness of the data with respect to exposure and toxicity to infants and children.

The development of the FQPA factors involves:

- examining the level of concern for special sensitivity/susceptibility of children
- assessing whether traditional uncertainty factors already incorporated into the development of RfDs and RfCs are adequate to protect infants and children
- assessing the adequacy of the exposure assessment database

Another important distinction is that the development of the FQPA safety factor takes place not during the dose-response assessment (i.e., during the development of RfD) but as part of risk characterization. The safety factor thus incorporates not only any uncertainty not built into the RfD itself but also any uncertainty that is part of hazard identification, exposure assessment, and risk characterization. By default, the factor is set to 10, but it can be smaller if justified during the "weight-of-evidence" assessment process.

Currently, USEPA uses the *benchmark dose* and *benchmark concentration* to derive some of its new RfDs and RfCs. The newer benchmark method addresses at least some of the disadvantages of using the experimentally derived NOAELs. For example, a NOAEL is established using a single data point and the rest of the dose-effect relationship is not used for establishing the apparent threshold; this reduces the reliability of the estimation of the threshold. The NOAEL is dependent on the degree of magnitude

Table 1.4 Use of uncertainty factors and modifying factors in deriving reference doses

Factors	USEPA Definitions
Uncertainty Factor (UF), comprising:	
• Variation of sensitivity in humans	Use a 10-fold factor when extrapolating from valid experimental results in studies using prolonged exposure to average healthy humans. This factor is intended to account for the variation in sensitivity among members of the human population and is referenced to as "10H."
• Extrapolation from animal data to humans	Use an additional 10-fold factor when extrapolating from valid results of long-term studies on experimental animals when results of studies of human exposure are not available or are inadequate. This factor is intended to account for the uncertainty involved in extrapolating from animal data to humans and is referenced to as "10A."
• Extrapolation from less-than-chronic NOAELs to chronic NOAELs	Use an additional 10-fold factor when extrapolating from less than chronic results on experimental animals when there are no useful long-term human data. This factor is intended to account for the uncertainty involved in extrapolating from less-than-chronic NOAELs to chronic NOAELs and is referenced to as "10S."
• Extrapolating from LOAEL instead of NOAEL	Use an additional 10-fold factor when deriving an RfD from a LOAEL instead of a NOAEL. This factor is intended to account for the uncertainty involved in extrapolating from LOAELs to NOAELs and is referenced to as "10L."
Modifying Factor (MF)	Use professional judgment to determine the MF, which is an additional uncertainty factor that is greater than zero and less than or equal to 10. The magnitude of MF depends upon professional assessment of scientific uncertainties of the study and database not explicitly accounted for (see above) e.g., completeness of the overall database and the number of species tested. The default value for MF is 1.

Source: USEPA 1993a

of the response ("signal") relative to the variability of the test ("noise"); thus, the various factors affecting "noise" also affect the magnitude of the NOAEL. In other words, a NOAEL is not a biological threshold but a value at or below the detection limit of a particular bioassay.

The following is the list of limitations USEPA has identified for using the NOAEL/LOAEL approach to develop RfDs and RfCs:

- The NOAEL/LOAEL is highly dependent on dose selection since the NOAEL/LOAEL is limited to being one of the doses included in a study.
- The NOAEL/LOAEL is highly dependent on sample size. The ability of a bioassay to distinguish a treatment response from a control response decreases as sample size decreases, so that the NOAEL for a compound (and thus the POD)[7] will tend to

be higher in studies with smaller numbers of animals per dose group.

- More generally, the NOAEL/LOAEL approach does not account for the uncertainty in the estimate of the dose-response, which is due to the characteristics of the study design.

- NOAELs/LOAELs do not correspond to consistent response levels for comparisons across studies/chemicals/endpoints and for use as PODs for the derivation of RfCs.

- The slope of the dose-response curve is not taken into account in the selection of a NOAEL or LOAEL, and is not usually considered unless the slope is very steep or very shallow.

- A LOAEL cannot be used to derive a NOAEL when a NOAEL does not exist in a study. Instead, a tenfold uncertainty factor has been routinely applied to the LOAEL to account for this limitation.

- While the NOAEL has typically been interpreted as a threshold (no-effect level), simulation studies (i.e., Leisenring and Ryan 1992) and reanalyses of developmental toxicity bioassay data (Allen et al. 1994a) have demonstrated that the rate of response above control at doses fitting the criteria for NOAELs, for a range of study designs, is about 5–20% on average, not 0%. (USEPA 2000a)

The benchmark approach avoids the above limitations. Determination of the benchmark dose starts with identifying the appropriate data set and fitting a model to the dose-response data, especially at the lower end of the observable dose-response range. The curves (or functions) representing both the best fit and the lower confidence limit (USEPA [2000a] recommends the use of 95 percent lower confidence limit) of the dose-effect relationship are obtained. A specified low level of response is selected, typically within the 1–10 percent range of excess health effect above background; the value of 5 percent is usually chosen. This value is referred to as the *benchmark response* or BMR (USEPA 2000a).[8]

Using the dose-effect relationship representing the best fit to the data, the dose corresponding to BMR is called *benchmark dose* or BMD. The dose corresponding to the lower confidence limit on the BMD is called BMDL. Visually, BMDL is the dose that corresponds to the 95 percent lower confidence limit of the benchmark response.

USEPA defines the terms in the following manner:

- *Benchmark Dose (BMD):* "An exposure due to a dose of a substance associated with a specified low incidence of risk, generally in the range of 1% to 10%, of a health effect; or the dose associated with a specified measure or change of a biological effect."

- *Benchmark Response (BMR):* "The response, generally expressed as in excess of background (see for example, Extra Risk), at which a benchmark dose or concentration is desired (see Benchmark Dose, Benchmark Concentration)."

- *BMDL:* "A lower one-sided confidence limit on the BMD" (USEPA 2000a).

The methods using either BMD or NOAEL have the disadvantage of not providing guidance about the magnitude of the risk once the threshold is exceeded, since this type of information is reflected only by the slope and shape of the dose-response curve. Since environmental exposures to several chemicals often exceed the threshold, the limitation of using approaches that define only the threshold and do not quantify the effect has practical importance. The following is a discussion on this subject by the IPCS:

> Exposure of humans at doses/intakes above that considered to be "without appreciable health risk" may show the potential to move some human subjects above the threshold. Individuals who may move above the threshold for the human population when the ADI or TDI was exceeded would be those for whom the full 10-fold factor for inter-individual variability was necessary and also on the assumption that the full 10-fold inter-species factor was also necessary for that chemical. Interpretation of risk above the threshold can only be made by reference back to the dose response relationship present in the animal database, which gave the threshold. Therefore, this has to be done on a case-by-case basis. (IPCS 2000)

RfD can be derived from BMDL in a manner comparable to the NOAEL/LOAEL approach. On average, the two values are numerically related. Based on the review of the literature by USEPA, the values of NOAEL and BMDL are generally within one-half to one order of magnitude, depending on the type of data used for comparison and the BMR selected for making the comparison (USEPA 2000a).

USEPA now uses the benchmark method to derive RfDs and RfCs, where possible. However, most of the existing RfDs and RfCs were developed using NOAELs/LOAELs, although some chemicals (e.g., benzene) were assessed using the benchmark approach.

Unit Risks and Slope Factors

USEPA traditionally assumed low-dose linearity for all carcinogens. Low-dose linearity is experimentally untestable, because the positive responses at low doses cannot be statistically distinguished from "no effect" responses. Nevertheless, the assumption of low-dose linearity is considered to be conservative, and the functions now in common use for dose-response assessment generate similar results. It is therefore useful to view the low-dose extrapolation as an accepted convention, rather than a proven fact.

The initial slope obtained from the best fit of a dose-response function to the dose-effect data is called the *maximum likelihood estimate* (MLE). MLE corresponds to the value of q_1 (i.e., slope) in the linearized multistage model. The slope of the function that represents the 95 percent upper confidence limit of the dose-response function is called the *upper confidence limit* (UCL). UCL corresponds to the q_1^* in the linearized multistage model. Unit risks and slope factors can be derived from either MLE or UCL. It is important to distinguish between potency estimates based on MLE and UCL because the two estimates may differ significantly; UCL is usually, but not always, the more conservative estimate of potency. MLE, UCL, and their differences are well described in an assessment report from the U.S. Office of Technology Assessment (1987).

At present, there does not appear to be a consensus on the preference for MLE or UCL. For some dose-response data, the difference may not be large, but in some cases the difference could be significant. In favor of MLE are the following arguments:

- MLEs are thought to represent the best fit to the available data and thus represent the most realistic estimate of the actual risk. As a result, the MLE value is most consistent with an assessment where the goal is to derive best estimate of risk. The UCL value may be excessively conservative, encouraging unnecessary risk management actions.

- In most cases, MLEs are lower than the UCLs because UCLs conservatively add a 95 percent confidence interval to the best fit represented by MLE. Under the condition of marginally applicable data where no low-dose effects have been determined, the initial linear component of MLE may not be present; thus, MLE could exceed the UCL value, because the UCL (but not the MLE) fit is forced into linearity.

The supporters of UCL see the advantage of the built-in conservatism and the enforced linearity with any data. They also point out that with MLE, small fluctuations in the underlying data at high doses in the bioassay can significantly change the maximum likelihood estimate at the low doses of interest. The UCL is a more stable number.

Both MLE and UCL are currently in use. USEPA has used both MLEs and UCLs for different contaminants, but on the whole tends to prefer UCL. In contrast, Occupational Safety and Health Administration (OSHA) assessments often present both MLEs and UCLs, but prefer the use of MLEs. A recent report by the U.S. National Research Council (NRC) based its unit risk for arsenic in drinking water solely on the MLE (National Research Council 2001b).

USEPA currently recommends the use of the more-conservative bound rather than the central tendency for several reasons:

> A lower limit rather than a central estimate is appropriate for several reasons. One considers the relative consequences of overestimating or underestimating risk and the Agency's choice to use methods that are not likely to underestimate risk. Another is that use of a bound, as opposed to a central estimate, accounts for the variability (i.e., the sampling error) in the experimental data. In addition, use of the lower bound is consistent with the goal of harmonization with the current practice for assessing effects other than cancer—also based on the lower limit on dose. (USEPA 2003c)

Whatever the model, the estimates of potency for most carcinogens are based primarily on experimental data, although human data have also been used.

In recent guidelines, USEPA (2003c) proposed the continued use of this approach for mutagenic carcinogens or carcinogens usually found in the environment above threshold quantities and as a conservative default. However, when there is sufficient evidence to conclude that the dose-response relationship is not linear at low doses and that the agent does not demonstrate mutagenic or other activity consistent with linearity at low doses, the USEPA recommends applying a nonlinear model. This is a welcome change to USEPA policy, as this approach reflects more closely the current scientific understanding of carcinogenic processes. The new USEPA approach also reduces bias toward excessively conservative estimates for nonmutagenic carcinogens. As a result of the new guidelines, RfCs and RfDs will be developed for some carcinogens, while slope factors and unit risks will continue to be developed for others.

Although USEPA has removed the administrative barrier to the use of nonlinear models, most of the USEPA assessments contained in the IRIS database have assumed linearity for carcinogens. In contrast, WHO (1999) and Health Canada's CEPA reports (Environment Canada and Health Canada 1993; Health Canada 2003a) often apply a nonlinear model in the dose-response assessment of nonmutagenic carcinogens. It may be preferable to use the WHO or CEPA assessment for a carcinogen that is not strongly mutagenic, rather than what is currently available in the IRIS database.

3.2.3   Current Estimated Potency Factors: Other Organizations

Health Canada

For noncancer effects, Health Canada develops doses or concentrations below which no adverse effect is expected. *Tolerable intakes* (TIs) or *tolerable concentrations* (TCs) are similar to RfDs and RfCs and are estimated using critical effect levels divided by default or substance-specific uncertainty factors. In the recent assessments, TIs and TCs have been derived using the benchmark approach. The critical effect level used in the development of TIs (TCs) is the Benchmark Dose$_{05}$ (Benchmark Concentration$_{05}$)[9] or its 95 percent lower confidence limit.

For cancer effects, Health Canada does not apply low-dose extrapolation to the dose-response data beyond the observed dose range. Instead, it develops Tumorigenic Dose$_{05}$ (TD$_{05}$) and Tumorigenic Concentration$_{05}$ (TC$_{05}$) as the measure for dose response (Health Canada 1994, 2003e).[10] These measures are considered to be more accurate than estimates of risks predicted based on extrapolation over many orders of magnitude from experimental studies in animals to the much lower levels of contaminants to which the general population is likely to be exposed, generally in the absence of information on the mode of action. In view of these considerable uncertainties, these risks are not specified in absolute terms of predicted incidence or numbers of excess deaths per unit of the population. Note that Health Canada also provides the 95 percent lower confidence limit on TD$_{05}$ and TC$_{05}$.

TD$_{05}$ and TC$_{05}$ can be converted to slope factor and unit risk as follows:

$$\text{Slope factor} = 0.05/\text{TD}_{05}$$
$$\text{Unit risk} = 0.05/\text{TC}_{05}$$

WHO

WHO develops air guidelines to provide a basis for protecting public health from adverse effects of air pollution and for eliminating, or reducing to a minimum, those air contaminants that are known or likely to be hazardous to human health and well-being. WHO (2000) defines a guideline as "any kind of recommendation or guidance on the protection of human beings or receptors in the environment from adverse effects of air pollutants. As such, a guideline is not restricted to a numerical value but might also be expressed in a different way, for example as exposure–response information or as a unit risk estimate."

A *guideline value* (GV) is a particular form of guideline available for some contaminants only:

> It has a numerical value expressed either as a concentration in ambient air or as a deposition level, which is linked to an averaging time. In

the case of human health, the guideline value provides a concentration below which no adverse effects or (in the case of odorous compounds), no nuisance or indirect health significance are expected, although it does not guarantee the absolute exclusion of effects at concentrations below the given value. (WHO 2000)

Some air guidance values represent total aggregate exposure from all media and not just from air.

WHO develops an air quality GV for *noncarcinogenic* effect in a manner similar to the way USEPA develops an RfC and Health Canada develops a TC. The GV is derived by applying an uncertainty factor to either a NOAEL or a LOAEL that has been identified for a critical effect (WHO 2000). The choice of NOAEL or LOAEL depends on the availability of the data and how representative the data are of the true NOAEL and LOAEL values. The working consensus has been that the level of exposure of concern in terms of human health is more easily related to the LOAEL, and this level therefore has been used whenever possible.

In the case of irritant and sensory effects on humans, it is desirable where possible to determine the no-observed-effect level (WHO 2000). Tolerable intake for ingestion is developed by a similar process.

Where dose-response information allows the identification of a BMD, a tolerable intake or tolerable concentration is developed by applying an uncertainty factor to the BMDL, the 95 percent lower confidence limit of the BMD. There is currently no consensus on the incidence of effect to be used as the basis for the BMD; however, it is generally agreed that the BMD should be comparable with a level of effect typically associated with the NOAEL or LOAEL (WHO 2000). Allen et al. (1992, 1993) have estimated that a BMDL at 5 percent is, on average, comparable to the NOAEL, whereas choosing a BMDL at 10 percent is more representative of a LOAEL.

WHO expresses GVs for most airborne carcinogens in terms of *incremental unit risk estimates*. The incremental unit risk estimate for an air pollutant is defined as "the additional lifetime cancer risk occurring in a hypothetical population in which all individuals are exposed continuously from birth throughout their lifetimes to a concentration of 1 $\mu g/m^3$ of the agent in the air they breathe" (WHO 2000).

For those substances for which appropriate human studies are available, the unit risk is derived using the "average relative risk model" to conduct linear extrapolation of the dose-response curve to zero (WHO 2000). On the other hand, several methods have been used to estimate potency based on animal data. Two general approaches have been proposed: a strictly linearized estimation method used by USEPA and a nonlinear approach. Nonlinear relations have been proposed for use when the data derived from animal studies indicate a nonlinear dose-response relationship or when there is evidence that the capacity to metabolize the polluting chemical to a carcinogenic form is of limited capacity (WHO 2000).

There is no clear consensus on appropriate methodology for the risk assessment of chemicals for which the critical effect may not have a threshold, such as genotoxic carcinogens and germ cell mutagens. Methods are evolving and a number of approaches based largely on characterization of dose-response have been adopted for assessment of such effects (WHO 2000):

- quantitative extrapolation by mathematical modeling of the dose-response curve to estimate the risk at likely human intakes or exposures (low-dose risk extrapolation)
- relative ranking of potencies in the experimental range
- division of effect levels by an uncertainty factor

Where data are available, attempts are made to incorporate the dose delivered to the target tissue into the dose-response analysis (physiologically based pharmacokinetic modeling)[11] (WHO 2000).

European Chemicals Bureau

The ECB reports LOAELs and NOAELs for noncarcinogens. It appears to be noncommittal on its approach to evaluate carcinogenic risk:

> Unless a threshold mechanism of action has been clearly demonstrated, it is considered prudent to assume that a threshold cannot be identified in relation to mutagenicity. As stated by IPCS (1994) "there is no clear consensus on appropriate methodology for the risk assessment of chemicals, for which the critical effect may not have a threshold, such as genotoxic carcinogens and germ cell mutagens." (European Chemicals Bureau 2003c)

3.2.4   Comparison among Jurisdictions

This section compares the dose-response assessment nomenclature and approaches of selected agencies to those of USEPA. USEPA terminology is used because it is generally well known by risk assessors and there is a degree of consistency in the terminology and the units used.

For noncancer effects, most jurisdictions present potencies as doses or concentrations below which no adverse effect is expected. It should be pointed out that although units from other jurisdictions may appear similar (see table 1.5), there are subtle differences. For example, while Health Canada's TC and TI are similar to RfC and RfD, respectively, there are also some obvious and subtle differences.

For non-threshold cancer effects, Health Canada does not apply low-dose extrapolation to the dose response data beyond the observed dose range. Instead, it defines potency in terms of TC_{05} and TD_{05} (Health Canada 1994, 2003e), whereas USEPA develops unit risks and slope factors. The obvious difference is that Health Canada presents estimated potency as the concentration or dose that induces a 5 percent increase in the incidence of, or deaths due to, tumors in a population associated with exposure, while USEPA presents estimated potency as the slope of a dose-response curve. The value of 5 percent is arbitrary.

A more subtle difference between Health Canada and USEPA is the usual (but not completely consistent) reporting by USEPA of the slope factor as a 95 percent upper bound on the calculated slope, while Health Canada reports the slope itself. Health Canada recommends the application of an appropriate factor (without specifying what is appropriate) to lower the concentration or the dose to achieve an acceptable measure of safety. USEPA does not make adjustment to the unit risks or slope factors, because conservatism is already incorporated in the use of the upper bound on the slope (see definitions in ITER 2003) All other things being equal, USEPA's slope factors

Table 1.5 Comparison of terminology and units used by different jurisdictions

	Threshold or nonlinear		Non-threshold or linear	
	Factor	Unit[g]	Factor	Unit[g]
USEPA (IRIS Database)[a]				
Inhalation	RfC	mg/m^3	Unit risk	$(g/m^3)^{-1}$
Ingestion, dermal exposure	RfD	mg/kg/day	Slope factor	$(mg/kg/day)^{-1}$
Drinking water	RfD	mg/kg/day	Unit risk	mg/L
Health Canada (CEPA)[b]				
Inhalation	TC	xg/m^3	TC_{05}	xg/m^3
Ingestion	TI	xg/kg/day	TD_{05}	xg/kg/day
WHO (Air Guidelines)[c]				
Inhalation	GV[e]	yg/m^3	Unit risk	$(g/m^3)^{-1}$
Ingestion	TI[f]			
European Chemicals Bureau[d]				
Inhalation	LOAEL or NOAEL	xg/m^3		
Ingestion	LOAEL or NOAEL	xg/kg/day		

[a] USEPA 2003f
[b] Health Canada 2003e
[c] WHO 1999, 2000
[d] European Chemicals Bureau 2003a, 2003b
[e] GV represents exposure from air only for some pollutants and total exposure from all environmental media for others
[f] TI (tolerable intake) = either TDI (tolerable daily intake) or ADI (acceptable daily intake)
[g] "xg" can be g, ng, or mg; "yg" can be g or mg

and unit risks would lead to more conservative assessments as compared to using slope factors and unit risks derived from Health Canada's TC_{05} and TD_{05} values.

The guideline values reported in the WHO *Air Quality Guidelines* (1999) differ from USEPA values. Some GVs are comparable to USEPA's RfCs, but other GVs represent total aggregate exposure from all media and not just from air.

The European Chemicals Bureau reports LOAELs and NOAELs, while USEPA makes regulatory decisions using RfDs and RfCs, which are made more stringent by applying uncertainty and modifying factors. Since USEPA also provides LOAELs or NOAELs, direct comparison can be made between USEPA and ECB values. The ECB does not use linear models, even for potent, direct-acting, and mutagenic carcinogens such as acrylamide (European Chemicals Bureau 2003b, 2002). Acrylamide is a well-known potent carcinogen, which USEPA has classified as B2 (USEPA 2003f), and is consistently mutagenic in various in vivo and in vitro systems and binds covalently to DNA and protein.

3.2.5 Dermal Potency Estimates

Most agencies have not estimated dermal potency factors. As a result, risk assessors usually need to extrapolate dermal factors from estimated oral potency factors. USEPA has demonstrated that route-to-route extrapolation of exposure limits from oral to inhalation exposure may lead to significant underestimation of the risk (USEPA 1996b). Although no similar evaluation is possible for extrapolation from oral to dermal exposure, it is probable that the estimated potency by oral and dermal route could be quite different in some cases. USEPA (2003c) recommends a case-by-case approach and "when route-to-route extrapolation is used, the assessor's degree of confidence in both the qualitative and quantitative extrapolation is discussed in the assessment and highlighted in the dose-response characterization."

Some assessors estimate dermal potency factors from oral dose-response factors by correcting for incomplete absorption through dermal exposure, following the initial USEPA guidance (1989). This practice implicitly assumes that health effects elicited are related only to the total uptake and not dependent on the route of exposure. This assumption may hold for some chemicals, such as lead and polychlorinated dibenzo-p-dioxins. However, some other substances tend to have different potencies and act at different organ sites depending on the route of exposure. Polycyclic aromatic hydrocarbons (PAHs) are good examples (Ontario Ministry of the Environment 1997b). They are bioactivated significantly in the skin via metabolism following dermal absorption prior to being delivered to the liver, where metabolism of most xenobiotics takes place. Some substances may cause skin irritation, which is a local action—a phenomenon that is not normally taken into consideration in developing an oral exposure limit. These factors make dermal exposure significantly different from oral exposure, and the use of oral dose-response parameters in evaluating health risk from dermal exposure may significantly underestimate the risk.

USEPA (1989, 1992a, 1998b, 2001b) discusses another important issue: RfDs and slope factors are developed and used differently for oral exposure as compared to dermal exposure. Oral dose-response parameters are usually developed from epidemiological or experimental studies in terms of the *administered* dose rather than the *absorbed* dose. The administered dose represents the total intake of the test substance, uncorrected for absorption. In contrast, the absorbed dose is the administered dose that has been corrected for uptake. The vehicle by which the test substance is delivered in the experimental studies is usually selected to minimally impede uptake. When the oral RfD or slope factor is applied to an environmental situation, the actual absorption may be less complete than in the experimental conditions, making the application of the uncorrected slope factor conservative. For example, absorption from a soil matrix may be less than from a solution matrix, which is commonly used to administer test substance in the laboratory.

While the estimated oral potency factors are usually reported in terms of intake, dermal exposure is generally presented in terms of the absorbed dose (uptake). Therefore, to extrapolate a dermal potency factor from an estimated oral potency factor requires converting the estimated oral potency expressed as intake into an expression of uptake. Most risk assessors simply assume a 100 percent oral uptake into the body tissues. In effect, the administered oral dose is assumed to be equal to the absorbed oral dose. These practices lead to an underestimation of the dermal uptake due to overcorrection for oral bioavailability; the magnitude of the underestimation is inversely proportional to the true oral absorption of the chemical in question. USEPA (1989, 1992a) therefore recommends correcting the estimated oral potency factor for bioavailability before extrapolation to dermal toxicity values, if defensible data allow for such a correction.

An oral RfD expressed as uptake ($RfD_{oral\ uptake}$) can be obtained by applying an oral bioavailability factor (BVF_{oral}) to the oral RfD expressed in terms of intake ($RfD_{oral\ intake}$). That is,

$$RfD_{oral\ uptake} = RfD_{oral\ intake} \times BVF_{oral}$$

Assuming that the same internal (or absorbed) dose obtained either through oral or dermal exposure would produce identical levels of the same systemic effect (i.e., $RfD_{dermal\ uptake} = RfD_{oral\ uptake}$), one can derive the dermal RfD as an uptake ($RfD_{dermal\ uptake}$):

$$RfD_{dermal\ uptake} = RfD_{oral\ intake} \times BVF_{oral}$$

Similarly, a dermal slope factor can be derived from an oral slope factor as shown:

$$Slope_{dermal\ uptake} = Slope_{oral\ intake} / BVF_{oral}$$

If correction for bioavailability is impractical, USEPA (1992a) recommends conducting route-to-route extrapolation from oral to dermal toxicity values only when *accompanied with a strong statement emphasizing the uncertainty involved.*

3.2.6 Complex Mixtures

Complex mixtures tend to be mixtures of chemicals generated simultaneously by a single source or process (e.g., coke oven emissions, diesel exhaust) or mixtures of chemically related compounds (e.g., PAHs, polychlorinated biphenyls [PCBs]). Assessment of these mixtures poses special challenges because the mixture can contain a large number of contaminants with different properties and the toxicity of many individual components may not be known. To monitor for all components of the mixture is also often not practical.

In principle, there are several approaches commonly used to assess the potency of a complex mixture. They could be classified into approaches that attempt to assess the complex mixture as a whole and approaches that attempt to integrate individual components' contributions to the toxicity of the mixture.

Assessing Toxicity of Mixtures as a Whole

It is possible to evaluate the risk of exposure to the whole mixture if the potency of the mixture or a sufficiently similar mixture has been previously estimated. For example, the potency of coke oven emissions as a PAH mixture has been estimated from epidemiological studies involving human exposure to coke oven emissions (USEPA 1984). USEPA (2000h) refers to this approach as the *whole-mixture dose-response assessment.*

While it is attractive to assume that the toxicity of coke oven emissions is the same regardless of the fuel and the combustion process that generate the emissions, and regardless of the type and effectiveness of pollution-reduction measures that are in place, this assumption may not always hold. Furthermore, the composition of a mixture can change over time in the environment as a result of transport and transformations. Also, complex mixtures from different sources, even if they are made up of similar chemical substances, likely differ in potency (e.g., coke oven emissions versus diesel engine exhaust; see Albert et al. 1983). This reality seriously limits the application of this approach, because a number of different sources can simultaneously contribute to the complex mixtures in the environment.

An alternative approach is the *comparative potency method* (see Albert et al. 1983; USEPA 2000h). This method is based on the assumption that the relative potencies of two similar complex mixtures are approximately the same in humans and in animal assays, as represented by the following equation:

$$\frac{\text{Potency}_{\text{Human Mixture1}}}{\text{Potency}_{\text{Experimental Mixture1}}} \cong \frac{\text{Potency}_{\text{Human Mixture2}}}{\text{Potency}_{\text{Experimental Mixture2}}}$$

The potency of an unknown mixture in humans can then be estimated by solving the following equation:

$$\text{Potency}_{\text{Human Mixture1}} \cong \frac{\text{Potency}_{\text{Human Mixture2}}}{\text{Potency}_{\text{Experimental Mixture2}}} \times \text{Potency}_{\text{Experimental Mixture1}}$$

The comparative potency method is applicable to more situations than the whole-mixture dose-response assessment, but it is solely dependent on the initial assumption of animal–human comparability being valid.

Another approach is the *surrogate approach*. This approach is usually applied not to the entire mixture but to a subset of contaminants in the mixture. For example, benzo[a]pyrene (B[a]P), a highly carcinogenic PAH, is often used as the surrogate for the PAH fraction of the complex mixture. The model assumes that the levels of chemicals in the mixture fraction vary proportionately with the level of surrogate in the fraction. In the case of the PAH fraction, the level of B[a]P is assumed to vary in proportion to the levels of carcinogenic PAHs in the PAH fraction.

In this model, the surrogate compound assumes the potency of the entire fraction of the complex mixture (e.g., B[a]P assumes the potency of all carcinogenic PAHs in the fraction, but not of any carcinogenic metals that may also be present in the mixture). Therefore, when the surrogate model is used, the actual potency of the surrogate is not important and the surrogate may not even be toxic itself. The surrogate substance is used solely as a measure of the concentration of the relevant fraction of the mixture. WHO (1999), the European Commission (2001), Great Britain (Expert Panel on Air Quality Standards 1999), the Netherlands (NIPHEP 1989), and Sweden (Boström et al. 2002) have adopted the surrogate approach to conduct risk assessment of PAHs.

This approach, like all other approaches that try to estimate the potency of the mixture as a whole, is critically dependent on the assumption that the potency of the complex mixture is proportional to the concentration of the surrogate (indicator) substance. To apply this approach, one has to determine the exposure to the surrogate substance that has been chosen and assign the toxicity of the mixture to this surrogate substance.

Assessing Toxicity of Mixtures by Integrating Toxicity of Individual Components

The simplest approach to assessing the potency of a complex mixture is to add up the risks attributable to individual components of the mixture as one would add the cancer risks from exposure to arsenic and to formaldehyde. For threshold effects, USEPA (2000h) recommends the use of a *hazard index* (see USEPA 2000h for an extensive discussion). In its most straightforward form, the hazard index for a mixture can be calculated by summing up the *hazard quotients* for individual components of the mixture. The hazard quotient for a single substance is E/AL, where E represents the exposure level and AL represents the acceptable exposure level for that substance. The hazard index (HI) for the mixture is therefore given by

$$HI = \sum_{i=1}^{n} \frac{E_i}{AL_i},$$

as the summation of hazard quotients (E/AL) for n chemicals in the mixture. In practice, RfD or RfC usually replaces the more generic AL.

USEPA also provides more elaborate formulas, which take into account interaction among different components in a mixture.

The hazard index approach can be applied to a mixture for which there are enough data to develop dose-response relationships for individual components in the mixture. This requirement is often not met, however, and instead the *relative potency approach* is often used. In the relative potency approach, the potencies of the components of the mixture are expressed as *toxic equivalencies* of an index substance selected because the dose-response database is strong for that substance. These unitless values are most often called *toxic equivalence factors* (TEFs).

In order to conduct the assessment, the exposure levels of the substances have to be converted to toxic equivalents of the index substance. For example, the concentrations of individual PAHs would have to be converted into B[a]P equivalents. The risk is then calculated by treating exposure to the equivalents as if they were actually the index substance. USEPA (2000h) has reviewed how the relative potency approach has been applied to polychlorinated dibenzo-p-dioxins and dibenzofurans, coplanar PCBs, and PAHs.

Whole-Mixture versus Component-by-Component Approaches

The approaches that focus on individual components of a mixture have two major disadvantages. First, it is difficult to assess any interaction that might be taking place among components in a complex mixture. At least for PAH mixtures, though, the weight of evidence suggests that the interactions may not play a role that would impact significantly on the outcome of the assessment (Ontario Ministry of the Environment 1997b). These approaches work well for mixtures of polychlorinated dibenzo-p-dioxins, polychlorinated dibenzofurans, and coplanar PCBs because the potency has been estimated for all congeners. For PAH complex mixtures, on the other hand, the potencies of most of the components are unknown. Experimental evidence suggests that in the case of PAHs, the assessment method that focuses on individual components underestimates the risk of the whole complex mixture (Ontario Ministry of the Environment 1997b). The underestimation may be because integrating risks from individual PAHs typically involves less than 20 compounds, while the complex mixture contains hundreds of PAHs and the toxicity of most of them is unknown.

When additivity of risk is assumed, the risk assessment is typically driven by one or two components of the mixture. For example, consider a mixture that has 10 distinct components. The difference between the risk from exposure to the most potent (single) contaminant and the sum of the risk from all 10 components combined is probably less than an order of magnitude, except in the unlikely event that the risks from all 10 individual components are exactly identical. Based on the experience of the authors, adding the risks from approximately 10 contaminants usually does not lead to more than a two- to threefold increase over the risk estimated from exposure to the one component that contributes the most risk.

One important advantage of approaches that focus on individual components is knowing the exact environmental levels of all the contaminants being assessed. In contrast, the environmental levels and toxicity of individual components and their relative contribution to the overall risk of the mixture may not be known when the mixture is assessed as a whole.

An advantage of whole-mixture approaches is that, as humans are usually exposed to the whole complex mixtures rather than individual components alone, assessing the potency of a complex mixture as a whole can better utilize human epidemiological data. Furthermore, any interaction among individual components is implicitly taken into consideration. Perhaps more importantly, all the contaminants in the mixture are being included in the assessment and not just the subset with known toxicity and environmental levels.

The major challenge and limitation of using the whole-mixture dose-response assessment and comparative potency methods is knowing when the mixture of interest is sufficiently similar to the standard reference mixture. Furthermore, the number of mixtures that are sufficiently similar to these standard reference mixtures could be small.

The surrogate model is more promising. It is dependent on one major assumption: that the environmental levels of individual components of an identifiable fraction of the mixture vary proportionately with the level of the surrogate in the fraction.

The surrogate approach and the relative potency approach are the two approaches that are most likely applicable to the majority of situations. In general, USEPA (2000h) recommends using approaches that involve the whole mixture and using the component-by-component approach only when the data are insufficient to assess the mixture as a whole.

3.2.7 New Trends

Dose-Response Factors for Children

Assessment of whether children are more vulnerable than adults to environmental contaminants is one of the new developments in dose-response assessment. The recent publication *Supplemental Guidance for Assessing Cancer Susceptibility from Early-Life Exposure to Carcinogens* (USEPA 2003l) represents USEPA's effort in this area. In this study, USEPA has separately examined mutagenic carcinogens (with assumed low-dose linear dose-response curves) and nonmutagenic carcinogens (with assumed low-dose nonlinear dose-response curves). For mutagenic carcinogens, USEPA recommends that:

- for infants up to 2 years old, the estimated carcinogenic potency derived for adults be increased 10 times

- for children aged 2 to 15 years, the estimated potency be increased 3 times

- individuals above the age of 15 be treated as adults

For carcinogens with a mode of action other than mutagenicity, USEPA recommends no default value, but rather an adjustment of the risk "if children may be susceptible to this mode of action" (USEPA 2003l).

USEPA recognizes that when exposure is fairly uniform over a lifetime, these adjustments will have relatively small impact on the estimated lifetime cancer risk, given the overall uncertainty of the assessment. In fact, if exposure is assumed to be constant over a lifetime, the adjustment for early-life exposures results in only a 40 percent increase in the risk estimate. Given that the uncertainty associated with potency estimates may often be about 10-fold, this 40 percent increase does not represent a large increase in conservatism. More importantly, USEPA reminds the readers that this guidance does not address childhood cancer. At present, there are not enough data to develop suitable guidance for this condition.

While USEPA has not yet published guidance on estimating potency for children for endpoints other than carcinogens, the Food Quality Protection Act of 1996 directs USEPA to use an additional 10-fold margin of safety in setting pesticide tolerances for the protection of infants and children. This FQPA safety factor accounts for the potential for pre- and postnatal toxicity and the completeness of the toxicology and exposure databases. The development of FQPA safety factors has been discussed in greater detail in section 3.2.2.

Change in Carcinogen Risk Assessment Approaches

USEPA has revised its risk assessment approach for carcinogens significantly. The process has been outlined in detail in section 3.2.2. Here, we summarize all the changes and discuss their implications.

In its recent guidelines, USEPA (2003c) has replaced the linearized multistage model with a two-step process for developing potencies for carcinogens. The first step involves curve-fitting to the data within the dose range for which data are available. The functions can be either biologically or nonbiologically based functions that fit the data well. Unlike previously, the model function is applied only up to an appropriate *point of departure* (POD) and not to zero. The POD "marks the beginning of extrapolation to lower doses. The POD is an estimated dose (expressed in human-equivalent terms) near the lower end of the observed range, without significant extrapolation to lower dose" (USEPA 2003c). Both the dose corresponding to the central estimate of POD and the lower bound on the dose representing the POD are determined.

The second step involves extrapolation below the range of doses used to establish the POD. This extrapolation may involve the extension of a biologically based model (from the first step) if there are sufficient data to support the use of this model. Other default approaches can be used if the use of a biologically based model is not appropriate, provided that these models are consistent with current understanding of the mode(s) of action of the chemical. These models may assume a linear or nonlinear dose-response relationship, or both.

When there is sufficient evidence for existence of a threshold for carcinogenic effect, the default approach to derive RfDs is to treat the lower bound on the POD in a manner analogous to BMDL. The remaining chemicals are assumed not to have thresholds and extrapolation models are selected accordingly. A biological model that is consistent with what is known about the contaminant's mechanism of action will be given preference. In the absence of such a biologically based model, a default model that represents a straight line between zero and the response corresponding to the lower bound of the POD can be used. Unit risks or slope factors are derived from this straight-line function in the same manner as from the linearized multistage models.

This new approach is not expected to change the level of conservatism for carcinogens that are assumed to have a linear dose-response relationship; however, the level of conservatism is reduced for carcinogens that are assumed to pose no cancer risk below a certain threshold. USEPA's revised carcinogen risk assessment guidelines are still at the "draft final" stage and it may be a while before a significant number of carcinogens will be assessed using this method and posted on the IRIS database (USEPA 2003c).

Assessment of Toxicity from Exposure Duration Shorter Than a Lifetime

At present, the majority of the estimated potency factors that have been developed are for chronic effects. This reality makes the assessment of risk from short-term exposure difficult. A practice that is common (but of uncertain validity) is to apply RfDs and RfCs, which are meant for continuous exposure over a lifetime, to a shorter exposure duration (see section 3.2.2). Various agencies have developed less-than-lifetime factors or methodologies to handle such scenarios. Some of these approaches have been reviewed (see, for example, USEPA 2002a).

The *minimal risk levels* (MRLs) published by ATSDR are the best-known estimated potency factors for exposure periods shorter than a lifetime. MRLs are developed from NOAELs or LOAELs with the application of uncertainty factors in a manner similar to the derivation of RfDs and RfCs. MRLs have been developed for oral and inhalation routes of exposure for three durations: acute (less than 14 days), intermediate (15–364 days), and chronic (365 days or more). The definition of acute, intermediate, and chronic exposure durations are the same for all species regardless of their lifespans, which of course vary from species to species. It is not certain if such an assumption has a sound biological basis. Thus, while MRLs could be an important resource for a risk assessor, ATSDR is quite explicit about the limitations regarding their applications: "It is important to note that MRLs are not intended to define clean-up or action levels for ATSDR or other Agencies" (ATSDR 2003c).

CalEPA (2000a) has developed a large number of acute *reference exposure levels* (RELs) for inhalation exposure from air. CalEPA describes RELs as follows:

> The concentration level at or below which no adverse health effects are anticipated for a specified exposure duration is termed the reference exposure level (REL). RELs are based on the most sensitive, relevant, adverse health effect reported in the medical and toxicological literature. RELs are designed to protect the most sensitive individuals in the population by the inclusion of margins of safety. Since margins of safety are incorporated to address data gaps and uncertainties, exceeding the REL does not automatically indicate an adverse health impact." (CalEPA 1999)

RELs are mostly developed for one-hour exposure duration (one-hour averaging time). However, RELs that are developed for reproductive and developmental effects are for four- to seven-hour exposure durations.

The National Research Council (2001a, 2002, 2003) has developed *acute exposure guidelines* (AEGLs) for some contaminants. However, these contaminants are not typically found in most situations where a human health risk assessment is called for.

USEPA has several programs that have been developing methodologies to estimate potency factors for exposure durations that are less than a lifetime. The Office of Air and Radiation manages USEPA's Acute Reference Exposure (ARE) program; the methodology is currently under review and values are not yet available. USEPA's Office of Pesticide Programs (USEPA 2000b) develops *acute reference dose* (aRfD) values using

NOAELs as the starting point and applying an uncertainty factor, which usually consists of an interspecies factor (10×) and an intraspecies factor (10×). The RfD is defined as the amount of toxicant (in mg/kg-day) a person can be safely exposed to for one day. Some of the *Specific Chemical Fact Sheets* (USEPA 2003h) contain aRfD values.

USEPA's Office of Water Programs publishes *Drinking Water Standards and Health Advisories* (USEPA 2002c), which include not only Lifetime Health Advisories, but also One-Day Health Advisories and Ten-Day Health Advisories. USEPA (2002c) defines a One-Day Health Advisory (HA) as "the concentration of a chemical in drinking water that is not expected to cause any adverse noncarcinogenic effects for up to one day of exposure. The One-Day HA is normally designed to protect a 10-kg child consuming 1 liter of water per day." A Ten-Day Health Advisory is similarly defined as "the concentration of a chemical in drinking water that is not expected to cause any adverse noncarcinogenic effects for up to ten days of exposure. The Ten-Day HA is also normally designed to protect a 10-kg child consuming 1 liter of water per day." Further details about the development of One-Day and Ten-Day Health Advisories are available in USEPA 2002c.

The authors propose that USEPA develop less-than-lifetime factors, which could be applied agency-wide and placed in the IRIS database. The advisories is not a guideline, but a review of existing situations and a tool for planning and scoping of further USEPA activities. Thus no immediate results in the form of estimated potency factors should be expected. In fact, the plans for an agency-wide program may affect existing programs in the different USEPA offices that are engaged in estimation of potency factors.

Physiologically Based Pharmacokinetic Modeling

Some of the major challenges in risk assessment are the extrapolation of observed dose-response relationships from those of animals to humans, from high to low exposure levels, and from one route of exposure to another. The most powerful and preferred tool for species extrapolation today is physiologically based pharmacokinetic (PBPK) modeling—whenever the needed pharmacokinetic information is available. (PBPK modeling and its application in chemical risk assessment were recently reviewed in Andersen 2003.)

Extrapolation using PBPK modeling is based on the assumption that the same *concentration of a toxicant* at the target site leads to the same level of toxic effect, regardless of the species or route of exposure. Some examples of a target site are DNA at the molecular level (formation of DNA adduct) in the case of cancer, cells in the case of cell death, or organs in the case of organ hypertrophy. Therefore, when extrapolating from a mouse to a human, one assumes that the level of response in the two species is the same as long as the concentration of the toxicant is the same at the target site. If the toxicant causes liver toxicity both by dermal and oral exposure in the same species, the assumption is that they would produce the same level of response, provided that the concentration in the liver is the same.

PBPK modeling, using a realistic description of mammalian physiology and biochemistry, attempts to describe the relationship between external measures of applied dose and internal measures of biologically active concentration (Clewell 1995). PBPK modeling integrates information on uptake, distribution, metabolic activation and deactivation, excretion from the body, and their kinetics to estimate the level of the biologically active agent, be it the parent substance or a metabolite in the target organ. For example, PBPK models allow the estimation of the concentration of the biologically active agent at the target site from an oral dose of the parent substance and then use this concentration to estimate the dermal dose of the parent substance. Any biological processes that affect the disposition and, therefore, the toxicity of the contaminant can be added to the model, and the model supports cross-species extrapolation (Clewell 1995).

The model is empirically built, that is, it is based on empirical data on pharmacokinetics. Factors that show species differences and are important for the expression of the toxic effect (pharmacodynamics) may be included into the model, including information on all important active metabolites, formation of DNA adducts, and efficiency of DNA repair. A lot of emphasis is placed on validation of the model against experimental data.

Physiologically based pharmacokinetic modeling provides important capabilities for improving the reliability of the extrapolations—not just across species but also across dose, exposure route, and exposure duration—that are generally required in chemical risk assessment regardless of the toxic endpoint being considered (Andersen et al. 1987; Clewell, Andersen, and Barton 2002). It has the capability of dealing with variation within the human population (Clewell 1995). PBPK modeling can also identify physiological and biochemical parameters that determine individual risk. The application of PBPK modeling to risk assessment of chemical mixtures is being actively explored (Yang et al. 1995; Liao et al. 2002).

Recently, there has been an increasing focus on harmonization of the cancer and noncancer risk assessment approaches used by regulatory agencies. Although the specific details of applying pharmacokinetic modeling within these two paradigms may differ, it is possible to identify important elements common to both. A framework has been developed to incorporate pharmacokinetic modeling to estimate tissue dosimetry into chemical risk assessment, whether for cancer or noncancer endpoints (Clewell, Andersen, and Barton 2002).

The key challenge for PBPK modeling is obtaining the necessary experimental data to evaluate model adequacy as to whether predictions are accurate (Clewell 1995). The limitations in the use of PBPK modeling in extrapolation across species, dose, exposure route, and exposure duration are well recognized. PBPK modeling can account for differences in pharmacokinetics but not differences in pharmacodynamics. Scientists are working on developing physiologically based pharmacodynamic (PBPD) models to model toxicity and the mode of action, and biologically based dose-response (BBDR) models (Setzer and De Woskin 2000). The integration of PBPK and PBPD modeling and the application of PBPK/PBPD modeling to risk assessment of chemicals are being actively pursued (Andersen 1995; Daston and Corley 2002).

3.2.8 *De Minimis* Risk

It is theoretically possible to reduce the environmental levels of threshold contaminants to below their toxicity thresholds. For contaminants that are believed not to have a safe level below which no toxic effect is expected, complete removal from the environment is usually impossible and often impractical. With non-threshold contaminants, health risk increases in direct proportion to the increase in exposure. Since there is no "safe" level, a specific level is often selected from the continuum of risk levels to represent an *acceptable* or *tolerable* risk level.

The decision on a "tolerable" risk level varies from chemical to chemical and from organization to organization depending on the circumstances and is often controversial. The opinion of what is or is not an acceptable risk is not really expert judgment, and different people may have a different perspective on what risk is tolerable. However, past U.S. regulatory records (Travis and Hattemer-Frey 1988) and present-day convention consider risks of less than one in a million (10^{-6}) negligible and no action is usually taken. Some regulatory action is usually taken if the risks from environmental exposures exceed one in 10 thousand (10^{-4}). When the risk is higher than 10^{-6} and lower than 10^{-4}, regulatory decisions to intervene are usually made on a case-by-case basis. This convention is described in the Canadian federal government guidance report published in support of the Canadian Environmental Protection Act:

> For assessment of "toxic" under paragraph 11(c) of CEPA for non-threshold toxicants, it is considered inappropriate to specify a concentration or dose associated with a negligible or de minimis level of risk (such as a lifetime cancer risk of 1 in 1 million) by low-dose extrapolation procedures, primarily since this would involve inclusion of considerations other than those based on science at this stage (i.e., making a societal judgement about what level constitutes de minimis risk). There is no single "correct" value which adequately characterizes de minimis risk associated with a concentration or dose below which risks are acceptable and above which they are not; rather, the risk at low doses or concentrations is assumed to be a continuum, with reduction of exposure leading to an incremental reduction of risk and increases in exposure leading to incremental increases in risk. In addition, in view of the considerable uncertainties of current low-dose extrapolation procedures, it is also considered inappropriate to specify risks in terms of predicted incidence or numbers of excess deaths per unit of the population.
>
> However, it is recognized that the incremental risks associated with exposure to low levels of such substances, although difficult to characterize, may be sufficiently small so as to be essentially negligible compared with other risks encountered in society and that on this basis, control action to reduce exposure may not be justified. Decisions concerning the need for, and development of, control strategies may be made only following a judicious balancing of the estimated risks against the associated costs and feasibility of controls, and/or benefits to society (i.e., in stages subsequent to assessment of "toxic" under the Act such as strategic options analysis).
>
> Obviating the establishment of a single de minimis risk level enables the assessment of "toxic" for "non-threshold toxicants" to be based to the extent possible on scientific considerations. This approach is also consistent with the objective that exposure to "nonthreshold toxicants" should be reduced to the extent possible. (Health Canada 1994)

The Canadian Federal Government does not specify a *tolerable* or *acceptable* risk level for environmental exposure. USEPA similarly does not specify an acceptable risk level:

> EPA uses the general 10^{-4} to 10^{-6} risk range as a "target range" within which the Agency strives to manage risks as part of a Superfund

cleanup. Once a decision has been made to make an action, the Agency has expressed a reference for cleanups achieving the more protective end of the range (i.e., 10^{-6}), although waste management strategies achieving reductions in site risks anywhere within the risk range may be deemed acceptable by the EPA risk manager. Furthermore, the upper boundary of the risk range is not a discrete line at 1×10^{-4}, although EPA generally uses 1×10^{-4} in making risk management decisions. (USEPA 1991b)

3.2.9 Application of Published Factors

Some agencies mandate the use of their own estimated potency factors. Even when this is not the case, many assessors choose to adopt values from a single convenient source, often without examining how the factors were derived or how applicable they are to the situation at hand and without comparing these values to those of other agencies. Risk assessors often justify this approach as being "conservative" or by saying that the organization that derived the factors is reputable. They often do not analyze how estimated potency factors differ among jurisdictions and do not provide a rationale for choosing a particular factor for their assessments. Sometimes estimated potency factors for the same substance can differ by more than one order of magnitude among organizations; as a result, the outcome of the assessment may well depend on which factor has been selected.

Although most estimated potency factors are designed to be conservative, they are not all equally appropriate for a particular application. Furthermore, even reputable organizations may have their estimated potency factors distorted by overarching organizational policies. For example, USEPA's traditional policy of assuming a non-threshold dose-response relationship for all carcinogens, regardless of their mechanism of action, means that assessments using USEPA's estimated potency factors tend to yield more conservative risk estimates than if WHO's or Health Canada's factors were used.

It is the authors' opinion that estimated potency factor selection is one of the key decisions that can significantly affect the outcome of a risk assessment. For this reason, risk assessors ought to put greater effort in selecting estimated potency factors. This issue is discussed in more detail in the uncertainty assessment section (see section 3.4).

Risk assessors are recommended to consider the following:

- It is important to distinguish between estimated potency factors (such as RfC and RfD) and health-based exposure limits (such as air standards). Development of environmental standards and other exposure limits usually involves considering a wide range of issues including health. In contrast, an RfD or RfC is a health benchmark; its derivation does not take into consideration nonhealth factors. Standards and other similar regulatory limits are therefore unsuitable for use in risk assessment.

- Risk assessors are advised to consider toxicity factors listed in only databases published by the agencies that developed the factors. Some secondary databases may not be updated regularly or may inadvertently alter the estimated potency factors during transcription. For example, USEPA Region 3, in its memo to Risk-Based Concentration

(RBC) users, warns: "As usual, updated toxicity factors have been used wherever available. However, because IRIS and provisional values are updated more frequently than the RBC Table, RBC Table users are ultimately responsible for obtaining the most up-to-date values. The RBC Table is provided as a convenience, but toxicity factors are compiled from the original sources and it is those original sources that should serve as the definitive reference" (USEPA 2003m).

• It is advisable to consider only those estimated potency factors that have undergone a thorough peer review and for which the derivation method is available and well documented.

• It is important to respect the intended application of the values. For example, ATSDR's MRLs are not intended to define cleanup or action levels.

• Risk assessors are advised to compare estimated potency values from different selected sources. After taking into account how these factors are derived (see sections 3.2.3 and 3.2.4, which compare estimated potency factors from different jurisdictions), risk assessors need to verify whether they would lead to reasonably similar risk estimates.

"Within about threefold" is suggested as a criterion for judging whether two values are "similar," as within threefold is about half an order of magnitude on the logarithmic scale. If estimates differ by more than threefold, it is important to understand why different jurisdictions arrive at different values. Greater effort ought to be applied in situations where the difference is bigger. A strong rationale must be developed for choosing one value over another value, and a discussion of this issue is a significant part of the dose-response assessment. In some instances, the risk assessment is conducted using estimated potency factors from more than one jurisdiction; the implication of using this approach is discussed as part of the uncertainty assessment and/or risk characterization.

If toxicity values among different jurisdictions do not differ significantly, the discussion and the rationale for using a specific value need not be developed to the same degree.

3.2.10 Key Sources of Information

Table 1.3 lists key sources of information for hazard identification. These sources are similar to those for dose-response assessment.

ATSDR's *Toxicological Profiles* provide comprehensive qualitative and quantitative summaries of dose-response information, including estimated potency factors for contaminants (Agency for Toxic Substances and Disease Registry 2003a); ATSDR's *Tox FAQs* (2003b) have only brief and general qualitative summaries. USEPA's *Chemical Summaries* (2003b) also provide detailed summaries of dose-response information. Both IPCS's *Environmental Health Criteria Monographs* (2003b) and *Concise International Chemical Assessment Documents (CICADs)* (2003a) provide comprehensive dose-response information, on par with the ATSDR's *Toxicological Profiles*.

The European Chemicals Bureau's *European Union Risk Assessment Reports* (2003a) and Health Canada's *Priority Substances List Assessment Reports* (2003a) all contain pertinent dose-response information. Health Canada guidance values are also published in *Health-Based Tolerable Daily Intakes/Concentrations and Tumorigenic Doses/Concentrations for Priority Substances* (Health Canada 2003e) and in *Environmental Carcinogenesis and Ecotoxicology Reviews* (*Journal of Environmental Science and Health* C12, no. 2 [1994], and C19, no. 1 [2001]). The reports discussed so far include different types of dose-response studies (i.e., inhalation, oral, dermal) of various durations (i.e., acute, subchronic, and chronic) in both humans and animals.

USEPA's well-organized Integrated Risk Information System website (2003f) is a comprehensive database of estimated potency factors (i.e., RfCs, RfDs) and contains detailed summaries of relevant dose-response studies for many contaminants USEPA has evaluated. The Risk Assessment Information System website (2003) also maintains a useful database of estimated potency factors that references various sources, including IRIS.

The Hazardous Substances Data Bank (2003), available via TOXNET, provides referenced summaries of human and animal dose-response information. The TSCATS database (Syracuse Research Corporation 2003c) contains abstracts and summaries of unpublished technical reports with dose-response data.

With regard to pesticide, both IPCS's *Pesticide Data Sheets (PDSs)* (2003d) and *Joint Meeting on Pesticide Residues (JMPR)* reports (2003c) contain summaries of dose-response studies. The *JMPR* is relatively more comprehensive in its discussion of dose-response data than are the *PDSs*. USEPA's reregistration eligibility decision (RED) reports (2003i) have detailed discussions on dose-response assessments for selected pesticides. The accompanying fact sheets provide only brief summaries.

Health Canada's *Proposed Regulatory Decision Documents (PRDD Series)* (2003c) and *Regulatory Notes (REG Series)* (2003d) also contain detailed discussions of dose-response data. Part A of the *Risk Assessment Guidance for Superfund (RAGS)* (USEPA 1989) provides a general outline of the toxicity assessment process.

3.3   Exposure Assessment

3.3.1   Description

Exposure assessment involves estimating the magnitude of exposure of subpopulations that come into contact with contaminants present in air, soil, water, sediment, food and beverages, consumer products, and other sources, usually via inhalation, ingestion, or dermal exposure.

Exposure assessment is rife with specialized terminology. In order to understand exposure assessments, it is necessary to understand some terms. For example, a *receptor* is a subpopulation sharing certain biological and lifestyle factors that affect the magnitude of exposure. Some examples of a receptor are "resident" of a contaminated site, "construction worker" working on a contaminated site, and "carpenter" exposed occupationally. *Biological factors* are those physiological factors that influence one's exposure (e.g., respiration rate, body weight). *Lifestyle factors* reflect the behavior of the subpopulation (e.g., time spent outdoors, consumption rate of different food groups). *Exposure scenario* is a collection of pathways by which a receptor can be exposed. For example, a resident on a contaminated site may be exposed through inhalation of contaminated dust and volatile contaminants or through ingestion of contaminated soil and produce grown in the contaminated soil. Exposure assessment typically assesses the magnitude of exposure of selected receptors for several selected exposure scenarios, assuming specific biological lifestyle factors for each receptor.

Despite apparent complexity, exposure calculation is relatively simple. A receptor's exposure to a contaminant in an environmental medium (e.g., soil, air, drinking water, food) via a particular pathway can be calculated as the product of the concentration of the contaminant in the medium and the amount of medium the receptor inhales, ingests, or is exposed to dermally. In cases of dermal and oral exposure, exposure is normalized in terms of body weight.

Estimated potency factors are developed assuming continuous exposure to a fixed contaminant concentration; in practice, humans are often exposed to different levels of contamination during the course of a day (as they move from work or school to home and from indoor or outdoor microenvironments), a week (working day vs. weekend), and different seasons or life stages (infant, child, adolescent, adult, retired). A key aspect of exposure assessment therefore involves determination of the average exposure over a lifetime (or over a shorter *averaging time*) by summing all exposures within the lifetime (averaging time frame) and dividing this total exposure by the lifespan (averaging time). The following section provides the basic set of equations that can be used to calculate exposure.

3.3.2   Examples of Exposure Equations

This section presents examples of equations commonly used in exposure assessment to calculate inhalation, oral, and dermal exposure from contaminated soils. The equations are applied to the resident receptor, which represents a subpopulation that may spend most of its life from infancy to adulthood on a specific contaminated site. The equations are modified from the equations contained in the Superfund document (USEPA 1989). There are many similar equations published in several guidance documents, including those listed in section 1.5.

Inhalation

For inhalation, exposure is expressed as the average concentration to which a receptor is exposed over the duration of the averaging time. This can be calculated using the following formula:

$$CA_{avg} = \sum_{1}^{j}\sum_{1}^{i} \frac{CA_j \times DEF \times IhR_{i,j} \times ET_{i,j} \times EF_{i,j} \times ED_{i,j}}{AR \times AT}$$

where:

CA_{avg} = average airborne concentration of a contaminant a receptor is exposed to over the averaging time ($\mu g/m^3$)

i = the number of life stages within the averaging period

j = the number of microenvironments (such as residential indoor air) to which the receptor is exposed

CA_j = the airborne concentration in microenvironment j ($\mu g/m^3$)

DEF = the (e.g., soil particle) matrix desorption factor[12]

$IhR_{i,j}$ = the inhalation rate at life stage i in microenvironment j (m^3/hr)

$ET_{i,j}$ = the daily exposure time during life stage i in microenvironment j (hrs/day)

$EF_{i,j}$ = the exposure frequency during life stage i in microenvironment j (days/year)

$ED_{i,j}$ = the exposure duration during life stage i in microenvironment j (years)

AR = the average respiration rate over the AT period (m³/hr)

AT = the averaging time (hours)

Some agencies, including USEPA (1989), calculate intake values instead of average inhalation rates and therefore must convert unit risk values expressed in terms of airborne concentrations into intake values. Such conversions should be avoided because expressing intake in terms of body weight implies that intake and the resulting adverse effect are proportional to body weight, which may not be the case.

Adjustment by body weight may be reasonable if the contaminant is readily absorbed from the lungs and the effects are systemic. This may be the case for some contaminants, such as dioxins, but may not apply for a number of metals, including chromium (VI), nickel, and cadmium, which when inhaled tend to have direct effect on the lung tissues. The same is true for polycyclic aromatic hydrocarbons (PAHs), which, when administered by inhalation, have an adverse effect on the lungs at exposure levels that do not result in tumors outside of the respiratory tract (for review, see Ontario Ministry of the Environment 1997b).

Inhalation rate itself is also not necessarily proportional to the body weight, as is clearly shown when plotting the ratio of recommended default inhalation rates and body weights for different age groups (see figure 1.1). Aerodynamics also change with body size and particulate is more likely to be deposited higher in the respiratory tract in individuals with smaller body size; this effect may not be dependent on body weight.

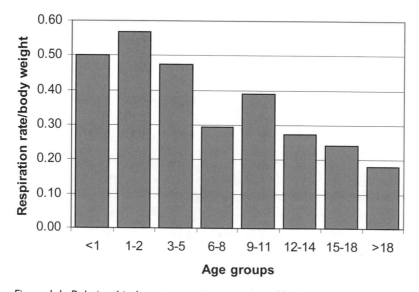

Figure 1.1 Relationship between respiration rate and body weight for different age groups
Note: Respiration rates and body weights are those recommended by USEPA based on results of population surveys (1997a).

Oral

Unlike inhalation, oral exposure is generally expressed as intake value:

$$I_{oral} = \sum_{1}^{i} \frac{CS \times DEF \times CF \times IgR_i \times EF_i \times ED_i}{BW_i \times AT_i}$$

where:

I_{oral} = intake over the averaging period (mg/kg-day)

i = the number of life stages within the averaging period

CS = the concentration of the contaminant in soil (µg/g)

DEF = the desorption factor, a unitless multiplier that accounts for desorption of the contaminant from the soil matrix into a bioavailable form, which can be absorbed by the digestive tract

CF = a unit conversion factor that has a value of 10^{-6}, which serves to convert the IgR unit from mg of soil/day to g of soil/day (because CS is expressed as µg/g of soil) and the result from µg/kg-day into mg/kg-day

IgR_i = the soil ingestion rate during life stage i (mg/day)

EF_i = the exposure frequency during life stage i (days/year)

ED_i = the exposure duration during life stage i (years)

BW_i = body weight during life stage i (kg)

AT_i = the averaging time during life stage i (days)

Dermal

While oral exposure is usually estimated in terms of intake, dermal exposure is calculated as uptake:

$$U_{dermal} = \sum_{1}^{j} CS \times DEF \times BVF \times CF \times ED_j \times \frac{(SA_{o,j} \times SL_{o,j} \times EF_{o,j}) + (SA_{i,j} \times SL_{i,j} \times EF_{i,j})}{BW_j \times AT_j}$$

where:

U_{dermal} = the dermal uptake over the averaging period (mg/kg-day)

j = the number of life stages within the averaging period

CS = the concentration of the contaminant in soil (µg/g)

DEF = the desorption factor, a unitless multiplier that accounts for desorption of the contaminant from the soil matrix into a bioavailable form, which can be absorbed by the digestive tract

BVF = the dermal bioavailability factor, another unitless multiplier

CF = a unit conversion factor that has a value of 10^{-6}, which serves to convert the SL unit from mg of soil/cm² to g of soil/cm² (because CS is expressed as µg/g of soil) and the result from µg/kg-day into mg/kg-day

ED_j = the exposure duration during life stage j (years)

SA_j = exposed skin area during life stage j (cm²)

SL_j = skin-soil load during life stage j (mg/cm²)

EF_j = the exposure frequency during life stage j (days/year)

BW_j = body weight during life stage j (kg)

AT_j = the averaging time during life stage j (days)

The subscripts o and i in the terms SA, SL, and EF signify outdoor and indoor microenvironments, respectively. EF_o and EF_i, the outdoor and indoor exposure frequencies, respectively, can be calculated using the two equations below:

$$EF_o = \frac{ES}{W} \times \frac{ES}{M} \times Y$$

and

$$EF_i = Y - EF_o$$

where W is a constant representing 7 days of the week, M represents 12 months of the year, and Y represents 365 days. EW represents the number of days exposed per week (days) and ES stands for exposure season, defined as the number of months in the year exposed (months/year). The equations contain a number of exposure parameters, which are discussed in the following section.

3.3.3   Exposure Parameters

The basic premise behind risk assessments is that they need to reflect the specific conditions experienced by the receptors. Ideally, exposure parameters should be derived for the specific situations the receptors are in. In practice, many or most exposure parameters used in typical assessments are default parameters obtained from published sources. In the case of biological factors (e.g., respiration rate), the use of default parameters is usually reasonable, unless there are marked differences in body sizes, physical conditions, and other racial differences between the populations for which the default parameters were derived and the population to which these default parameters are applied.

When there are surprisingly large differences in values for the same biological factors between jurisdictions, they are often due to methodological differences and not because of any real difference in population characteristics. For example, it cannot be true that Canadians breathe 59 percent more air every day than the neighboring Americans, as suggested by Health Canada's recommendation for adult inhalation rate (compare the Health Canada [1995] estimate for inhalation rate of 23 m³/day to a more reasonable USEPA [1997a] default value of 14.5 m³/day). Similarly, it is unlikely that U.S. adult residents consume two and a half times more soil daily than their Canadian counterparts (20 mg/day in Canada versus 50 mg/day in the United States).

Greater caution needs to be exercised in selecting lifestyle factors that best describe the population of interest. Lifestyle factors can be influenced by cultural differences, climate, and other factors. Using the default exposure factors produced by the jurisdiction in which the assessment is being conducted is not a guarantee of best match to the particular situation one wants to evaluate. It is prudent to examine the recentness of the data, the comprehensiveness of the database used to derive the default factors, and the absence of consistent bias in their derivation. In cases where the quality of data is questionable, it may be preferable to use data from a nearby jurisdiction that publishes better quality default exposure parameters.

3.3.4   Time Averaging

One of the difficulties in conducting a human health risk assessment is how to address less-than-lifetime exposures. For example, it is often necessary to assess the health risk construction workers may experience from conducting subsurface work at a contaminated site for a few weeks. The workers may be significantly exposed from direct contact with contaminated soil, but only during working hours and only for a few weeks. Both estimated cancer and noncancer potency factors (unit risks, slope factors, RfDs, and RfCs) assume a lifetime exposure or an exposure close to a lifetime. The number of less-than-lifetime estimated potency factors available from various agencies is small. Even the one potentially most appropriate source for less-than-lifetime exposure toxicity factors, the MRLs prepared by ATSDR, "are not intended to define clean-up or action levels" (Agency for Toxic Substances and Disease Registry 2003c). Risk assessors therefore have several options:

*   One can calculate the health risk assuming lifetime exposure, even for situations involving only short-term exposure. Averaging short-term exposures over a lifetime effectively means that even very large short-term exposures would be found "acceptable." This approach is not recommended because it likely underestimates risk.

*   One can refuse to assess risk from short-term exposures and provide the rationale why such an assessment cannot be undertaken. This approach is scientifically defensible, but not very useful to risk managers.

*   An imperfect solution is to use MRLs where available, while documenting ATSDR's position on this practice and providing the rationale for their use.

*   Another imperfect solution, but perhaps preferable to the option above, is to apply RfCs and RfDs to situations with exposure averaged over less than a lifetime.

The last solution has the advantage of being conservative. Most chronic adverse effects observed only after a lifetime of exposure tend to occur at lower dose levels than acute or short-term effects. Furthermore, USEPA applies a larger UF (the uncertainty factor used in the development of RfDs and RfCs) when the RfDs or RfCs are developed based on shorter-than-lifetime studies, and this adjustment leads to more-conservative estimated potency factors. For example, to evaluate the health risk of workers working in the subsurface on a site five days a week for a few weeks, one can determine the average daily exposure for the weeks they work by averaging the total weekly exposure over seven days. One can also determine the average exposure for each life stage separately and assess the risk for the most highly exposed life stage. This approach is designed to be protective of children, who are sometimes more exposed than adults. However, this approach applies RfCs and RfDs in a way not intended by USEPA.

3.3.5   Absorption Factors

The way absorption factors are selected and used can significantly affect the outcome of a risk assessment by as much as an order of magnitude or more. Despite the impact, regulatory agencies pay relatively little attention to absorption factors in drafting site-specific risk assessment guidelines. They also do not pay enough attention to evaluating existing absorption factors or derivation methods of absorption factors. Even when absorption factors are available, how the soil or particulate matrix affects absorption is often not discussed. Problems in the proper use of absorption factors are some of the most common limitations of site-specific risk assessments conducted. Paustenbach (2001) has authored an excellent review of issues pertaining to absorption of contaminants.

Intake and Uptake

Under normal circumstances, people are exposed to environmental chemicals by inhalation, oral, or dermal exposure. In order to get to the lungs, GI (gastrointestinal) tract, or skin, chemicals must be inhaled, ingested, or applied to the skin. The lungs, GI tract, and skin are body boundaries that need to be crossed in order for the chemicals to reach the tissues and exert their effects. The rate at which these chemicals come into contact with these boundaries is called *intake*. Paustenbach (2001) defines intake as "the process by which a substance crosses the outer boundary of an organism without passing an absorption barrier (e.g., through ingestion or inhalation)."

In contrast, *uptake* is the process by which substances that are inhaled, ingested, or applied to the skin cross the absorption barrier (lungs, GI tract, skin) and enter the living tissues. The actual toxic effect is due solely to that portion of the total dose that reaches the tissues. In experimental situations, *administered dose, external dose,* or *applied dose* is the dose delivered to the body boundaries—but not necessarily absorbed. In contrast, *absorbed dose, internal dose,* or *uptake* is the fraction of the administered dose that crosses the body boundaries through the uptake mechanism. USEPA (1989) also defines the *delivered dose* as the "the amount transported to an individual organ, tissue, or fluid of interest" and the *biologically effective dose* as "the amount that actually reaches cells, sites, or membranes where adverse effects occur."

Bioavailability, Bioaccessibility, and Desorption

Pure substances, such as benzene and trichloroethene gases or liquids, are in a form that can be readily absorbed through the lungs, GI tract, or skin. The *bioavailability factor* represents the ratio of the absorbed dose to the administered dose. This term should be reserved for situations when the applied dose is delivered in a bioavailable form. Substances can be administered either in an absorbable form or in a matrix that requires initial desorption before the substances can become available for absorption. For example, benzo[a]pyrene (B[a]P) is largely bound to particulate matter such as fine lipophilic carbonaceous particles in the natural environment. Absorption is only partial and slow, because B[a]P first needs to desorb from its soil matrix to be available for uptake. Experimentally, however, B[a]P can be administered to animals adsorbed on fine sodium chloride particles, which are likely to dissolve upon contact with the lung tissue. Under these experimental conditions, B[a]P is readily available for uptake.

The *bioaccessibility factor* represents the proportion of a chemical present in a matrix (e.g., soil) that can cross the body barrier. For example, in the case of ingestion of B[a]P-contaminated soil, the bioaccessibility factor of B[a]P is the proportion of B[a]P present in the soil matrix that will be absorbed into the body. Thus, uptake of a contaminant from the soil matrix is equal to the product of intake from soil and the bioaccessibility factor, or alternatively equal to the product of intake from soil, desorption factor, and bioavailability factor:

$$Uptake = Intake \times Bioaccessibility\ Factor$$

or

$$Uptake = Intake \times Desorption\ Factor \times Bioavailability\ Factor$$

The *desorption factor* can be derived from the bioavailability and bioaccessibility factors. For example, if it is known that 50 percent of a pure solubilized chemical is absorbed but only 5 percent is absorbed from soil, then the bioavailability factor of this chemical would be 0.5 (50%), its bioaccessibility factor from soil would be 0.05 (5%), and the desorption factor from soil would be 0.1 (5% / 50%).

Most information sources discuss "absorption factors" without clearly differentiating whether they refer to bioavailability, bioaccessibility, or a combination of both. It is up to the risk assessor to determine what type of information they are dealing with.

Adjusting Estimated Potency Factors for Bioaccessibility

It is important to have a good grasp of this issue, because agencies generally do not distinguish between estimated potency factors derived from human studies and factors derived using animal data. In epidemiological studies, exposure often involves particle-bound contaminants, whereas experimental studies are often conducted using contaminants in a bioavailable form. Estimated oral potency factors are generally reported on a per-administered-dose (intake) basis, while estimated inhalation potency factors are generally expressed in terms of airborne concentrations, from which intake can be calculated. As a result, exposure for the two exposure routes should be expressed in terms of intake or average airborne concentration.

In presenting estimated potency factors, agencies generally do not distinguish between different degrees of bioaccessibility. For example, WHO's guidelines (1999) for PAHs are based on occupational exposure to coke oven emissions. However, B[a]P, which was used as a surrogate for other PAHs, was mostly particle-bound in coke oven emissions, and WHO did not adjust for bioaccessibility when it developed its guideline value. In contrast, Health Canada's CEPA program (Health Canada 1994), CalEPA (1993), and USEPA (1994a, now withdrawn) derived B[a]P cancer potency from hamster inhalation studies, which administered to the animals B[a]P adsorbed on fine sodium chloride particles. Thus, the B[a]P was delivered in a bioavailable form in the hamster studies. The difference has practical implication.

When WHO's estimated potency value for B[a]P is used, no adjustment is necessary, because environmental exposure generally involves B[a]P in a particle-bound form, just as in coke oven emissions that were used to develop WHO guideline value. In contrast, when the CEPA's or CalEPA's estimated potency factors for B[a]P are used, a bioaccessibility factor should be applied to account for the use of a more bioaccessible form of B[a]P in the experimental studies. The same principle applies to oral exposure.

Exposure estimates should be adjusted for bioaccessibility when necessary to ensure that exposure is expressed in terms of a contaminant form similar to the contaminant form involved in the toxicity data used to derive the estimated potency factor.

Such adjustments are often not conducted in routine site-specific risk assessments. The need for such adjustments is also not always emphasized by regulatory or quasi-regulatory agencies, even though the application of these factors may have significant impact on the outcome of the risk assessment. The regulatory agencies also do not publish bioaccessibility factors.

Oral Intake versus Dermal Uptake

Since estimated oral potency factors are usually developed as intake, oral bioavailability factors are usually not needed to estimate oral exposure. One exception involves the use of oral bioavailability factors to derive bioaccessibility factors when dealing with ingestion of contaminants from soil. In contrast, oral bioavailability factors are needed for estimating dermal potency factors from oral factors because, unlike estimated oral potency factors, estimated dermal potency is defined in terms of uptake. For further detail on converting oral factors into dermal factors, see section 3.2.5.

Development of Bioavailability and Bioaccessibility Factors

Research is ongoing to obtain data for the derivation of bioavailability and bioaccessibility factors. Methods are also being developed for the derivation of bioaccessibility factors for contaminants in soil for both generic and site-specific applications. Some methods involve elution from a soil matrix, some use in vitro experiments that measure permeability through isolated animal or human skin, and others involve in vivo experiments. The discussion of these methods is outside the scope of this review, but some discussions and references on this subject can be found elsewhere (see, for example, Grøn and Andersen 2003; Paustenbach 2001; U.S. Navy 2000a, 2000b).

3.3.6   Key Sources of Information

Section 1.5 provides a listing of general risk assessment guidance documents. Most of these reports provide extensive guidance and information related to exposure assessment. These reports will not be discussed further in this section. Other key sources one can consult when conducting exposure assessment are listed in table 1.6. Extensive guidance is available in USEPA's *Guidelines for Exposure Assessment* (1992). Health Canada's *Handbook of Exposure Calculations* provides Canadian guidance. The Australian enHealth Council has released *Exposure Scenarios and Exposure Settings* (2001a), which provides guidance on exposure assessments and outlines exposure settings and assumptions that are appropriate for Australia.

USEPA's *Exposure Factors Handbook*, a series of three volumes (1997a, 1997b, 1997c), is one of the most comprehensive resources available for human exposure factors. The handbook includes point estimates and (when available) distributional data for general factors (e.g., drinking water intake, soil ingestion), food ingestion factors (e.g., fruit and vegetable intake, meat intake), and activity factors (e.g., activity patterns, population mobility). Also available from USEPA is the *Child-Specific Exposure Factors Handbook (Interim Report)* (2002d), which includes exposure data more relevant to children. USEPA's *Supplemental Guidance for Dermal Risk Assessment* (2001b) offers updated and detailed discussions on conducting dermal exposure assessment. In addition, USEPA maintains a website for the Exposure Factors Program, which provides references to recent studies that could provide new information about exposure factors and generally keeps the reader up to date on USEPA's initiatives in the area of exposure factors.

Table 1.6 Key sources of information for exposure assessment

Sources	Links
CalEPA (OEHHA)—*Air Toxics Hot Spots Program Risk Assessment Guidelines,* part 4: *Technical Support Document for Exposure Assessment and Stochastic Analysis*	http://www.oehha.org/air/hot_spots/finalStoc.html
enHealth Council—*Exposure Scenarios and Exposure Settings*	http://www.health.gov.au/pubhlth/publicat/document/metadata/env_exposure.htm
Health Canada—*Investigating Human Exposure to Contaminants in the Environment: A Handbook for Exposure Calculations*	http://www.hc-sc.gc.ca/ehp/ehd/catalogue/bch_pubs/95ehd193.htm
ODEQ—*Guidance for Use of Probabilistic Analysis in Human Health Risk Assessments*	http://www.deq.state.or.us/wmc/cleanup/hh-intro.htm
Richardson—*Compendium of Canadian Human Exposure Factors for Risk Assessment*	http://irr.uwaterloo.ca/B_Descrip.htm#Anchor-Compendium-59218
Risk Assessment Information System (RAIS)	http://risk.lsd.ornl.gov/rap_hp.shtml
USEPA—*Guidelines for Exposure Assessment*	http://cfpub.epa.gov/ncea/cfm/recordisplay.cfm?deid = 15263
USEPA—*Exposure Factor Handbook/Child-Specific Exposure Factors Handbook*	http://cfpub.epa.gov/ncea/cfm/recordisplay.cfm?deid = 20563
USEPA—*Exposure Factors Handbook*	http://cfpub.epa.gov/ncea/cfm/recordisplay.cfm?deid = 12464

The Risk Assessment Information System website (2003) provides some human health risk exposure models for different exposure scenarios in different environmental media (e.g., residential exposure to contaminants in soil). The Oregon Department of Environmental Quality's *Guidance for Use of Probabilistic Analysis in Human Health Risk Assessments* (1998) presents human exposure factors and models. Continuous distributions are provided for exposure factors. Similarly, California's Office of Environmental Health Hazard Assessment produced the *Technical Support Document for Exposure Assessment and Stochastic Analysis*, part 4 of its *Air Toxics Hot Spots Program Risk Assessment Guidelines* (CalEPA 2000b), which provides evaluated and usable exposure factors, including those associated with dermal exposure assessment. In Canada, Health Canada has published *A Handbook of Exposure Calculations*. This document contains Canadian guidance for conducting exposure assessment and also provides Canadian exposure factors. Richardson has authored a *Compendium of Canadian Human Exposure Factors for Risk Assessment* (1997), which contains lognormal distributions for several exposure factors.

3.4   Uncertainty Assessment

3.4.1   Description

It is useful to distinguish between the terms *variability* and *uncertainty*. USEPA describes variability thus:

> Variability arises from true heterogeneity in characteristics such as dose-response differences within a population, or differences in contaminant levels in the environment. The values of some variables used in an assessment change with time and space, or across the population whose exposure is being estimated. Assessments should address the resulting variability in doses received by members of the target population. Individual exposure, dose, and risk can vary widely in a large population. Central tendency and high-end individual risk descriptors capture the variability in exposure, lifestyles, and other factors that lead to a distribution of risk across a population. (USEPA 2000c)

In contrast, the same source describes uncertainty as follows:

> Uncertainty represents lack of knowledge about factors such as adverse effects or contaminant levels which may be reduced with additional study. Generally, risk assessments carry several categories of uncertainty, and each merits consideration. Measurement uncertainty refers to the usual error that accompanies scientific measurements—standard statistical techniques can often be used to express measurement uncertainty. An amount of uncertainty is often inherent in environmental sampling, and assessments should address these uncertainties. There are likewise uncertainties associated with the use of scientific models, e.g., dose-response models, models of environmental fate and transport. (USEPA 2000c)

It is useful to think of variability as natural heterogeneity inherent in a specific parameter. For example, individuals differ in their body weights and these differences are not a result of imprecise measurement. In contrast, uncertainty indicates a lack of knowledge, which may result from imprecise or incomplete testing (which can be expressed using statistical methods). Uncertainty can also arise from incomplete knowledge of relevant science, which requires the use of assumptions in the place of hard facts. The latter type of uncertainty is much less amenable to the use of statistical methods and the estimation of magnitude of this type of uncertainty calls for expert judgment.

Uncertainty assessment involves collection and interpretation of information about variability and uncertainty in order to assist in the scoping and planning of the assessment and in interpreting quantitatively exposure and risk estimates. Methods have been developed to conduct quantitative uncertainty analysis in environmental health assessment. More discussion on this topic is provided later in section 4.3. The publication *An Introductory Guide to Uncertainty Analysis in Environmental and Health Risk Assessment* (Hammonds et al. 1994) is a good resource for this subject.

3.4.2   Uncertainty as a Tool for Planning Risk Assessment

Risk assessment can be seen as a process of reducing uncertainty in risk estimation. The overall uncertainty is reduced to various degrees as sources of uncertainty are addressed, some to a greater degree than others. As more accuracy is needed, more data and more sophisticated analyses are required.

A perfect assessment would address all sources of uncertainty, whether they contribute slightly or significantly to the overall uncertainty. However, as most risk assessments are conducted within a limited budget and time frame, it is usually not possible for risk assessors to deliver such a perfect assessment. Under these circumstances, it is important to focus one's effort on addressing the biggest sources of uncertainty and to deal with smaller ones if resources are available.

This section discusses the relative magnitude of uncertainties based on some simple sensitivity analysis and on the experience of the authors.

Sensitivity analysis is the study of how variations in the inputs (data and judgment) affect the output (risk estimate) of a risk assessment. The variation of inputs arises from uncertainty and variability associated with the data and judgment used. As noted above, variability refers to differences that exist among individuals in a population, while uncertainty results from a lack of knowledge. Sensitivity analysis involves a number of analytical tools that can be applied in both deterministic and probabilistic risk assessments. Factors are identified and ranked in terms of their contributions to variability and uncertainty (for review, see USEPA 2001d).

A simple analysis is presented here to provide readers a perspective of sensitivity analysis. Take the exposure equations from section 3.3.2 as an example; if addition of different life stages is disregarded, they involve only multiplications and divisions. Equal weight is given to all the parameters in the equation. As a result, changing any of the parameters threefold will result in a threefold change in the exposure or risk estimate. Examination of the variability or uncertainty of each parameter can therefore provide a direct indication of the parameter's impact on the outcome of the exposure assessment or risk assessment. Uncertainty associated with different components of risk assessment will be discussed in the subsequent sections.

3.4.3 Assessment of Uncertainty: Estimated Potency Factors

USEPA (2003e) states that the uncertainty associated with RfDs and RfCs spans *perhaps an order of magnitude.* The term "span" suggests a range or a confidence interval. It is noted, however, that an uncertainty factor of two to three orders of magnitude is not rare in the derivation of RfDs and RfCs. Similarly, unit risks and slope factors have a built-in conservatism by using the 95 percent upper confidence limit on the initial slope of the dose-response curve. The difference between the upper confidence limit on the slope and the slope with the best fit is about an order of magnitude, at least in the case of coke oven emissions (discussed elsewhere in this chapter).

Furthermore, despite best efforts, the potency estimates have to be revised over time as new data become available. For example, the risk from ingesting arsenic as estimated by the NRC in 2001 (National Research Council 2001b) (converted by the authors to slope factor) is between 2.1 and 5.2 $(mg/kg/day)^{-1}$ for bladder cancer and 1.8–4.7 $(mg/kg/day)^{-1}$ for lung cancer. These values contrast with the earlier USEPA (1998) assessment of 1.5 $(mg/kg/day)^{-1}$ based on skin cancer rates. Not only are the slope factors higher but the more recent NRC assessment also based the slope factors on two types of cancer with high fatality, instead of on the low-fatality skin cancer (squamous cell carcinoma) used in the earlier USEPA assessment. USEPA routinely updates and revises its potency estimates as new data and new assessments become available.

One source of uncertainty arises from the degree to which different agencies differ in their estimation of potency factors. Table 1.7 shows examples of significantly different potency estimations for the same chemical by three different reputable jurisdictions. This table is not meant as an indication of flaws in the assessment, but rather as an illustration of uncertainty involved in potency estimation. While a fivefold or greater difference in potency estimates is quite common, the estimates can differ by as much as two to three orders of magnitude. The biggest difference in potency estimates appears to involve weak carcinogens or nongenotoxic carcinogens, and complex mixtures.

3.4.4 Assessment of Uncertainty: Environmental Levels of Contaminants

One important issue regarding exposure assessment is the selection of an appropriate metric for contaminant levels in an environmental medium. USEPA and other agencies may specify in their guidelines that the highest measured level, the 95th percentile, the 95th upper confidence limit on the mean (95 percent UCL), or the average level be used in exposure estimation.

A detailed review has not been conducted; however, to illustrate the situation, we examined the concentrations of 13 contaminants on four contaminated sites (a total of 20 site-contaminant combinations). Contaminant concentrations generally vary spatially in soil and the use of a different metric for contaminant concentrations may affect the outcome of a risk assessment. The effect can be illustrated in figure 1.2, which shows a frequency distribution of the ratios of the highest measured to the mean (max/mean) contaminant levels, 95 percent UCL to the mean, and 95th percentile to the mean contaminant levels at the four sites. The figure shows that maximal values are often more than one order of magnitude higher than the mean, while most of the 95th percentile values are within a factor of five of the mean. The 95 percent UCL is usually within a factor of two of the mean and always within a factor of three.

These findings may not apply to all sites, but they do provide readers with a general idea of the issue. Sediments are expected to vary spatially in a similar fashion to soils. On the other hand, the levels of contaminants in groundwater, surface water, and air vary to a lesser extent spatially within a short distance and may vary over time if the measurements are taken near a major intermittent source at different time intervals or near a major point source.

Variation in ambient air levels can be illustrated based on the measurements obtained from Environment Canada National Air Pollution Surveillance (NAPS) monitoring network for air toxics (Environment Canada 2003b). The ambient air levels are reported for a number of contaminants as percentiles, means, and standard deviations. The number of samples collected at stationary monitoring stations across Canada is also reported. The mean, 90th percentile, and maximum ambient air levels for 2002 were extracted for benzene, toluene, tetrachloroethene, dichloromethane, 1,1,1-trichloroethane, and 1,3-butadiene. Data where the mean was not reported or where the mean was zero were excluded from the calculations. The analysis was conducted on a total of 156 monitoring station–contaminant combinations. The results, as illustrated in figure 1.3, show that the 90th percentile levels are mostly within 2.5 times the mean levels and the maximal levels are mostly an order of magnitude higher than the mean. These results may not be representative of all situations and all contami-

Table 1.7 Comparison of estimated potency factors from three agencies

Max/Min	Chemicals	Agencies[a]			Reference
		USEPA	Health Canada[b]	WHO	
8.4	1,4-Dichlorobenzene Inhalation—noncancer	**RfC: 0.8 mg/m³** *Based on a rat multigeneration reproductive study by the Chlorobenzene Producers Association*	**RfC: 0.095 mg/m³** *Based on a study that did not find increased liver and kidney weight, urinary protein, and coproporphyrin in rats at a concentration of 450 mg/m³*		CEPA 1996 USEPA 1996c
300	Cadmium Inhalation—cancer	Unit risk: 1.8E–3 $(\mu g/m^3)^{-1}$ **RSC[c]: 0.006 $\mu g/m^3$** *Based on epidemiological human data, lung cancer*	Unit risk: 1.0E–2 $(\mu g/m^3)^{-1}$ **RSC: 0.001 $\mu g/m^3$** *Based on estimated carcinogenic potency by inhalation in long-term bioassays*	**RfC: 0.3 $\mu g/m^3$** *Based on an occupational study, showing renal effects*	CEPA 1994a USEPA 1994c WHO 2000
6.3	Chromium (VI) Inhalation—cancer	**Unit risk: 1.2E–2 $(\mu g/m^3)^{-1}$** *Occupational study, lung cancer*	**Unit risk: 7.6E–2 $(\mu g/m^3)^{-1}$** *Occupational study, lung cancer*	**Unit risk: 4.0E–2 $(\mu g/m^3)^{-1}$** *Based on several epidemiological studies and a range of values that has been established*	CEPA 1994b USEPA 1998c WHO 2000
143	2,3,7,8-Tetrachlorodibenzo-p-dioxin (TCDD) Inhalation	**Inhal. cancer slope factor:** 1.5×10^5 $(mg/kg/day)^{-1} =$ 1.5×10^{-4} $(pg/kg/day)^{-1}$ **RSC: 0.07 pg/kg/day** *Occupational inhalation studies, wide range of cancer types, sites*	**RfD: 10 pg/kg/day** *Based on cancer effects in animal studies; TI is intended as total exposure from all routes*	**RfD: 1–4 pg/kg/day** *Based on reproductive and developmental effects in animal studies; TDI is intended as total exposure from all routes*	CEPA 1993 USEPA 1994d, 2000f, 2003n WHO 2000

(continues)

Table 1.7 Continued

Max/Min	Chemicals	Agencies[a]			Reference
		USEPA	Health Canada[b]	WHO	
15.4	Nickel Oral—Noncancer	**RfD: 2.0E–2 mg/kg/day** (soluble nickel as a group, one of the two studies used nickel chloride) *Based on animal studies in rats and dogs, loss of body weight*	**RfD: 1.3E–3 mg/kg/day** (nickel chloride)		CEPA 1996b USEPA 1996d
3.8	Styrene Inhalation—noncancer	**RfC: 1 mg/m³** *Based on an epidemiological study on neuro-psychological function*		**RfC: 0.26 mg/m³** *Based on an epidemiological study (or studies; no reference was given)*	USEPA 1993b WHO 2000
14.4	Toluene Inhalation—noncancer	**RfC: 0.4 mg/m³** *Based on an occupational study using neurobehavioral tests*	**RfC: 3.75 mg/m³** *Based on a clinical study on volunteers (decrease in neurological function, increase in neurological symptoms, and irritation of the respiratory tract)*	**RfC: 0.26 mg/m³** *Based on an occupational study (critical effects were central nervous system effects)*	CEPA 1992 USEPA 1994e WHO 2000
6.3	Toluene Oral—noncancer	**RfD: 0.2 mg/kg/day** *Based on an animal study (increased relative liver, kidney weight in males)*	**RfD: 1.25 mg/kg/day** *Based on an animal study (increased relative liver, kidney weight in males)*		CEPA 1992 USEPA 1994c

[a] For the sake of clarity, Health Canada and WHO estimated potency factors are relabeled using the closest matching terms from USEPA.

[b] Health Canada reports TC_{05} for cancer potency. To allow comparison of estimated potency factors between different jurisdictions, Health Canada's estimated potency factors have been converted to unit risk assuming linear dose-response relationship between TC_{05} and the origin.

[c] RSC = Risk Specific Concentration, the concentration of a chemical associated with a specific excess lifetime cancer risk level. In this table, the specific risk level is 10^{-5}. At the risk level of 10^{-6}, the RSC would be 10 times lower.

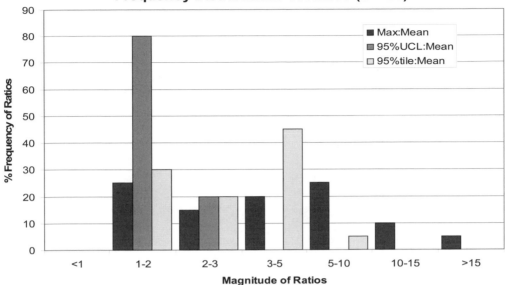

*Figure 1.2 Frequency distribution of contaminant levels in soil at a contaminated site
Ratio of maximum soil level to mean, 95th UCL level to mean and 95th percentile to mean level.*

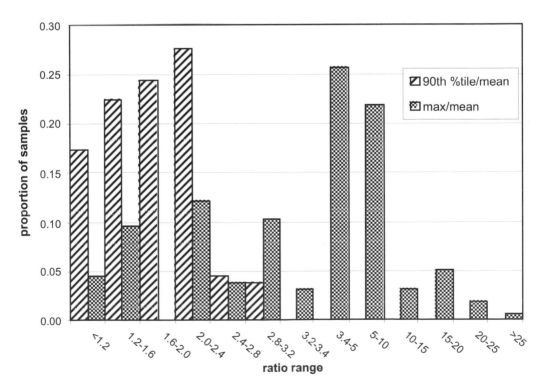

*Figure 1.3 Frequency distribution of contaminant levels in ambient air
Ratio of maximum air level to mean level, 90th percentile level to mean level. (Data from Environment
Canada 2003b)*

nants, but they do provide the typical variability of air contaminant levels at a given location and assist in comparison with variability from other sources.

3.4.5   Assessment of Uncertainty: Bioavailability and Bioaccessibility

Bioavailability factor is another important source of uncertainty. Hrudey and colleagues tabulated published inhalation, oral, and dermal bioavailability factors for a number of substances, including metals, solvents, PAHs, dioxins, benzene, toluene, ethylbenzene, and xylenes (Hrudey, Chen, and Rousseaux 1996). Inspection of the table indicates that factors for the same substance often vary by as much as fivefold, and in some instances tenfold, from one source to another, even when variables such as chemical form and solubility of chemical species are taken into consideration. It is therefore important to select absorption factors carefully, as they may have significant impact on the outcome of the risk assessment.

Bioaccessibility factors are less available than bioavailability factors. Great care must be exercised to avoid confusion in the use of the two factors.

3.4.6   Lifestyle and Biological Factors

There seems to be general agreement on the values of many lifestyle and biological factors, despite some amusing differences between Canadian and U.S. respiration and soil ingestion rates (see section 3.3.3). Health Canada and USEPA differ significantly with respect to guidance on food intakes.

Health Canada (1994) recommends food intake estimates that were derived from an unpublished *Nutrition Canada Survey*, which took place from 1970 to 1972. Health Canada acknowledges that this study is dated and that the food intake of Canadians has changed over the years. On the other hand, the U.S. Department of Agriculture (1997) provides detailed and up-to-date analysis of food intake in the United States. Furthermore, Canada sometimes relies on its estimation of contaminants in food that is based on relatively limited and dated studies. In contrast, the *Total Diet Study* statistics compiled by the U.S. Food and Drug Administration (2000) represent a large-scale and ongoing food-testing program in the United States. People in the United States and Canada generally have similar lifestyles and are expected to consume similar food. Health Canada's guidance on exposure assessment needs updating, at least in those areas that affect significantly the outcome of risk assessments.

One of the most important lifestyle factors is the duration of residency on a given property. Lifetime residence (nominally 75 years) is sometimes assumed, and it may be a reasonable worst case in some rural communities where long-term residency is a norm. However, a much shorter-term residency is assumed in most cases. USEPA's (1997c) recommendation for typical residency is nine years and for a reasonable worst case is 33 years.

3.4.7   Soil Loading

A wide range of soil loadings has been recommended for different age groups and occupations. Depending on the assumption an individual risk assessor makes, one can expect a two- to threefold difference between assessments conducted by different assessors, all using appendix C of the *Supplemental Guidance for Dermal Risk Assessment* (USEPA 2001b) as the reference. Soil loading values can range more than an order of magnitude.

3.4.8 Built-in Bias toward Conservatism in Dose-Response Assessment

One of the goals of many risk assessments is to provide an unbiased "best-guess" estimate of risk. This goal is difficult to achieve as long as one uses estimated potency factors, which have an inherent bias toward conservatism. This bias applies to both contaminants that are thought to act via a threshold and those with a non-threshold mode of action.

Assessments Assuming Threshold

USEPA (1993a, 1994b), Health Canada's CEPA program (Health Canada 1994), WHO (2000), Baars et al. (2001), CalEPA (2000c), and others apply an uncertainty factor and sometimes a modifying factor to derive RfDs and RfCs (and conceptually similar values) in order to account for specific factors that have not been well delineated. These assessments are therefore biased toward conservatism.

Risk management decisions are made based on the *hazard quotient* (HQ) in the case of USEPA; some investigators also make use of *exposure ratio* (ER). Both HQ and ER are calculated as the ratio of the estimated exposure to the reference exposure level (estimated potency factor adjusted for uncertainty, such as RfD, TDI, ADI, TC, or RfC).[13] Values of HQ or ER below one are considered safe. Levels higher than one are indicative of possible concern. Use of *uncertainty* or *modifying factors* biases the HQ and ER toward a conservative outcome even when an attempt is made to estimate the exposure realistically. Since estimated potency factors such as RfDs are derived from various databases, this type of uncertainty may not always be presented in risk characterization. However, this information is important and should be available to help risk managers in making decisions.

The same type of bias is not inherent in the European approach (European Chemicals Bureau 2003c), which reports the *margin of safety* (MoS) for human populations. MoS is calculated as the ratio of the NOAEL to the exposure (i.e., MoS = NOAEL / Exposure).[14] If the MoS value is judged to be sufficiently large, the corresponding contaminant levels are considered to be "of no concern"; if MoS is small, a *risk to health cannot be excluded*. The ECB guidance is interesting because of its use of MoS for regulatory purposes, since MoS is not modified by any factor.

To achieve environmental protection goals, the ECB expresses risk characterization as the ratio of *predicted environmental concentration* (PEC) to *predicted no-effect concentration* (PNEC). PNEC is created using both a NOAEL and a *safety factor*. PNEC therefore appears similar to ADI or RfD. However, PNEC is defined in terms of not exposure per day but environmental concentrations (e.g., concentrations in soils, sediments, or groundwater). PECs are predicted (modeled) rather than measured environmental concentrations. When the predicted no-effect concentration is exceeded—i.e., PEC / PNEC > 1—a risk is considered to be present. The details of the process involving PNEC and PEC as applied to human health risk assessment are not clear, as this part of the ECB guidance is still under development.

USEPA (1993a) introduced the *margin of exposure* (MOE) concept in 1993. MOE can be calculated in much the same way as MoS:

$$MOE = NOAEL \text{ (experimental dose)} / EED \text{ (human dose)},$$

where EED represents the estimated (calculated or measured) exposure dose. EED should include all sources and routes of exposure involved. It is expressed in the same units as NOAEL; MOE therefore is unitless.

USEPA states that when the MOE is equal to or greater than UF × MF, "the need for regulatory concern is likely to be small." To the best of the authors' knowledge, USEPA rarely uses MOE to make regulatory decisions to mitigate human health risk. In contrast, ECB does make decisions based on MoS values.

ECB's approach is not less conservative than the USEPA and Health Canada approaches. The difference is the requirement by ECB to explicitly discuss and address uncertainty issues associated with toxicity potency estimation in the risk characterization component of the assessment. This approach, in the authors' opinion, is preferable to burying decisions related to the management of uncertainties associated with potency estimation in the dose-response assessment process itself, as is the case with USEPA and Health Canada. ECB states:

> Where the exposure estimate is less than the N(L)OAEL, the risk assessor will need to decide which of the possible results applies. For this step, the magnitude by which the N(L)OAEL exceeds the estimated exposure (i.e., the "margin of safety") needs to be considered taking account of the following parameters:
>
> From Regulation 1488/94 and Directive 93/67/EEC:
> * the uncertainty arising, among other factors, from the variability in the experimental data
> * and intra- and interspecies variation;
> * the nature and severity of the effect;
> * the human population to which the quantitative and/or qualitative information on exposure applies.
>
> Other factors:
> * the differences in exposure (route, duration, frequency and pattern);
> * the dose-response relationship observed;
> * the overall confidence in the database.
>
> Expert judgement is required to weigh these individual parameters on a case-by-case basis. The approach used should be transparent and a justification should be provided by the assessor for the conclusion reached. (European Chemicals Bureau 2003c)

Health Canada is closer to the ECB approach. Although no formal guidance has been published, a recently released evaluation of malathion by the Pesticide Management Regulatory Agency of Health Canada (Health Canada 2003f) established the NOAEL and LOAEL for the critical adverse health effect. Different target MOEs were selected for the protection of different target subpopulations. Decisions on managing the risk to the specific target subpopulation would be based on whether the estimated MOE was greater than the target MOE. Similarly, priorities for risk management actions for priority substances under the CEPA program are established based on a comparison of the estimated exposure level to the NOAEL, LOAEL, or benchmark dose on which the TI or TC is based (Health Canada 1994).

Assessments Assuming Non-threshold

The assumption of a linear dose-response relationship (non-threshold assumption) for mutagenic carcinogens has been a controversial issue for a long time. Some scientists continue to argue for threshold assumption (i.e., a nonlinear dose-relationship; see section 2.2); however, most jurisdictions have adopted the more conservative non-threshold assumption. ECB is the exception and appears to be noncommittal on this issue:

> Unless a threshold mechanism of action has been clearly demonstrated, it is considered prudent to assume that a threshold cannot be identified in relation to mutagenicity. As stated by IPCS (1994) "there is no clear consensus on appropriate methodology for the risk assessment of chemicals, for which the critical effect may not have a threshold, such as genotoxic carcinogens and germ cell mutagens." Risk characterisation has therefore to be conducted on a pragmatic basis in which account is taken of the pattern and extent of human exposure, what is known or can be deduced about the relevance for humans of the effect observed in the studies (e.g., from toxicokinetics data), the potency of the substance and the risk reduction measures already in place. (European Chemicals Bureau 2003c)

Using the slope of the best fit to the data (MLE) or the 95 percent confidence interval on this slope (UCL) to assess cancer risk is also often a policy decision. Health Canada uses MLE, while USEPA mostly uses UCL. Health Canada (2003e) estimated that the use of MLE versus UCL results in a twofold difference in the dose corresponding to the same risk level. Assuming that the slope of the dose response curve is linear at low dose, the slope factors developed by the two jurisdictions would therefore differ by a factor of two. However, a twofold difference seems to underestimate the "real" difference, at least in some cases. For example, USEPA's MLE and UCL potency estimates for coke oven emissions differ by about an order of magnitude. Therefore, one has to be mindful that USEPA's (and some other organizations') unit risks and slope factors have a built-in bias toward conservatism.

Like USEPA, Health Canada assumes the risk to be a continuum at low doses or concentrations. Where Health Canada differs from USEPA is the recognition of the difficulty to characterize risk at low dose; instead of conducting low-dose extrapolation beyond the observed dose range, Health Canada (1994) expresses potency in terms of TC_{05} or TD_{05}. Guidance on priority to develop exposure reduction options is provided based on the magnitude of the *exposure/potency index* (EPI) (Health Canada 1994). For example, a "low" priority for control action to reduce exposure is assigned when estimated exposure is only a very small proportion of the TC_{05} or TD_{05}.[15] This approach avoids the numerous uncertainties associated with low-dose extrapolation and does not introduce a built-in bias toward conservatism into the dose-response assessment. Conservatism can be discussed in the risk characterization component and incorporated explicitly into decision making, taking into consideration all uncertainty issues associated with risk estimation.

Decisions regarding whether or not to assume a threshold for a particular health effect, for which chemicals, and how to develop the dose-response relationships are policy decisions made by individual regulatory agencies for territories under their jurisdiction. A conservative bias is often applied to assess health effects that have a threshold and those that do not. Applying conservatism in risk assessment is a policy decision and is appropriate. The authors argue, however, that a risk assessment would be more transparent and better understood if this conservatism were built into the risk characterization section after the calculation of the risk has been completed. The dose-response assessment section is not always reviewed or understood by decision makers who may not have strong grounding in environmental risk assessment. Risk managers usually focus their attention on the risk characterization section and it is in this section that any bias toward conservatism is best introduced and discussed.

Some regulatory agencies' risk assessment processes do not apply conservatism at the dose-response level, but rather at the risk characterization level. This approach is more easily achievable for threshold effects because NOAELs are generally readily available. It is more difficult for cancer risk assessments, as most potencies are defined in terms of slope factors and unit risks as UCL estimates. Most guidance documents and regulatory policies force risk assessors to introduce a conservative bias at the dose-response assessment stage.

3.4.9   Conclusions

Uncertainty assessment is an important component of any risk assessment because it assists risk managers in evaluating the level of confidence they can place on the outcome of the risk assessment. A good understanding of the variability and uncertainty inherent in different data sets and judgments can help in scoping a risk assessment so that the greatest emphasis is devoted to those elements that are associated with the greatest uncertainty. It is the authors' opinion that the greatest sources of uncertainty are associated with dose-response assessment, application of bioavailability and bioaccessibility factors, and assessment of the dermal exposure route. The authors recommend that risk assessors and regulators pay greater attention to these topic areas.

3.5   Risk Characterization

3.5.1   Description

Risk characterization is the final step of risk assessment. Risk characterization summarizes the quantitative and qualitative findings, as well as the data, methods, and assumptions used in the assessment. Risk characterization is described in numerous risk assessment guidance documents, but the most comprehensive and current publication on the subject is probably the USEPA *Risk Characterization Handbook* (2000c).

As defined in the *Risk Characterization Policy* (USEPA 1995b; USEPA 2000c, appendix A), "Risk characterization integrates information from the preceding components of the risk assessment and synthesizes an overall conclusion about risk that is complete, informative, and useful for decision makers." In essence, risk characterization conveys the risk assessor's judgment on the existence (or lack of) and nature of human health risk.

Risk characterization is an instrument by which a risk assessor communicates the findings of the risk assessment to the risk manager or decision maker. It is therefore important that this component of the assessment contain information the risk manager needs—presented in a readily understandable manner—to allow her or him to make an informed decision. For the communication to be effective, a high-quality risk characterization, as defined by USEPA, should be transparent, clear, consistent with other similar USEPA assessments, and reasonable (referred to as the TCCR principles).

Risk characterization must contain sufficient detail to permit another expert to reconstruct what has been done in the assessment. All uncertainties and assumptions made should be included in the discussion. The risk manager needs to know the major elements that affect the characterization of risk, including:

- Key information
- Context
- Sensitive subpopulations
- Scientific assumptions
- Policy choices
- Variability
- Uncertainty
- Bias and perspective
- Strengths and weaknesses
- Key conclusions
- Alternatives considered
- Research needs

USEPA stresses that the risk characterization section is more than just a place where the calculated risks are tabulated:

> Every risk characterization has a fundamental, irreducible set of information consisting of the *key* findings that must be conveyed to every audience to adequately characterize the risk; again, it is more than just a number. (USEPA 2000c)

USEPA (2000c) recommends addressing a list of issues (see table 1.8) in risk characterization. Inclusion of all the information listed may be useful and appropriate in

Table 1.8 Issues to be covered in risk characterization

a What studies are available and how robust they are (e.g., have the findings been repeated in an independent laboratory?)

b The estimates of the major risks, the assumptions and the extrapolations made during risk calculation, and the residual uncertainties and their impact on the range of plausible risk estimates. Description of the risk estimate should indicate what receptors are being assessed (e.g., individual, population, ecosystem) and include such things as the high-end and central tendency estimates.

c Use of defaults, policy choices, and risk management decisions made (e.g., refer the reader to an agency risk assessment guidance, guideline, or other easily obtainable reference source that explains the terminology used)

d Whether the key data used for the assessment are considered experimental, state of the art, or generally accepted scientific knowledge

e The meaning of quantitative data in an easily understandable form—the use of tables and graphics may be helpful

f Variability

Source: USEPA 2000c

large-scale assessments involving sophisticated risk management teams. However, evaluation of studies considered in the dose-response assessment (item a), as suggested, may be excessive for some assessments of modest size. For example, builders with limited experience and understanding of environmental risk management issues could be making decisions respecting contaminated soil based on routine site-specific risk assessments.

In summary, the primary role of the risk characterization component is to provide risk managers (decision makers) all the key information required for making an informed decision. While the intended audience for the other components of a risk assessment may be expert readers, such as peer reviewers, risk characterization needs to be prepared with decision makers in mind. Risk assessors need to anticipate decisions that have to be made and present the needed information in a format that would facilitate making these decisions.

Experience and training of risk managers differ from project to project. More detail may be appropriate when risk assessments are conducted for regulatory agencies or large industrial firms with experienced risk managers. However, many assessments are conducted for decision makers who do not have much environmental risk management experience. Writing for this audience should still cover all the key issues but include fewer details so as not to overwhelm the readers. Clear interpretation of the situation, outline of key risk management options, and recommendations are particularly important in these situations.

3.5.2 Adding Risks

This section discusses adding risks from multiple pathways, multiple routes of exposures, and multiple toxic agents in the risk characterization component of the risk assessment. The discussion presupposes that a decision has been made at the dose-response assessment stage not to treat the whole array of toxic agents as a complex mixture. Approaches to the evaluation of the dose-response relationships of chemical mixtures have been discussed in section 3.2.6.

Many assessors attempt to estimate the overall risk using traditional methods wherein the exposure of a given receptor is examined pathway by pathway, route by route, and contaminant by contaminant, summing the risks or exposures. The usual justification for this approach is that it is "conservative." Although such an approach can be justified in some instances, it should not be applied routinely. It can be argued that adding individual risks has minimal impact on risk management decisions in most situations. For example, consider three separate risks: 10^{-4}, 10^{-6}, and 10^{-8}. The sum of the risks is essentially the same as the largest of the three risks (10^{-4}). Therefore, adding risks that differ significantly results in a total value that is nearly identical to the largest risk value. The only instance when summing two individual risks matters is when the two individual risks are of the same order of magnitude, in which case the risk only doubles. Perhaps more important, it is rare for two independent risks to be almost identical. In the real world, risks are sufficiently different. Unless some of the many risks are of the same magnitude as the largest risk, summing risks would not produce a different outcome and is not generally appropriate.

Adding Risks or Exposures from Multiple Pathways

A single route of exposure may involve several exposure pathways. For example, oral exposure to a contaminant is possible from ingestion of supermarket food, homegrown food, drinking water, and soil. It is reasonable to sum exposures or risks resulting from all these exposure pathways, although one exposure pathway usually dominates. The total exposure from all other pathways may not differ significantly from the exposure via the dominant pathway.

Adding Exposures from Multiple Routes of Exposure

This type of addition has a number of issues. As shown in section 3.3.2, inhalation is expressed in terms of average airborne contaminant level, ingestion in terms of intake, and dermal exposure in terms of uptake. In order to calculate the total dose, however, all exposures need to be expressed in terms of either intake or uptake. Also, in calculating risk, the estimated potency factors must be adjusted to intake or uptake as dictated by the exposure calculation.

A bigger issue lies with the fact that many contaminants produce different effects depending on the route of exposure; these effects may occur at dose levels that are much lower than the dose levels that elicit systemic effects common to other exposure routes. For example, several metals are carcinogenic when exposure is via inhalation, but not necessarily so via other routes of exposure. In the case of PAHs, the doses needed to induce tumors near the site of entry or absorption are significantly lower than the doses needed to induce similar effects by other distal exposure routes.

As in the case of adding exposure from multiple pathways, a single route of exposure usually dominates. Therefore, adding exposures from multiple routes of exposure rarely adds any significant degree of conservatism. Only under a rare condition where exposures via two exposure routes are equal does the combined exposure (and risk) become twice the exposure (and risk) from the most important exposure route. In most situations, the difference will be much less than double. On the whole, addition of exposures from different routes of exposure is not recommended.

Adding Cancer Risks of Multiple Contaminants

Adding cancer risks from exposures to different contaminants by different pathways and routes is scientifically defensible, provided the site and type of tumors are identical. For risk management purposes, it is sometimes useful to add the risks, but the fatality of different types of cancer should be mentioned when cancer risks with widely different fatality are added. For example, skin tumors (squamous cell carcinoma) have a very low fatality, while lung tumors have a very high fatality (see table 1.9 for fatality of different tumors).

3.5.3 Key Sources of Information

Risk characterization is usually discussed in risk assessment guidance documents, which are listed in section 1.5. Table 1.10 lists two additional key sources that provide specific guidance on risk characterization. In 1992, USEPA published *Guidance on Risk Characterization for Risk Managers and Risk Assessors* (1992b). This document contains guidelines that ensure health risks are properly characterized using information from different stages in the risk assessment process. USEPA released an update to the 1992 document in 2000; the *Risk Characterization Handbook* (USEPA 2000c) provides updates on critical elements that should be considered in characterizing risk.

Table 1.9 Estimated new cases and deaths for cancer sites by gender

	Ratio of Deaths to Cases		
	Total	Male	Female
All Cancers	0.48	0.50	0.46
Lung	0.89	0.90	0.88
Breast	0.25	0.31	0.25
Prostate	0.22	0.22	—
Colorectal	0.46	0.45	0.47
Non-Hodgkin's Lymphoma	0.44	0.44	0.45
Bladder	0.31	0.29	0.34
Kidney	0.36	0.35	0.36
Melanoma	0.22	0.25	0.18
Body of Uterus	0.19	—	0.19
Leukemia	0.61	0.62	0.61
Pancreas	0.99	0.99	0.98
Oral	0.35	0.35	0.36
Stomach	0.69	0.65	0.75
Ovary	0.61	—	0.61
Brain	0.66	0.68	0.63
Thyroid	0.08	0.11	0.08
Multiple Myeloma	0.68	0.66	0.70
Esophagus	1.08	1.11	0.99
Cervix	0.31	—	0.31
Larynx	0.42	0.42	0.39
Hodgkin's Disease	0.14	0.14	0.14
Testis	0.05	0.05	—
All Other Sites	0.74	0.84	0.66

Note: Based on Canadian statistics for 2003 (National Cancer Institute of Canada 2003)

Mortality to incidence ratio

	Ratio of Deaths to Cases		
	Total	Male	Female
Non-melanoma	0.013	0.014	0.012

Source: National Cancer Registry (Ireland) 1997

4   Issues in Risk Assessment

4.1   Use of Default Assumptions

Section 1.2 acknowledges that all risk assessments are associated with a degree of uncertainty, partly because of the lack of needed information, which is replaced by

Table 1.10 Key sources of information for risk characterization

Sources	Links
USEPA—*Guidance on Risk Characterization for Risk Managers and Risk Assessors*	http://www.epa.gov/superfund/programs/risk/habicht.htm
USEPA—*Risk Characterization Handbook*	http://www.epa.gov/osp/spc/2riskchr.htm

expert judgment or applicable regulatory policy. Default assumptions are often used and may be justifiable when the objective of the risk assessment is to exclude the presence of health risk or to compare various types of risks using the same set of assumptions. Using default assumptions is also practical for risk assessments conducted for a party potentially responsible for remediation in a legal, quasi-legal, or public proceeding. Use of well-established conservative default assumptions could defuse arguments that the risk assessor is biased toward more permissive risk estimates. However, some risk assessors tend to use default assumptions even in situations where useful and scientifically defensible data could be found if they were to analyze the available information more critically. They often adopt this approach because they can save time and money, and they justify it as being *conservative*.

There has been an ongoing debate between using a scientifically defensible, case-specific approach where possible and relying on default assumptions from which risk assessors deviate only when deemed inappropriate. The current thinking tilts toward relying on default assumptions only when one fails to identify scientifically defensible case-specific solutions and is a welcome trend. The following are excerpts from USEPA cancer guidelines.

> NRC [National Research Council 1994] reaffirmed the use of default options as "a reasonable way to cope with uncertainty about the choice of appropriate models or theory" (p. 104). It saw the need to treat uncertainty in a predictable way that is "scientifically defensible, consistent with the agency's statutory mission, and responsive to the needs of decision-makers" (p. 86). Accordingly, default options have a science component and a policy component.
>
> NRC discussed two opposing principles for governing departures from default options. One suggested principle would evaluate a departure in terms of whether "it is scientifically plausible" and whether it "tends to protect public health in the face of scientific uncertainty" (p. 601). An opposing principle "emphasizes scientific plausibility with regard to the use of alternative models" (p. 631). Reaching no consensus on a single approach, NRC recognized that developing criteria for departures is an EPA policy matter.
>
> With increasing understanding becoming available, these guidelines adopt a view of default options that is consistent with EPA's mission to protect human health. Rather than viewing default options as the starting point from which departures may be justified by new scientific information, these guidelines view a critical analysis of the avail-

able information as the starting point from which a default option may be invoked if needed to address uncertainty or the absence of critical information. The primary goal of EPA actions is public health protection; accordingly, as an Agency policy, any default options used in the absence of scientific data to the contrary should be health protective. (USEPA 2003c)

4.2   Best Risk Estimate or Central Tendency

Best risk estimate is what risk assessors perceive as the most realistic risk estimate that is most representative of the population at risk. The goal is to estimate the risk associated with *central tendency* exposure (CTE) in a population using central tendency input parameters. In practice, arithmetic means are used to represent central tendency for most input parameters, especially with regard to contaminant concentrations in soil. In some circumstances, the median or geometric mean may be a more appropriate measure of central tendency.

USEPA (1999a) defines CTE as "a risk descriptor representing the average or typical individual in a population, usually considered to be the mean or median of the distribution."

Best estimates usually receive the greatest attention from risk managers. It is therefore important for risk assessors to examine each step in their assessments to assure each assumption made is realistic for the particular population for which the risk is being determined. Some assessors prefer to use a probabilistic, rather than a deterministic, risk assessment to calculate the best estimates of risk, although probabilistic risk assessment is not always necessary or appropriate. A discussion on probabilistic and deterministic risk assessments follows.

4.3   Probabilistic versus Deterministic Risk Assessment

A probabilistic risk assessment (PRA) is "a risk assessment that uses probability distributions to characterize variability or uncertainty in risk estimates. In a PRA, one or more variables in the risk equation are defined as a probability distribution rather than a single number. Similarly, the output of a PRA is a range or probability distribution of risks experienced by the receptors" (USEPA 2001d).

The traditional deterministic (point estimate) approach assigns a single value to each parameter in the risk model to derive a single risk value. In contrast, the probabilistic (stochastic) approach inputs the whole distribution (a range of values) for one or more parameters in the model and uses statistical methods to propagate the variance of the input parameters through the model. The output variable is a distribution, described by its range, arithmetic mean, standard deviation, and different quartiles and is presented as a probability density function (PDF) or a cumulative distribution function (CDF). Thus, the probabilistic approach makes greater use of the available data than the point estimate approach and provides more information on the range of potential risks and their likelihood of occurrence in a population. Although both deterministic and probabilistic assessments are compatible with and can incorporate sensitivity analysis, some sensitivity analysis methods are available only with the probabilistic approach, making it easier to identify those parameters that drive the risk. Such information allows the risk manager to better prioritize risks and make better-informed decisions.

The authors' intent is not to provide a detailed review of this method but rather to discuss the advantages and disadvantages of the method and to provide readers with some key references that may be helpful in evaluating and performing probabilistic risk assessments.

4.3.1   Advantages and Disadvantages of Probabilistic Risk Assessment

Probabilistic risk assessment has both advantages and disadvantages. The views of individual European risk assessors and risk managers are contained in the report written by Jager (1998) for the Dutch National Institute of Public Health and the Environment (RIVM).[16] According to the USEPA (2001d), the primary advantage of PRA is that it is seen as a good way to quantify uncertainty and variability. USEPA also claims that it gives the risk manager greater flexibility to select the level of uncertainty in the continuum of uncertainties in defining a desirable risk management level. PRA is seen in some ways as an antidote to deterministic assessment, which in the past has been—but does not need to be—biased toward excessive conservatism. Proponents of PRA also point to the possibility of conducting a more sophisticated sensitivity analysis than is possible with deterministic assessment.

There are five major disadvantages of probabilistic risk assessment. First, the data to construct some distributions are at present incomplete and the data gaps are replaced by distributions that are sometimes built on conjecture rather than real data. This practice raises questions about the reliability of the output distributions of a PRA. Second, the uncertainty related to expert judgment (also called model uncertainty) is difficult to capture in terms of distributions, yet this type of uncertainty may be greater than data variability and uncertainty. Third, development of potency factors and absorption factors is usually not included in the PRA, yet these factors are major sources of uncertainty, which in all likelihood dwarf the variability and uncertainty associated with exposure assessment. To the authors, it is difficult to understand the benefit of using detailed probabilistic distributions of exposure factors when uncertainty factors of three orders of magnitude are applied to estimate potency factors. The fourth and perhaps most important disadvantage follows from the disadvantages already discussed: the graphical output from probabilistic assessment creates a false sense of reliability and precision. Rather than making the risk assessment more transparent, there is a greater tendency that data and knowledge gaps are hidden behind attractive graphs. The final drawback is that PRA takes more time and costs more to prepare than a deterministic risk assessment.

Like PRA, a deterministic risk assessment can also provide risk managers with a range of risk estimates for a range of scenarios using a range of different assumptions for biological and lifestyle exposure factors. A PRA does not always provide more information than a deterministic risk assessment. Before launching a probabilistic risk assessment, risk assessors are advised to consider carefully whether the potential benefits of a PRA justify the extra time and cost it entails.

4.3.2   Key Sources of Information

Guidelines and distributions relating to PRAs are available from USEPA (1998d, 2000g, 2001d), the Oregon Department of Environmental Quality (1998), CalEPA (2000), and Jager et al. (2000). Some of the relevant software is shown in table 1.11.

Table 1.11 Probabilistic risk assessment software

Product	Company	Link
@Risk, V.4.5	Palisade Corporation	http://www.palisade.com/
Crystal Ball 2000	Decisioneering	http://www.decisioneering.com/products.html
Analytica 3	Lumina	http://www.lumina.com/

4.4   Multiple Exposure Pathways, Sources, and Stressors

Initially, most environmental risk assessments were source focused and chemical focused, that is, the assessment focused on considering the effects of a single source—and often of a single chemical—on various subpopulations (receptors). Exposures via different pathways were treated as independent events—in other words, one individual exposed to one substance via a single pathway at a time. However, in the real world, exposures do not occur as single events, but rather as a series of sequential or simultaneous events that are linked in time and place. To correct for this inadequacy, USEPA (1999e) has developed guidance for conducting aggregate exposure and risk assessment specifically for the use of pesticides, although the same approach applies to other substances:

> The approach to aggregate exposure and risk assessment focuses on the potential exposure to a single chemical by multiple routes to individuals in a population. Exposures to an individual in a population: (1) may occur by more than one route (i.e., oral, dermal, and/or inhalation); (2) may originate from more than one source and/or pathway (i.e., food, drinking water, and residential); (3) occur within a time frame such that the chemical exposure overlaps correspond to the effective period of the adverse toxicological effect; (4) occur at a spatially relevant set of locations that correspond to an individual's potential exposure; and (5) be demographically consistent. (USEPA 1999e)

Though an improvement, aggregate risk assessment ignores the fact that in the environment, people are exposed to complex mixtures that may include hundreds of potentially interacting contaminants. This issue was recognized early on and USEPA (1986b, 2000d) has developed guidelines for the assessment of mixtures. USEPA (2003f) has also estimated potencies for several mixtures (e.g., PCBs, coke oven emissions, diesel engine exhaust). The World Health Organization (WHO 1999) has developed guidelines for diesel exhaust and PAHs.[17] Other agencies also have estimated potencies for complex mixtures (Ontario Ministry of Environment and Energy 1997b).

Perhaps even more important, a given population may be exposed to multiple sources and may experience aggregate effect from these sources. USEPA describes the difference between the traditional single source approach and the newer multiple source approach as follows:

> A chemical-focused assessment may look at several populations affected by exposure to the chemical but not at other chemicals. A population-focused assessment looks at one population for perhaps many stressors but not at other populations. Consequently, for traditional,

chemical-focused assessments, we say we conduct a "risk assessments for a certain chemical." In contrast, the essence of a cumulative risk assessment is that the assessment is conducted "for a certain population." (USEPA 2003d)

Concerns for exposure from multiple sources lead to the development of community-based risk assessment and eventually cumulative risk assessment methods. The methodology for community-based assessment was developed partly by the Total Exposure Assessment Methodology (TEAM) program (USEPA 1987a) and the National Human Exposure Assessment Survey (NHEXAS) (Sexton, Kleffman, and Callahan, 1995). Community-based assessments attempt to estimate the effect of multiple sources on a community. The overall risk and the relative risk from individual sources for different segments of the community are usually assessed, taking into consideration the location of residence and the types of activity in which residents are engaged. In the United States, these projects are tracked by USEPA in the *Air Toxics Community Assessment and Risk Reduction Projects Database* (2003a). The database provides a description of the individual projects and their status of completion; copies of reports are also available. Currently, the database contains well over a hundred studies.

Cumulative risk assessment evolved from community-based risk assessment, as there was a perceived need to consider factors other than chemical or radiation exposure, such as the quality of health care, socioeconomic status, and other factors not traditionally included in a human health risk assessment. These factors, which can separately or together induce adverse effects, are called *stressors*. USEPA describes a stressor as

> a physical, chemical, biological, or other entity that can cause an adverse response in a human or other organism or ecosystem. Exposure to a chemical, biological, or physical agent (e.g., radon) can be a stressor, as can the lack of, or destruction of, some necessity, such as a habitat. The stressor may not cause harm directly, but it may make the target more vulnerable to harm by other stressors. A socioeconomic stressor, for example, might be the lack of needed health care, which could lead to adverse effects. Harmful events, such as automobile crashes, could also be termed stressors. (USEPA 2003d)

USEPA (2003d) defines *cumulative risk* as "the combined risks from aggregate exposures to multiple agents or stressors" and *cumulative risk assessment* as "an analysis, characterization, and possible quantification of the combined risks to health or the environment from multiple agents or stressors."

Several programs within USEPA have been working relatively independently to develop methodology for cumulative risk assessment. An agency-wide framework was published recently (USEPA 2003d). Readers who are particularly interested in additional references regarding pesticides and ecological risk assessments are advised to consult this report. USEPA is not the only agency that is developing cumulative risk assessment.

Cumulative risk assessment aims at assessing scenarios involving simultaneous exposure of heterogeneous populations to multiple interacting stressors in the past, present, and future. This exercise can lead to extremely complex assessments that are expensive and time-consuming and demand considerable commitment on the part of the assessors and managers. Furthermore, during an external peer reviewer meeting to discuss the USEPA cumulative risk assessment report (2003d), it became clear that the methods for conducting cumulative human health risk assessments are still very much under development. In many instances, analysis of many stressors can be at best qualitative and the results may need to be interpreted with caution, as substantial knowledge gaps persist. Although some concepts, thinking, and elements of cumulative risk assessments could be readily incorporated into pragmatic risk assessments, one has to consider carefully whether the additional benefits of conducting an ambitious cumulative assessment justify the large commitment of time and resources that could otherwise be used to address other environmental problems.

Cumulative risk assessments, community-based assessments, and the integrated decision-making process (discussed in section 5.3) have a lot in common. They analyze the impact of multiple sources of contamination on a given population and its subpopulations. Therefore, any one of the three approaches could be considered when an impact on a given community or an area needs to be assessed. Community-based assessments can be relatively simple and generate data that are relatively easy to understand and use for developing management solutions. It could be argued, however, that this type of assessment oversimplifies the situation and management decisions made on the basis of such an assessment could be overly simplistic. In contrast, cumulative risk assessment is intended to consider more stressors and takes into account the interactions of the various effects these stressors may have on human health.

Both the integrated environmental decision-making (IED) process and cumulative risk assessment can consider the impact of a broad range of stressors on a given population. The two processes have a lot of similarities (see USEPA 2003d); however, there is one important difference. The ultimate goal of the IED process is to achieve optimal risk reduction with the resources available. The IED process includes risk management decision making and evaluation of the effectiveness of the risk management steps. In contrast, cumulative risk assessment as designed by USEPA explicitly stays away from incorporation of any risk management process (USEPA 2003d). Thus, the outcome of this resource- and time-consuming assessment may not be focused on providing answers to the questions that risk managers need to address to achieve effective risk reduction.

The differences between the three types of assessment may be to some degree arbitrary. Elements of cumulative risk assessment can be built into community-based assessment, and similarly cumulative risk assessment could be pared down to resemble a community-based assessment. Cumulative risk assessment could be designed with risk management and decision making in mind. The goals and the objectives of the assessment need to be made clear at the beginning of the project, so that the type of assessment suited to achieve preset goals can be designed.

5 Applications of Risk Assessment

Risk assessment is not a single standard spreadsheet-based process that accepts certain inputs and provides fixed outputs. Rather, it is a tool that needs to be adjusted to reflect the purpose for which the assessment is undertaken; in particular, the level of conservatism and the level of complexity must meet the purpose of the assessment. For example, if the goal is to exclude any safety concern, the risk assessment can be simple and conservative. If the purpose is to estimate realistically the risk facing a population, the assessment has to be detailed and must minimize consistent bias toward either permissiveness or conservatism.

Although risk assessment is an imprecise tool, it has important risk management applications. The following sections describe the common applications of risk assessment and the special consideration for each application.

5.1   Meeting Regulatory Criteria

Most risk assessments are conducted according to particular regulatory guidelines. Examples of such guidelines are provided in section 1.5. Guidelines usually provide a general direction for conducting risk assessment and define the level of risk that can be considered "sufficiently low" from a regulatory perspective. The estimated risk is in effect compared to numerical values (e.g., standards, guidelines, or criteria). If the assessment is conducted according to the guidelines and the risk is determined to be below regulatory concern, little or no risk reduction is usually required. Different regulatory bodies may tolerate modest exceedances (up to two- or threefold), which often can be justified by the conservatism that is built into a typical risk assessment. But as the exceedance becomes bigger, risk managers face increasing pressure to implement risk reduction steps.

In many cases, a risk assessment for the purpose of meeting regulatory criteria is better viewed as a comparison of risk to a fixed prescribed risk level rather than an estimation of the true risk. Nonetheless, such calculation has its applications. For example, it can be used to prioritize different carcinogens for action or different risk-reduction approaches for the same carcinogen. But it is generally not recommended for estimating the number of residents in city X who will develop cancer due to exposure to a given carcinogen in the ambient air. The risk assessment is associated with so much uncertainty that any estimate of cancer cases may be unreliable.

5.2   Screening for Absence of Potential Health Risk

A risk manager may wish to have a simple risk assessment conducted to exclude the possibility of danger to humans from exposure to some contaminants. The purpose is not to estimate the risk accurately, but rather to arrive at a conservative risk estimate representing a worst-case scenario. If the assessment indicates that the risk is below the level of regulatory concern, it is unlikely that risk management actions would be needed and the risk manager would have more latitude in implementing risk-reduction steps.

When conducting this type of assessment, it is important to build in a reasonable degree of conservatism. The assessment needs to be conservative enough that the actual risk to the target population is unlikely to be higher than the estimated risk. At the same time, the risk assessment should not be so conservative as to grossly overestimate the risk.

The degree of conservatism in a risk assessment depends on how conservative the assumptions are regarding population exposure and contaminant toxicity. In an assessment for risk screening, the maximum possible values for exposure and dose that can conceivably happen are used in the calculation. The estimate is generally referred to as the "worst case," defined by USEPA as follows:

> A semiquantitative term referring to the maximum possible exposure, dose, or risk, that can conceivably occur, whether or not this exposure, dose, or risk actually occurs or is observed in a specific population. Historically, this term has been loosely defined in an ad hoc way in the literature, so assessors are cautioned to look for contextual definitions when encountering this term. It should refer to a hypothetical situation in which everything that can plausibly happen to maximize exposure, dose, or risk does in fact happen. This worst case may occur (or even be observed) in a given population, but since it is usually a very unlikely set of circumstances, in most cases, a worst-case estimate will be somewhat higher than occurs in a specific population. As in other fields, the worst-case scenario is a useful device when low probability events may result in a catastrophe that must be avoided even at great cost, but in most health risk assessments, a worst-case scenario is essentially a type of bounding estimate. (USEPA 2002e)

Most risk assessments designed to exclude any possibility of health risk to an exposed population do not use the worst-case scenario but rather a "reasonable worst-case" (sometimes called "plausible worst-case") scenario. Again a good description of the term is provided by the USEPA:

> Reasonable worst case: A semiquantitative term referring to the lower portion of the high end of the exposure, dose, or risk distribution. The reasonable worst case has historically been loosely defined, including synonymously with maximum exposure or worst case. As a semiquantitative term, it is sometimes useful to refer to individual exposures, doses, or risks that, while in the high end of the distribution, are not in the extreme tail. For consistency, it should refer to a range that can conceptually be described as above the 90th percentile in the distribution, but below about the 98th percentile. (USEPA 2002e)

It is important to understand that risk assessments based on worst-case or even reasonable worst-case scenarios are not suitable for making quantitative estimates of risk to human health. If the risk is found to be high using this method, it does not imply that the studied population is at risk. Instead, a more comprehensive assessment using more realistic exposure scenarios and more representative estimates needs to be conducted.

5.3 Risk Comparison and Risk Reduction

Comparative risk assessment is used expressly to compare risks for the purposes of planning and budgeting. It is used as an environmental management–planning tool, as a risk communication tool, and generally as an aid to environmental decision making. This type of risk assessment is often called *comparative risk assessment* (CRA).

In 1987, USEPA published its pioneering CRA report (USEPA 1987b) to help realign the allocation of USEPA's resources to maximize risk reduction. In the preface of that report, former USEPA administrator Lee M. Thomas wrote: "Although EPA's mission enjoys a broad public support, our agency nonetheless must operate on finite resources. Therefore, we must choose our priorities carefully so that we apply those resources as effectively as possible" (USEPA 1987b).

CRA has also become recognized as a communication tool. The Science Advisory Board (SAB) of USEPA concluded that the concept of comparative environmental risk assessment "helps people discuss disparate environmental problems with a common language. It allows many environmental problems to be measured and compared in common terms, and it allows different risk reduction options to be evaluated from a common basis. Thus the concept of environmental risk can help an organization to develop environmental policies in a consistent and systematic way" (USEPA 1987b).

A considerable number of comparative risk assessments have been undertaken (for references, see Green Mountain Institute for Environmental Democracy 1998), and USEPA has developed a draft comparative risk framework methodology (Boutin et al. 1998). More recently, SAB proposed incorporating CRA into an integrated environmental decision-making (IED) process, where risk assessment plays a role in prioritizing environmental problems and potential approaches to address these problems and serves as a common language for discussing environmental problems and communicating the outcome of the assessment (USEPA 2000e).

Although many comparative risk assessments are conducted by large jurisdictions, smaller municipalities have also undertaken this type of assessment. A streamlined version can be applicable at the industrial plant or at the corporate level (for more details, see Muller and Nicholson 2000).

Once environmental issues have been prioritized, it becomes necessary to find ways to achieve maximal risk reduction with the available resources. Risk reduction tools could involve market forces, science and technology, information, and product specifications. Risk reduction options can be prioritized. The final selection of risk reduction options may be made based on some preestablished criteria.

USEPA's Science Advisory Board has been the leader in this type of work. SAB introduced the basic concepts in 1990 (USEPA 1990). In spring 1999, SAB released a sequel to that report, entitled *Integrated Environmental Decision-making in the Twenty-First Century* (USEPA 1999d), in which it describes the framework in greater detail than in the 1990 report. The reference in the title to integrated environmental decision-making reflects the emphasis on incorporating and integrating into the decision-making framework multiple factors that impact the health of the environment and the various economic consequences of risk reduction efforts. In the final report (USEPA 2000e), many of the details contained in the 1999 draft were omitted, and it has become more of a framework as compared to the more detailed draft.

6 Definition of Terms

Absorption Factor (f): The proportion of a chemical in contact with the body outer surface (skin or lumen of the GI tract or respiratory tract) that would enter the blood compartment.

Average Respiration Rate (AR): The average rate of respiration over the averaging time.

Averaging Time (AT): The period over which exposure is averaged (in days).

Benchmark Dose (BMD) or *Benchmark Concentration (BMC):* An exposure due to a dose or concentration (of a substance) associated with a specified low incidence of risk, generally in the range of 1–10 percent, of a health effect; or the dose or concentration associated with a specified measure or change of a biological effect.

Benchmark Response (BMR): The response, generally expressed as in excess of background, at which a *benchmark dose* or *benchmark concentration* is desired.

Bioaccessible Fraction: The proportion of a chemical present in a matrix (e.g., soil) that can cross the body barrier (skin, respiratory, or digestive tracts).

Bioavailable Fraction: The proportion of a chemical, free from any interaction with a matrix (e.g., soil), that is available for absorption across the body barrier (skin, respiratory, or digestive tracts).

BMDL: The lower one-sided confidence limit on the *benchmark dose.*

Dermal Absorption Factor (ABS): The proportion of an agent in contact with the skin's surface that enters the blood system.

Exposure Duration (ED): The total time span of exposure (in years).

Exposure Frequency (EF): How often exposure occurs (in days/year).

Exposure Ratio (ER): The ratio of estimated exposure level to the reference exposure level over the same time period for a substance.

Exposure Time (ET): The number of hours per day during which exposure takes place.

Guideline Value (GV): For noncarcinogenic effects, WHO develops guideline values in a manner similar to *reference concentration, reference dose, tolerable concentration,* and *tolerable daily intake.* For carcinogenic effects, GV represents an incremental unit risk estimate (i.e., the additional lifetime cancer risk occurring in a hypothetical population in which all individuals are exposed continuously from birth throughout their lifetimes to a concentration of 1 g/m^3 of the agent in the air they breathe).

Hazard: A potential source of harm.

Human Equivalent Concentration (HEC) or *Human Equivalent Dose (HED):* The human concentration (for inhalation exposure) or dose (for other routes of exposure) of an agent that is believed to induce the same magnitude of toxic effect as the experimental animal species concentration or dose. This adjustment may require incorporation of toxicokinetic information on the particular agent, if available, or use of a default procedure such as assuming that daily oral doses experienced over a lifetime are proportional to the body weight raised to the 0.75th power.

Human Health Risk Assessment: The determination of potential adverse health effects from exposure to chemicals, including both qualitative and quantitative expressions of risk. The process of risk assessment involves four major steps: hazard identification, dose-response assessment, exposure assessment, and risk characterization.

Inhalation/Ingestion Rate (IR): Amount of contaminated medium inhaled/ingested per unit time or event.

Lowest-Observed-Adverse-Effect Level (LOAEL): The lowest exposure level at which there are statistically or biologically significant increases in frequency or severity of adverse effects in the exposed population and its appropriate control group. It is also referred to as the lowest-effect level (LEL).

Lowest-Observed-Effect Level (LOEL): The lowest dose or exposure level in a study at which a statistically or biologically significant effect is observed in the exposed population compared to an appropriate unexposed control group.

Maximum Likelihood Estimate (MLE): A statistical method for estimating model parameters. The MLE generally provides a mean or central tendency estimate, as opposed to a confidence limit on the estimate.

Modifying Factor (MF): A factor used in the derivation of a *reference dose* or *reference concentration.* The magnitude of the MF reflects the scientific uncertainties of the study and database not explicitly treated with standard *uncertainty factors* (e.g., the completeness of the overall database). An MF is greater than zero and less than or equal to 10, and the default value for the MF is 1.

"Non-threshold" Chemicals: Chemicals having a linear dose-response relationship at low concentrations.

No-Observed-Adverse-Effect Level (NOAEL): The highest exposure level at which there are no statistically or biologically significant increases in the frequency or severity of adverse effect between the exposed population and its appropriate control; some effects may be produced at this level, but they are not considered adverse, nor precursors to adverse effects.

No-Observed-Effect Level (NOEL): An exposure level at which there are no statistically or biologically significant increases in the frequency or severity of any effect between the exposed population and its appropriate control.

Point of Departure (POD): The dose-response point that marks the beginning of a low-dose extrapolation. This point is most often the upper bound on an observed incidence or on an estimated incidence from a dose-response model. The POD is an estimated dose (expressed in human-equivalent terms) near the lower end of the observed range, without significant extrapolation to lower doses.

Reference Concentration (RfC): An estimate (with uncertainty spanning perhaps an order of magnitude) of a continuous inhalation exposure to the human population (including sensitive subgroups) that is likely to be without an appreciable risk of deleterious effects during a lifetime. It can be derived from a *no-observed-adverse-effect level, lowest-observed-adverse-effect level,* or *benchmark concentration,* with *uncertainty factors* generally applied to reflect limitations of the data used. RfC is generally used in USEPA's noncancer health assessments.

Reference Dose (RfD): An estimate (with uncertainty spanning perhaps an order of magnitude) of a daily oral exposure to the human population (including sensitive subgroups) that is likely to be without an appreciable risk of deleterious effects during a lifetime. It can be derived from a *no-observed-adverse-effect level, lowest-observed-adverse-effect level,* or *benchmark dose,* with *uncertainty factors* generally applied to reflect limitations of the data used. RfD is generally used in USEPA's noncancer health assessments.

Risk: The probability of injury, disease, or death from exposure to a chemical agent or a mixture of chemicals.

Risk-Specific Concentration (RsC): The airborne concentration of a chemical (in mg/m^3) that is associated with a specified excess lifetime cancer risk, often one in a million (10^{-6}).

Risk-Specific Dose (RsD): The dose of a chemical (in mg/kg-day) that is associated with a specified excess lifetime cancer risk, often one in a million (10^{-6}).

Seasonal Exposure (ES): Duration of a season over which a given exposure rate takes place.

Semi-Volatile Organic Compound (SVOC): A substance that evaporates slowly at 20°C and 1 atm pressure.

Site-Specific Risk Assessment (SSRA): A form of risk assessment where site-specific conditions are used in the identification of the potential risks facing the users of a given site as a result of its environmental conditions.

Slope Factor: The upper bound, approximating a 95 percent confidence limit, on the increased cancer risk per unit exposure from a lifetime exposure to a carcinogenic agent. This estimate, usually expressed as the proportion (of an exposed population) affected per mg/kg/day, is generally reserved for use in the low-dose region of the dose-response relationship, that is, for exposures corresponding to risk less than one in 100.

"Threshold" Chemicals: Chemicals having a nonlinear dose-response relationship at low concentrations.

Tolerable Concentration (TC): An airborne concentration to which it is believed a person can be exposed continuously over a lifetime without deleterious noncarcinogenic effect.

Tolerable Daily Intake (TDI): The tolerable daily intake, expressed on a body weight basis (e.g., mg/kg/day), is the total intake by ingestion to which it is believed a person can be exposed daily over a lifetime without deleterious effect. TDI is based on noncarcinogenic effects and is usually calculated by applying an uncertainty factor to the *no-observed-adverse-effect level* or *lowest-observed-adverse-effect level.* Maximum tolerable intakes per day can be developed for various age groups by multiplying the TDI by the average body weight of the age group under consideration. It should be noted, however, that exceedance of such a calculated intake by a particular age group for a small proportion of a person's lifespan does not necessarily imply that the exposure constitutes an undue risk to health.

Toxic Equivalency Factor (TEF): A measure of the relative toxic potency of a chemical as compared to a well-characterized reference chemical.

Tumorigenic Concentration (TC$_{05}$): Concentration associated with a 5 percent increase in incidence or mortality due to tumor.

Tumorigenic Dose (TD$_{05}$): Total intake associated with a 5 percent increase in incidence or mortality due to tumor.

Uncertainty Factor (UF): One of several, generally 10-fold factors, used in deriving the *reference dose* and *reference concentration* from experimental data. UFs are intended to account for (1) the variation in sensitivity among members of the human population, i.e., interhuman or intraspecies variability; (2) the uncertainty in extrapolating animal data to humans, i.e., interspecies variability; (3) the uncertainty in extrapolating from data obtained in a less-than-lifetime exposure study to lifetime exposure, i.e., extrapolating from subchronic to chronic exposure; (4) the uncertainty associated with derivation of an RfC/RfD from a *lowest-observed-adverse-effect level* rather than from a *no-observed-adverse-effect level*; and (5) the uncertainty associated with extrapolation from animal data when the database is incomplete.

Unit Risk: The upper-bound excess lifetime cancer risk resulting from continuous exposure over one's lifetime to an agent at a concentration of 1 g/liter in the water one drinks, or 1 μg/m^3 in the air one breathes. Unit risk can be interpreted as follows: a drinking water unit risk of 2.0 E–6 per μg/L means that two persons out of a population of 1 million exposed daily to 1 g of the chemical in 1 liter of drinking water over a lifetime are expected to develop cancer in their lifetime.

Upper Confidence Limit (UCL): The upper bound of a confidence interval around any calculated statistical parameter, most typically the "average." For example, the 95 percent confidence interval for an average is the range of values that will contain the true average value (i.e., the average of the full statistical population of all possible data) 95 percent of the time. In other words, we can say with 95 percent certainty that the "true" average will exceed the UCL only 2.5 percent of the time. In the face of uncertainty, to avoid underestimating the true unit risk estimate, USEPA and others have based most unit risk estimates on the upper confidence limit of the response data or fitted dose-response curves.

Volatile Organic Compound (VOC): Any organic compound that evaporates readily to the atmosphere. VOCs contribute significantly to photochemical smog production and certain health problems.

7 References and Recommended Resources

Agency for Toxic Substances and Disease Registry (ATSDR). 2001. *A primer on health risk communication principles and practices.* Available at www.atsdr.cdc.gov/HEC/primer.html (accessed 22 February 2004).

———. 2003a. *Toxicological profiles.* Available at http://www.atsdr.cdc.gov/toxpro2.html (accessed 28 November 2003).

————. 2003b. *ToxFAQs*. Available at http://www.atsdr.cdc.gov/toxfaq.html (accessed 28 November 2003).

————. 2003c. *Minimal risk levels (MRLs) for hazardous substances*. Available at http://www.atsdr.cdc.gov/mrls.html (accessed 28 November 2003).

Albert, R. E., J. Lewtas, S. Nesnow, T. W. Thorslund, and E. Anderson. 1983. Comparative potency method for cancer risk assessment: Application to diesel particulate emissions. *Risk Analysis* 3:101–17.

Allen, B. C., R. J. Kavlock, C. A. Kimmel, and E. M. Faustman. 1993. Comparison of quantitative dose response modeling approaches for evaluating fetal weight changes in segment II developmental toxicity studies. *Teratology* 47:41. Cited in WHO 2000.

————. 1994a. Dose-response assessment for developmental toxicity, part 2: Comparison of generic benchmark dose estimates with NOAELs. *Fund. Appl. Toxicol.* 23:487–95. Cited in USEPA 2000a.

Allen, B. C., C. Van Landingham, R. B. Howe, R. J. Kavlock, C. A. Kimmel, and E. M. Faustman. 1992. Dose-response modeling for developmental toxicity. *Toxicologist* 12:300. Cited in WHO 2000.

American Chemical Society. 1998. *Understanding risk analysis: A short guide for health, safety, and environmental policy making*. Washington, D.C.: American Chemical Society.

Andersen, M. E. 1995. Development of physiologically based pharmacokinetic and physiologically based pharmacodynamic models for applications in toxicology and risk assessment. *Toxicol. Lett.* 79:35–44.

————. 2003. Toxicokinetic modeling and its applications in chemical risk assessment. *Toxicol. Lett.* 138:9–27.

Andersen, M. E., M. G. MacNaughton, H. J. Clewell, and D. P. Paustenbach. 1987. Adjusting exposure limits for long and short exposure periods using a physiological pharmacokinetic model. *Amer. Industrial Hygiene Assoc. Journal* 48:335–43.

Anderson, E. L., and U.S. Environmental Protection Agency Carcinogen Assessment Group. 1983. Quantitative approaches in use to assess cancer risk. *Risk Analysis* 3:277–95.

Association of State and Territorial Health Officials (ASTHO). 2002. *Communication in risk situations: Responding to the communication challenges posed by bioterrorism and emerging infectious diseases*.

Baars, A. J., R. M. C. Theelen, P. J. C. M. Janssen, J. M. Hesse, M. E. van Apeldoorn, M. C. M. Meijerink, L. Verdam, and M. J. Zeilmaker. 2001. *Re-evaluation of human-toxicological maximum permissible risk levels*. RIVM 711701025. Available at http://www.rivm.nl/bibliotheek/rapporten/711701025.html (accessed 26 December 2003).

Boström, C., P. Gerde, A. Hanberg, B. Jernström, C. Johansson, T. Kyrklund, A. Rannug, M. Törnqvist, K. Victorin, and R. Westerholm. 2002. *Cancer risk assessment, indicators and guidelines for polycyclic aromatic hydrocarbons (PAH) in the ambient air.* Stockholm, Sweden: Institute of Environmental Medicine, Karolinska Institutet.

Boutin, B., M. B. Brown, R. Clark, J. C. Lipscomb, R. Miltner, P. Murphy, L. R. Papa, D. Reasoner, and R. Rheingans. 1998. *Comparative risk framework methodology and case study.* Washington, D.C.: U.S. Environmental Protection Agency, Office of Research and Development, National Center for Environmental Assessment. Available at http://cfpub.epa.gov/ncea/cfm/recordisplay.cfm?-deid = 12465 (accessed 28 November 2003).

Calabrese, E. J., and L. A. Baldwin. 2000a. Chemical hormesis: Its historical foundations as a biological hypothesis. *Human & Experimental Toxicology* 19:2–31.

———. 2000b. Tales of two similar hypotheses: The rise and fall of chemical and radiation hormesis. *Human & Experimental Toxicology* 19:85–97.

CalEPA (California Environmental Protection Agency). 1993. Health assessment. Part B of *Benzo[a]pyrene as a toxic contaminant.*

———. 1997. *Final statement of reasons for rulemaking: Staff report/Executive summary: Proposed identification of inorganic lead as a toxic air contaminant.* Available at http://www.oehha.org/air/toxic_contaminants/leadrpt.html#** (accessed 23 February 2004).

———. 1999. The determination of acute reference exposure levels for airborne toxicants. Part 1 of *Air Toxics Hot Spots Program risk assessment guidelines.* Available at http://www.oehha.ca.gov/air/acute_rels/acuterel.html (accessed 28 November 2003).

———. 2000a. *All acute reference exposure levels developed by OEHHA as of May 2000.* Available at http://www.oehha.ca.gov/air/acute_rels/allAcRELs.html (accessed 28 November 2003).

———. 2000b. Technical support document for exposure assessment and stochastic analysis. Part 4 of *Air Toxics Hot Spots Program.* Oakland, Calif. Available at http://www.oehha.org/air/hot_spots/finalStoc.html (accessed 28 November 2003).

———. 2000c. Technical support document for the determination of noncancer chronic reference exposure levels. Part 3 of *Air Toxics Hot Spots Program.* Oakland, Calif. Available at http://www.oehha.ca.gov/air/chronic_rels/51702 chrel.html (accessed 28 November 2003).

———. 2002. *The Air Toxics Hot Spots Program: Guidance manual for preparation of health risk assessments.* Public review draft, June 2002. Available at http://www.oehha.ca.gov/air/hot_spots/index.html (accessed 28 November 2003).

Canadian Institute for Environmental Law and Policy. 1999. *Ontario's environment and the "common sense revolution": A four-year report.* Toronto: Canadian Institute for Environmental Law and Policy. Available at http://www.cielap.org/infocent/pub/reports/csr.html (accessed 28 November 2003).

Canadian Standards Association (CSA). 2002. *Risk management: Guideline for decision makers.* CAN/CSA-Q850–97 (R2002).

CCME (Canadian Council of Ministers of the Environment). 1996. *Guidance manual for developing site-specific soil quality remediation objectives for contaminated sites in Canada.* Winnipeg, Man.: Canadian Council of Ministers of the Environment Secretariat. Available at http://www.ec.gc.ca/ceqg-rcqe/English/Pdf/soil_sitespecific.pdf (accessed 28 November 2003).

———. 1997. *Guidance document on the management of contaminated sites in Canada.* Winnipeg, Man.: Canadian Council of Ministers of the Environment Secretariat.

———. 2002. *Canadian environmental quality guidelines.* Winnipeg, Man.: Canadian Council of Ministers of the Environment Secretariat. Available at http://www.ccme.ca (accessed 28 November 2003).

Clewell, H. J., III. 1995. The application of physiologically based pharmacokinetic modeling in human health risk assessment of hazardous substances. *Toxicol. Lett.* 79:207–17.

Clewell, H. J., III, M. E. Andersen, and H. A. Barton. 2002. A consistent approach for the application of pharmacokinetic modeling in cancer and noncancer risk assessment. *Environ. Health Perspect.* 110:85–93.

Committee on Environment and Natural Resources. 1997. *Integrating the nation's environmental monitoring and research networks and programs: A proposed framework.* Washington, D.C.: National Science and Technical Council, Executive Office of the President.

Covello, V., and F. Allen. 1988. *Seven cardinal rules of risk communication.* Washington, D.C.: U.S. Environmental Protection Agency, Office of Policy Analysis.

Crump, K. S. 1996. The linearized multistage model and the future of quantitative risk assessment. *Human & Experimental Toxicology* 15:787–98.

Daston, G. P., and R. A. Corley. 2002. The potential for biological modeling to improve children's risk assessment. *Toxicologist* 66:55.

Delhagen, D., J. Dea, and K. Kramer. 1996. *Comparative risk at the local level: Lessons from the road.* Boulder, Colo.: Western Center for Environmental Decision Making.

Dourson, M. L., and J. F. Stara. 1983. Regulatory history and experimental support of uncertainty (safety) factors. *Regulatory Toxicology and Pharmacology* 5:224–38.

Draize J. H., G. Woodward, and H. O. Calvery. 1944. Methods for the study of irritation and toxicity of substances applied topically to the skin and mucous membranes. *Journal of Pharm. and Exper. Therapeutics* 78:458–63.

enHealth Council. 2001a. *Exposure scenarios and exposure settings.* 3rd ed. Australia: Queensland Department of Health. Available at http://www.health.gov.au/pubhlth/publicat/document/metadata/env_exposure. htm (accessed 28 November 2003).

———. 2001b. *Health-based soil investigation levels.* 3rd ed. Australia: Queensland Department of Health. Available at http://www.health.gov.au/pubhlth/publicat/document/metadata/env_soil.htm (accessed 28 November 2003).

———. 2001c. *Health impact assessment guidelines, September 2001.* Australia: Commonwealth Department of Health and Aged Care. Available at http://www

.health.gov.au/pubhlth/publicat/document/metadata/env_impact.htm (accessed 28 November 2003).

———. 2002. *Environmental health risk assessment: Guidelines for assessing human health risks from environmental hazards, June 2002.* Australia: Commonwealth Department of Health and Ageing. Available at http://www .health.gov.au/pubhlth/publicat/document/metadata/env_hra.htm (accessed 28 November 2003).

———. 2003. *National environmental health monographs.* Available at http://enhealth.nphp.gov.au/council/pubs/ecpub.htm (accessed 28 November 2003).

Environment Canada. 2002. *Priority Substance Assessment Program: Assessment reports.* Available at http://www.ec.gc.ca/substances/ese/eng/psap/final/ main.cfm (accessed 28 November 2003).

———. 2003a. *CEPA Environmental Registry: General information.* Available at http:// www.ec.gc.ca/ceparegistry/gene_info (accessed 28 November 2003).

———. 2003b. *National Air Pollution Surveillance (NAPS) network annual data summary for 2002.* December. EPS 7/AP/35. Available at http://www .etcentre.org/publications/napsreports_e.html.

Environment Canada and Health Canada. 1992. *The Canadian Environmental Protection Act Priority Substances List assessment report: Toluene.* Environment Canada and Health Canada.

———. 1993. *Priority Substances List assessment report: Polychlorinated dibenzodioxins and polychlorinated dibenzofurans.* Environment Canada and Health Canada.

———. 1994a. *Priority Substances List assessment report: Cadmium and its compounds.* Environment Canada and Health Canada.

———. 1994b. *Priority Substances List assessment report: Chromium and its compounds.* Environment Canada and Health Canada.

———. 1994c. *Priority Substances List assessment report: Nickel and its compounds.* Environment Canada and Health Canada.

Environmental Defence Fund (EDF). 2003a. *Scorecard: About the chemicals.* Washington, D.C.: EDF. Available at http://www.scorecard.org/chcmical-profiles (accessed 28 November 2003).

———. 2003b. *Scorecard: Setting priorities: What is comparative risk analysis?* Washington, D.C.: EDF. Available at http://www.scorecard.org/comp-risk/def/ comprisk_explanation.html (accessed 28 November 2003).

European Chemicals Bureau (ECB). 2002. *European Union risk assessment reports: Acrylamide.* Available at http://ecb.jrc.it/existing-chemicals (accessed 28 November 2003).

———. 2003a. *European Union risk assessment reports.* Available at http://ecb.jrc.it/ existing-chemicals (accessed 28 November 2003).

———. 2003b. *Risk assessment principles.* Available at http://ecb.jrc.it/existing-chemicals (accessed 28 November 2003).

————. 2003c. *Technical guidance document on risk assessment.* 2nd ed. Available at http://ecb.jrc.it/existing-chemicals (accessed 28 November 2003).

European Commission. 1996. *Technical guidance documents in support of the Commission Directive 93/67/EEC on risk assessment for new notified substances and Commission Regulation (EC no. 1488/94 on risk assessment for existing substances).* Luxembourg: European Commission.

————. 2001. *Ambient air pollution by polycyclic aromatic hydrocarbons (PAH).* Position paper prepared by the Working Group on Polycyclic Aromatic Hydrocarbons (PAH). Luxembourg: European Communities.

Expert Panel on Air Quality Standards. 1999. *Polycyclic aromatic hydrocarbons.* London: Department of the Environment.

Flowers, L., and G. Hsu. 2004. Personal communication, February 3, 2004. U.S. Environmental Protection Agency.

Gaylor, D. W. 1989. Preliminary estimates of the virtually safe dose for tumors obtained from the maximum tolerated dose. *Regulatory Toxicology and Pharmacology* 9:1–18.

Golden, K. M. 1996. *Relative risk ranking evaluation for the DOE Oak Ridge operations.* Washington, D.C.: U.S. Department of Energy, Office of Environmental Management. Available at http://risk.lsd.ornl.gov/homepage/rap_tmn.shtml (accessed 28 November 2003).

Goodman, L. S., A. G. Gilman, L. E. Limbird, and J. G. Hardman, eds. 2001. *Goodman and Gilman's the pharmacological basis of therapeutics.* New York: McGraw-Hill.

Green Mountain Institute (GMI) for Environmental Democracy, and Western Center for Environmental Decision (WCED) Making. 1998. *PGMI and WCED Project Summaries.* Vermont.

Grøn, C., and L. Andersen. 2003. *Human bioaccessibility of heavy metals and PAH from soil.* Environmental Protection Agency (Denmark), Technology Programme for Soil and Groundwater Contamination. Environmental Project no. 840, 2003. Available at http://www.mst.dk/udgiv/publications/2003/87-7972-877-4/html/default_eng.htm (accessed 28 November 2003).

Guess H., and K. S. Crump. 1976. Low-dose extrapolation of data from animal carcinogenesis experiments: Analysis of a new statistical technique. *Mathematical Biosciences* 32:15–36.

Hammonds, J. S., F. O. Hoffman, and S. M. Bartell. 1994. *An introductory guide to uncertainty analysis in environmental and health risk assessment.* Oak Ridge, Tenn.: Oak Ridge National Laboratory.

Hazardous Substances Data Bank (HSDB). 2003. Available at http://toxnet.nlm.nih.gov/cgi-bin/sis/htmlgen?HSDB (accessed 28 November 2003).

Health Canada. 1994. *Canadian Environmental Protection Act: Human health risk assessment for priority substances.* Ottawa, Ont.: Health Canada. Available at http://www.hc-sc.gc.ca/hecs-sesc/exsd/psap3.htm (accessed 28 November 2003).

———. 1995. *Investigating human exposure to contaminants in the environment: A handbook for exposure calculations.* Ottawa, Ont.: Health Canada. Available at http://www.hc-sc.gc.ca/ehp/ehd/catalogue/bch_pubs/95ehd193.htm (accessed 28 November 2003).

———. 1996. *Health-based tolerable daily intakes/concentrations and tumourigenic doses/concentrations for priority substances.* Ottawa, Ont.: Health Canada.

———. 2003a. *Priority Substances Assessment Program: First Priority Substances Lists (PSL1) assessments.* Ottawa, Ont.: Health Canada. Available at http://www.hc-sc.gc.ca/hecs-sesc/exsd/psl1.htm (accessed 28 November 2003).

———. 2003b. *Proposed acceptability for continuing registration (PACR series).* Ottawa, Ont.: Health Canada, Pest Management Regulatory Agency. Available at http://www.hc-sc.gc.ca/pmra-arla/english/pubs/pacr-e.html (accessed 28 November 2003).

———. 2003c. *Proposed regulatory decision documents (PRDD series).* Ottawa, Ont.: Health Canada, Pest Management Regulatory Agency. Available at http://www.hc-sc.gc.ca/pmra-arla/english/pubs/prdd-e.html (accessed 28 November 2003).

———. 2003d. *Regulatory notes (REG series).* Ottawa, Ont.: Health Canada, Pest Management Regulatory Agency. Available at http://www.hc-sc.gc.ca/pmra-arla/english/pubs/reg-e.html (accessed 28 November 2003).

———. 2003e. *Health-based tolerable daily intakes/concentrations and tumorigenic doses/concentrations for priority substances.* Available at http://www.hc-sc.gc.ca/hecs-sesc/exsd/publications/tumorigenic_doses/toc.htm (accessed 28 November 2003). Note: this is an updated version of Health Canada 1996.

———. 2003f. *Proposed acceptability for continuing registration: Re-evaluation of malathion.* Ottawa, Ont.: Health Canada, Pest Management Regulatory Agency. PACR2003–07. Available at http://www.hc-sc.gc.ca/pmra-arla/english/pdf/pacr/pacr2003–10-e.pdf (accessed 3 January 2004).

Health Canada and Ontario Ministry of Health. 1997. *The health and environment handbook for health professionals.* Ottawa, Ont.: Health Canada and Ontario Ministry of Health.

Hrudey, S. E., W. Chen, and C. G. Rousseaux. 1996. *Bioavailability in environmental risk assessment.* Boca Raton, Fla.: CRC Press.

International Agency for Research on Cancer (IARC). 2003. *IARC monographs programme on the evaluation of carcinogenic risks to humans.* Available at http://monographs.iarc.fr/ (accessed 28 November 2003).

IPCS (International Program on Chemical Safety). 1994. *Environmental health criteria 170: Assessing human health risks of chemicals: The derivation of guidance values for health-based exposure limits.* Geneva: World Health Organization. Available at http://www.inchem.org/documents/ehc/ehc/ehc170.htm (accessed 28 March 2004).

———. 1999. *Environmental health criteria 210: Principles for the assessment of risks to human health from exposure to chemicals.* Available at http://www.inchem.org/documents/ehc/ehc/ehc210.htm (accessed 28 November 2003).

———. 2000. *General scientific principles of chemical safety*. IPCS training module no. 4. Available at http://www.who.int/pcs/training_material/module4/table_ of_contents.htm (accessed 28 November 2003).

———. 2003a. *Concise international chemical assessment documents (CICADs)*. Available at http://www.inchem.org/pages/cicads.html (accessed 28 November 2003).

———. 2003b. *Environmental health criteria (EHC) monographs*. Available at http:// www.inchem.org/pages/ehc.html (accessed 28 November 2003).

———. 2003c. *Joint Meeting on Pesticide Residues (JMPR): Monographs and evaluations*. Available at http://www.inchem.org/pages/jmpr.html (accessed 28 November 2003).

———. 2003d. *Pesticide data sheets*. Available at http://www.inchem.org/pages/pds .html (accessed 28 November 2003).

———. 2003e. *Screening information dataset (SIDS) for high production volume chemicals*. Available at http://www.inchem.org/pages/sids.html, http://www .chem.unep.ch/irptc/sids/sidspub.html, or http://www.chem.unep.ch/irptc/ Publications/sidsidex/sidsidex.htm (accessed 28 November 2003).

ITER. 2003. *ITER definitions*. Available at http://iter.ctcnet.net/publicurl/glossary .htm (accessed 28 November 2003).

Jager, T. 1998. *Uncertainty analysis of EUSES: Interviews with representatives from EU member states and industry*. October. Bilthoven, The Netherlands: National Institute of Public Health and the Environment (RIVM). RIVM 679102047. Available at http://www.rivm.nl/bibliotheek/rapporten/679102047.html (accessed 26 December 2003).

Jager, T., H. A. den Hollander, G. B. Janssen, P. van der Poel, M. G. J. Rikken, and T. G. Vermeire. 2000. *Probabilistic risk assessment for new and existing chemicals: Example calculations*. Bilthoven, The Netherlands: National Institute of Public Health and the Environment (RIVM). RIVM 679102049. Available at http://www.rivm.nl/bibliotheek/rapporten/679102049.pdf (accessed 26 December 2003).

Janssen, P. J. C. M., and G. J. A. Speijers. 1997. *Guidance on the derivation of maximum permissible risk levels for human intake of soil contaminants*. Bilthoven, The Netherlands: National Institute of Public Health and the Environment (RIVM). RIVM 711701006. Available at http://www.rivm.nl/bibliotheek/ rapporten/711701006.pdf (accessed 28 November 2003).

Kalant, H., W. H. Roschlau, and H. E. Roschlau, eds. 1997. *Principles of medical pharmacology*. New York: Oxford University Press.

Kamrin, M. A., D. J. Katz, and M. L. Walter. 2000. *Reporting on risk: A journalist's handbook on environmental risk assessment*. Available at http://www.facsnet .org/tools/ref_tutor/risk/index.php3 (accessed 28 November 2003).

Kevin, E. 2002. Scientific anomaly and regulatory policy: The case of chemical hormesis. Draft prepared for the Next Generation Leaders in Science and Technology Policy Symposium, November 2002, Washington, D.C. Available at http:// www.cspo.org/nextgen/Elliott.pdf (accessed 2 January 2004).

Krewski, D., D. W. Gaylor, A. P. Soms, and M. Szyszkowicz. 1993. Correlation between carcinogenic potency and the maximum tolerated dose: Implications for risk assessment. In *Issues in Risk Assessment*, National Research Council, 111–71. Washington, D.C.: NAS Press.

Krieger, W. C. 2001. Paracelsus dose response. In *The Handbook of Pesticide Toxicology*, ed. W. C. Krieger. Academic Press. Available at http://www.mindfully .org/Pesticide/Paracelsus-Dose-ToxicologyOct01.htm (accessed 2 January 2004).

Lehman, A. J., and O. G. Fitzhugh. 1954. One-hundredfold margin of safety. *Assoc. Food Drug Off. U.S.Q. Bull.* 18:33–35.

Leisenring, W., and L. Ryan. 1992. Statistical properties at the NOAEL. *Regulatory Toxicology and Pharmacology* 15:161–71. Cited in USEPA 2000a.

Liao, K. H., I. D. Dobrev, J. E. Dennison, Jr., M. E. Andersen, B. Reisfeld, K. F. Reardon, J. A. Campain, W. Wei, M. T. Klein, R. J. Quann, and R. S. Yang. 2002. Application of biologically based computer modeling to simple or complex mixtures. *Environ. Health Perspect.* 110 (Suppl. 6): 957–63.

Liteplo, R. 2004. Personal communication, 5 February 2004. Health Canada.

McColl, S., J. Hicks, L. Craig, and J. Shortreed. 2000. *Environmental health risk management: A primer for Canadians.* Waterloo: University of Waterloo, NERAM.

Meek, M. E., R. Newhook, R. G. Liteplo, and V. C. Armstrong. 1994. Approach to assessment of risk to human health for priority substances under the Canadian Environmental Protection Act. *Journal of Environmental Science and Health* C12 (2): 105–34.

Moolenaar, R. J. 1994. Carcinogen risk assessment: International comparison. *Regulatory Toxicology and Pharmacology* 20:302–36.

Moolgavkar, S. H., and A. G. Knudson, Jr. 1981. Mutation and cancer: A model for human carcinogenesis. *Journal of the National Cancer Institute* 66:1037–52.

Muller, P., and J. Nicholson. 2000. Integrated environmental decision-making: The next step in risk management: Legal brief. *Hazardous Materials Management* 12:29–30.

National Academy of Sciences (NAS). 1977. *Drinking Water and Health.* Washington, D.C.: National Academy Press.

National Cancer Institute of Canada (NCIC). 2003. *Canadian cancer statistics, 2003.* Available at http://www.cancer.ca/vgn/images/portal/cit_776/61/38/561 58640niw_statseen.pdf (accessed 28 November 2003).

National Cancer Registry (Ireland). 1997. *Incidence, mortality, treatment and survival: Annual report, 1997.* Available at http://www.ncri.ie/pubs/report-1997.shtml (accessed 28 November 2003).

National Industrial Chemicals Notification and Assessment Scheme (NICNAS). 2003. *Public chemical assessment reports.* Australia: Commonwealth Department of Health and Ageing. Available at http://www.nicnas.gov.au/publications/ CAR/ (accessed 28 November 2003).

National Research Council (NRC). 1983. *Risk assessment in the federal government: Managing the process.* Committee on the Institutional Means for Assessment of Risks to Public Health, Commission on Life Sciences. Washington, D.C.: National Academy Press.

———. 1994. *Science and judgment in risk assessment.* Washington, D.C.: National Academy Press.

———. 2001a. *Acute exposure guideline levels for selected airborne chemicals.* Vol. 1. Washington, D.C.: National Academy Press. Available at http://www.nap.edu/books/0309072948/html (accessed 28 November 2003).

———. 2001b. *Arsenic in drinking water, 2001 update.* Washington, D.C.: National Academy Press.

———. 2001c. *Food safety policy, science, and risk assessment: Strengthening the connection: Workshop proceedings, 2001.* Washington, D.C.: National Academy Press.

———. 2002. *Acute exposure guideline levels for selected airborne chemicals.* Vol. 2. Washington, D.C.: National Academy Press. Available at http://www.nap.edu/books/030908511X/html (accessed 28 November 2003).

———. 2003. *Acute exposure guideline levels for selected airborne chemicals.* Vol. 3. Washington, D.C.: National Academy Press. Available at http://www.nap.edu/books/0309088836/html (accessed 28 November 2003).

NIPHEP. 1989. *Integrated criteria document: PAHs.* RIVM 758474011. Bilthoven, Netherlands: RIVM.

Omenn, G. S. 1997. *Report on the accomplishments of the Commission on Risk Assessment and Risk Management.* Presidential/Congressional Commission on Risk Assessment and Risk Management (PCRARM). Available at http://www.risk world.com/Nreports/1997/risk-rpt/miscinfo/nr7mi002.htm (accessed 28 November 2003).

Ontario Ministry of the Environment (OMOE). 1999. *Reviewer's checklist for risk assessments.* Available at http://www.ene.gov.on.ca/envision/decomm/checkra .pdf (accessed 28 November 2003).

Ontario Ministry of the Environment and Energy (OMOEE). 1994. *Scientific criteria document for multimedia environmental standards development: Lead.* Toronto: Ontario Ministry of Environment and Energy.

———. 1997a. *Guidelines for use at contaminated sites in Ontario.* Toronto: Ontario Ministry of Environment and Energy, Standards Development Branch.

———. 1997b. *Scientific criteria document for multimedia standards development: Polycyclic aromatic hydrocarbons (PAH).* February. Toronto: Ontario Ministry of Environment and Energy, Standards Development Branch.

Oregon Department of Environmental Quality (ODEQ). 1998. *Guidance for use of probabilistic analysis in human health risk assessments.* Interim final draft. Portland, Ore. Available at http://www.deq.state.or.us/wmc/cleanup/hh-intro.htm (accessed 28 November 2003).

Paustenbach, D. J. 2001. The practice of exposure assessment. In *Principles and Methods of Toxicology*, 4th ed., ed. A. Wallace Hayes. Philadelphia: Taylor & Francis.

PubMed. 2003. Available at http://toxnet.nlm.nih.gov/cgi-bin/sis/htmlgen?Multi (accessed 28 November 2003).

Registry of Toxic Effects of Chemical Substances (RTECS) Database. 2003. Available at http://ccinfoweb.ccohs.ca/rtecs/search.html (accessed 28 November 2003).

Richardson, G. M. 1997. *Compendium of Canadian human exposure factors for risk assessment.* Toronto: O'Connor Associates Environmental.

Risk Assessment Information System (RAIS). 1999. *Guidance for conducting risk assessments and related risk activities for the DOE-ORO Environmental Management Program.* BJC/OR-271. Available at http://rais.ornl.gov/homepage/bjcor271/or271.shtml (accessed 28 November 2003).

———. 2003. *Welcome to the Risk Assessment Information System.* Available at http://risk.lsd.ornl.gov/rap_hp.shtml (accessed 28 November 2003).

Sandman, P. 1986. *Explaining environmental risk.* Washington, D.C.: U.S. Environmental Protection Agency, Office of Toxic Substances.

Schierow, L-J. 1998. *Risk analysis: Background on Environmental Protection Agency mandates.* Congressional Research Service Issue Brief for Congress. Washington, D.C.: National Council for Science and the Environment. Available at http://www.ncseonline.org/NLE/CRSreports/Risk/rsk-12.cfm?& CFID – 11299621&C FTO KEN = 28374922 (accessed 28 November 2003).

———. 2001. *IB94036: The role of risk analysis and risk management in environmental protection.* Congressional Research Service Issue Brief for Congress. Washington, D.C.: National Council for Science and the Environment. Available at http://www.ncseonline.org/NLE/CRSreports/Risk/rsk-1.cfm?& CFID = 11299597&C FTOK EN = 84634859 (accessed 28 November 2003).

Setzer, R. W., and R. S. De Woskin. 2000. Incorporating a validated PBPK pregnant rat model into a BBDR model for the embryotoxicity of 5-fluorouracil. *Toxicologist* 54:93.

Sexton, K., D. E. Kleffman, and M. A. Callahan. 1995. An introduction to the National Human Exposure Assessment Survey (NHEXAS) and related phase I field studies. *Journal of Exposure Analysis and Environmental Epidemiology* 5:229–33.

Sips, A. J. A. M., and J. C. H. van Eijkeren. 1996. *Oral bioavailability of heavy metals and organic compounds from soil: Too complicated to absorb? An inventarisation of factors affecting bioavailability of environmental contaminants from soil.* Bilthoven, The Netherlands: National Institute of Public Health and the Environment (RIVM). RIVM 711701002. Available at http://www.rivm.nl/bibliotheek/rapporten/711701002.pdf (accessed 28 November 2003).

Spurway, P. 2001. *Media relations training seminar.* Presentation. Public Works and Government Services Canada.

Syracuse Research Corporation (SRC). 2003a. Environmental Fate Data Base (EFDB). Available at http://esc.syrres.com/efdb.htm (accessed 28 November 2003).

————. 2003b. *Interactive LogKow (KowWin) demo.* Available at http://esc.syrres.-com/interkow/kowdemo.htm (accessed 28 November 2003).

————. 2003c. *TSCATS search.* Available at http://esc.syrres.com/efdb/TSCATS.htm (accessed 28 November 2003).

TOXLINE. 2003. Available at http://toxnet.nlm.nih.gov/cgi-bin/sis/htmlgen?Multi (accessed 28 November 2003).

ToxProbe. 1999. *Screening risk assessment on air contaminants in Teplice and Prachatice.* Toronto: ToxProbe.

Travis, C. C., and H. A. Hattemer-Frey. 1988. Determining an acceptable level of risk. *Environ. Sci. Technol.* 22:873–76.

Trevan, J. W. 1927. The error of determination of toxicity. *Proceedings of the Royal Society (London),* series B, 101:483–514.

U.S. Department of Agriculture (USDA). 1997. Data tables: Results from USDA's 1994–1996 Continuing Survey of Food Intakes by Individuals and 1994–1996 Diet and Health Knowledge Survey. Washington, D.C.: Agricultural Research Service.

U.S. Department of Defense. 1997. *Relative risk site evaluation primer.* Washington, D.C. Available at http://www.dtic.mil/envirodod/Policies/Cleanup/relrisk_toc.htm (accessed 28 November 2003).

U.S. Department of Energy. 1995. *Environmental restoration risk-based prioritization work package planning and risk ranking methodology.* Prepared by Lockheed Martin Energy Systems. Washington, D.C.: U.S. Department of Energy.

————. 1997a. *Business performance systems: Environment, safety and health risk-based prioritization.* Washington, D.C.: U.S. Department of Energy, Office of Environment, Safety, and Health. Available at http://tis-nt.eh.doe.gov/bps/eshplan/rpm.htm (accessed 28 November 2003).

————. 1997b. *Guidelines for risk-based prioritization of DOE activities.* Washington, D.C.: U.S. Department of Energy.

————. 1998. Risk-based priority model (RPM). In *Budget year 1998 guidance manual.* Washington, D.C.: U.S. Department of Energy, Office of Environment, Safety, and Health. Available at http://tis-nt.eh.doe.gov/bps/eshplan/rpmdesc.pdf (accessed 28 November 2003).

U.S. Department of Health and Human Services. 2002. *Communicating in a crisis: Risk communication guidelines for public officials.* Rockville, Md.: Public Health Service.

USEPA (U.S. Environmental Protection Agency). 1980. *Comprehensive Environmental Response, Compensation, and Liability Act (CERCLA or Superfund).* Washington, D.C. Available at http://www4.law.cornell.edu/uscode/42/ch103.html (accessed 3 January 2004).

————. 1984. *Carcinogen assessment of coke oven emissions.* EPA/600/6–82–003F. Washington, D.C.: Office of Health and Environmental Assessment.

————. 1986a. Guidelines for carcinogenic risk assessment. *Federal Register* 51 (185): 33992–34003.

———. 1986b. Guidelines for the health risk assessment of chemical mixtures. EPA/630/R-98–002. *Federal Register* 51 (185): 34014–25.

———. 1987a. *The Total Exposure Assessment Methodology (TEAM) study.* EPA/600/6–87–002. Washington, D.C.: Office of Acid Deposition, Environmental Monitoring, and Quality Assurance, Office of Research and Development.

———. 1987b. *Unfinished business: A comparative assessment of environmental problems.* EPA Overview Report. Washington, D.C.: USEPA.

———. 1989. *Human health evaluation manual.* Interim final draft. Part A of vol. 1 of *Risk assessment guidance for Superfund (RAGS).* EPA/540/1–89–002. Washington, D.C.: USEPA. Available at http://www.epa.gov/superfund/programs/risk/ragsa (accessed 28 November 2003).

———. 1990. *Reducing risk: Setting priorities and strategies for environmental protection.* EPA/SAB/EC-90–021. Washington, D.C.: Science Advisory Board.

———. 1991a. *Risk assessment for toxic air pollutants: A citizen's guide.* EPA/450/3–90–024. Washington, D.C.: USEPA.

———. 1991b. *Role of the baseline risk assessment in Superfund remedy selection decisions.* OSWER Directive 9355.0–30. Available at http://www.epa.gov/superfund/programs/risk/baseline.htm (accessed 26 February 2004).

———. 1992a. *Dermal exposure assessment: Principles and applications.* Interim report. EPA/600/8–91–011b. Washington, D.C.: USEPA.

———. 1992b. *Guidance on risk characterization for risk managers and risk assessors.* Available at http://www.epa.gov/superfund/programs/risk/habicht.htm (accessed 28 November 2003).

———. 1993a. *Reference dose (RfD): Description and use in health risk assessments.* Background Document 1A. Available at http://www.epa.gov/iris/rfd.htm (accessed 28 November 2003).

———. 1993b. Styrene. In IRIS database. Available at http://www.epa.gov/iris/subst/0104.htm (accessed 26 December 2003).

———. 1994a. Benzo[a]pyrene. In IRIS database. Available at http://www.epa.gov/iris/subst/0136.htm (accessed 28 November 2003).

———. 1994b. *Methods for derivation of inhalation reference concentrations and application of inhalation dosimetry.* October. EPA/600/8–90–066F. Available at http://www.epa.gov/cgi-bin/claritgw?op-Display&document=clserv:ORD:327;&rank=2&template=epa (accessed 26 December 2003).

———. 1994c. Cadmium. In IRIS database. Available at http://www.epa.gov/iris/subst/0141.htm (accessed 26 December 2003).

———. 1994d. *Estimating exposure to dioxin-like compounds.* External review draft. EPA/600/6–88–005Ca-c. Washington, D.C.: USEPA.

———. 1994e. Toluene. In IRIS database. Available at http://www.epa.gov/iris/subst/0118.htm (accessed 26 December 2003).

———. 1995a. *Region III technical guidance manual: Risk assessment: Assessing dermal exposure from soil.* EPA/903/K-95–003. Washington, D.C.: USEPA.

———. 1995b. *U.S. Environmental Protection Agency policy for risk characterization.* March. Cited in USEPA 2000c, appendix A.

———. 1996a. *The Food Quality Protection Act (FQPA) background.* Available at http://www.epa.gov/oppfead1/fqpa/backgrnd.htm (accessed 2 January 2004).

———. 1996b. *Soil screening guidance: Technical background document (TBD).* EPA/540/R-95–128. Washington, D.C.: USEPA.

———. 1996c. 1,4-Dichlorobenzene. In IRIS database. Available at http://www.epa.gov/iris/subst/0552.htm (accessed 26 December 2003).

———. 1996d. Nickel, soluble salts. In IRIS database. Available at http://www.epa.gov/iris/subst/0271.htm (accessed 26 December 2003).

———. 1997a. *Exposure factors handbook.* Vol. 1, *General factors.* EPA/600/P-95–002Fa. Washington, D.C.: Office of Research and Development. Available at http://cfpub.epa.gov/ncea/cfm/recordisplay.cfm?deid = 12464.

———. 1997b. *Exposure factors handbook.* Vol. 2, *Food ingestion factors.* EPA/600/P-95–002Fb. Washington, D.C.: Office of Research and Development. Available at http://cfpub.epa.gov/ncea/cfm/recordisplay.cfm?deid = 12464.

———. 1997c. *Exposure factors handbook.* Vol. 3, *Activity factors.* EPA/600/P-95–002Fc. Washington, D.C.: Office of Research and Development. Available at http://cfpub.epa.gov/ncea/cfm/recordisplay.cfm?deid = 12464.

———. 1998a. *Guidelines for ecological risk assessment.* EPA/630/R-95–002F. Washington, D.C.: USEPA.

———. 1998b. *Region 9 preliminary remediation goals (PRGs).*

———. 1998c. *Toxicological review of hexavalent chromium (CAS No. 18540–29–9) in support of summary information on the Integrated Risk Information System (IRIS).* August. Washington, D.C.: USEPA.

———. 1998d. *Guidance for submission of probabilistic human health exposure assessments to the Office of Pesticide Programs.* Draft, 4 November. Available at http://www.epa.gov/fedrgstr/EPA-PEST/1998/November/Day-05/6021.pdf (accessed 26 December 2003).

———. 1999a. *Integrated environmental decision-making in the twenty-first century: Summary recommendations.* Peer review draft. Washington, D.C.: Science Advisory Board. Available at http://www.epa.gov/science1/pdf/intenv20.pdf (accessed 28 November 2003).

———. 1999b. *Review of revised sections of the proposed guidelines for carcinogen risk assessment.* EPA/SAB/EC-99–015. Washington, D.C.: Science Advisory Board. Cited in USEPA 2003c. Available at http://www.epa.gov/ncea/raf/cancer.htm.

———. 1999c. *Community involvement in Superfund risk assessments.* Supplement to *Human health evaluation manual,* part A of vol. 1 of *Risk assessment guidance for Superfund.* March. Washington, D.C.: Office of Solid Waste and Emergency Response.

———. 1999d. *Integrated environmental decision-making in the twenty-first century (IED)*. Peer review draft. Washington, D.C.: Science Advisory Board.

———. 1999e. *Guidance for performing aggregate exposure and risk assessments*. Washington, D.C.: Office of Pesticide Programs. October. Available at http://www.epa.gov/fedrgstr/EPA-PEST/1999/November/Day-10/6043.pdf (accessed 22 February 2004).

———. 2000a. *Benchmark dose technical guidance document*. EPA/630/R-00–001. External review draft. Washington, D.C.: USEPA. Available at http://cfpub .epa.gov/ncea/cfm/recordisplay.cfm?deid=20871 (accessed 28 November 2003).

———. 2000b. Choosing a percentile of acute dietary exposure as a threshold of regulatory concern. Available at www.epa.gov/pesticides/trac/science/trac2b054 .pdf (accessed 28 November 2003).

———. 2000c. *Risk characterization handbook*. EPA/100/B-00/002. Washington, D.C.: Office of Research and Development. Available at http://www.epa.gov/ osp/spc/2riskchr.htm (accessed 28 November 2003).

———. 2000d. *Supplementary guidance for conducting health risk assessment of chemical mixtures*. EPA/630/R-95–002F. Washington, D.C.: Office of Research and Development, Risk Assessment Forum.

———. 2000e. *Toward integrated environmental decision-making*. Washington, D.C.: Science Advisory Board. EPA/SAB/EC-00–011. Available at http://www.epa .gov/sab/pdf/ecirp011.pdf (accessed 28 November 2003).

———. 2000f. *Draft exposure and human health reassessment of 2,3,7,8-tetrachlorodibenzo-p-dioxin (TCDD) and related compounds*. Washington, D.C.: Office of Research and Development.

———. 2000g. *Options for development of parametric probability distributions for exposure factors*. July. EPA/600/R-00–058. Available at http://www.epa.gov/ ncea/ (accessed 26 December 2003).

———. 2000h. *Supplementary guidance for conducting health risk assessment of chemical mixtures*. August. EPA/630/R-00–002. Washington, D.C.: Office of Research and Development, Risk Assessment Forum.

———. 2001a. *Standardized planning, reporting, and review of Superfund risk assessments*. Part D of vol. 1, *Human health evaluation manual*, of *Risk assessment guidance for Superfund*. Publication 9285.7–47. Washington, D.C.: Science Advisory Board. Available at http://www.epa.gov/superfund/programs/risk/ ragsd/tara.htm (accessed 28 November 2003).

———. 2001b. *Supplemental guidance for dermal risk assessment*. Part E of vol. 1, *Human health evaluation manual*, of *Risk Assessment Guidance for Superfund*. Interim review draft. EPA/540/R-99–005. Washington, D.C.: Office of Emergency and Remedial Response. Available at http://www.epa.gov/superfund/ programs/risk/ragse/ (accessed 28 November 2003).

———. 2001c. *Road map to understanding innovative technology options for brownfields investigation and cleanup*. 3rd ed. EPA/542/B-01–001. Washington,

D.C.: USEPA. Available at http://clu-in.org/roadmap (accessed 28 November 2003).

———. 2001d. *Process for conducting probabilistic risk assessment.* Part A of vol. 3 of *Risk Assessment Guidance for Superfund.* EPA/540/R-02–002. Washington, D.C.: Office of Emergency and Remedial Response.

———. 2002a. *A review of the reference dose and reference concentration processes.* EPA/630/P-02–002F. Washington, D.C.: USEPA. Available at www.epa.gov/iris/RFD_FINAL%5B1%5D.pdf (accessed 28 November 2003).

———. 2002b. *Determination of the appropriate FQPA safety factor(s) in tolerance assessment.* February. Washington, D.C.: Office of Pesticide Programs. Available at http://www.epa.gov/oppfead1/trac/science/determ.pdf (accessed 22 February 2004).

———. 2002c. *Drinking water standards and health advisories.* EPA/822/R-02–038. Washington, D.C.: USEPA. Available at http://www.epa.gov/waterscience/drinking (accessed 28 November 2003).

———. 2002d. *Child-specific exposure factors handbook.* Interim report. EPA/600/P-00–002B. Washington, D.C.: Office of Research and Development.

———. 2002e. *Working draft technical document for characterizing and presenting summary chemical exposure assessment results.* Washington, D.C.: Office of Pollution Prevention and Toxics.

———. 2003a. *Air toxics community assessment and risk reduction projects database.* Available at http://yosemite.epa.gov/oar/CommunityAssessment.nsf/Community%20Assessment %20L ist?OpenForm (accessed 28 November 2003).

———. 2003b. *Chemicals in the environment: OPPT chemical fact sheets.* Washington, D.C.: Office of Pollution Prevention and Toxics. Available at http://www.epa.gov/chemfact/ (accessed 28 November 2003).

———. 2003c. *Draft final guidelines for carcinogen risk assessment.* External review draft. February. NCEA-F-0644A. Washington, D.C.: Office of Research and Development, Risk Assessment Forum. Available at http://oaspub.epa.gov/eims/eimscomm.getfile?p_download_id=36765 (accessed 28 November 2003).

———. 2003d. *Framework for cumulative risk assessment.* EPA/630/P-02–001F. Washington, D.C.: USEPA. Available at www.pestlaw.com/x/guide/2003/EPA-20030500B.pdf (accessed 28 November 2003).

———. 2003e. *Glossary of IRIS terms.* Available at http://www.epa.gov/iris/gloss8.htm (accessed 28 November 2003).

———. 2003f. *Integrated Risk Information System (IRIS).* Available at http://epa.gov/iris/index.html (accessed 28 November 2003).

———. 2003g. *National Center for Environmental Assessment.* Available at http://cfpub.epa.gov/ncea/cfm/nceapubtopics.cfm?ActType=PublicationTopics (accessed 28 November 2003).

———. 2003h. *Pesticides: Topical and chemical fact sheets.* Available at http://www.epa.gov/pesticides/factsheets/chemical_fs.htm (accessed 28 November 2003).

———. 2003i. *Pesticides: Reregistration.* Available at http://cfpub.epa.gov/oppref/ rereg/status.cfm?show=rereg (accessed 28 November 2003).

———. 2003j. *Superfund.* Available at http://www.epa.gov/superfund/health/risk/ index.htm (accessed 28 November 2003).

———. 2003k. *Superfund risk assessment glossary.* Available at http://epa.gov/ superfund/programs/risk/glossary.htm (accessed 28 November 2003).

———. 2003l. *Supplemental guidance for assessing cancer susceptibility from early-life exposure to carcinogens.* External review draft. EPA/630/R-03–003. Washington, D.C.: USEPA. Available at www.epa.gov/ncea/raf/cancer2003.htm (accessed 28 November 2003).

———. 2003m. Cover memo from Jennifer Hubbard to RBC table users, October 15, 2003. Available at http://www.epa.gov/reg3hwmd/risk/cov1003.htm (accessed 28 November 2003).

———. 2003n. 2,3,7,8-tetrachlorodibenzo-p-dioxin (2,3,7,8-TCDD). Available at http://www.epa.gov/ttn/atw/hlthef/dioxin.html (accessed 26 December 2003).

———. 2003o. *Risk communication.* February. Part of the Superfund Community Involvement Toolkit. Available at http://www.epa.gov/superfund/tools/pdfs/ 37riskcom.pdf (accessed 22 February 2004).

U.S. Food and Drug Administration (FDA). 1949. Procedures for the appraisal of the toxicity of chemicals in foods. *Food, Drug, and Cosmetic Law Quarterly* (September): 412–34. Cited in P. B. Hutt, *The Historical Development of Animal Toxicity Testing from Its Ancient Origins to the Middle of the Twentieth Century.* Cambridge: Harvard Law School, 1997. Available at http://leda.law.harvard .edu/leda/data/196/ngertler2.html (accessed 28 November 2003).

———. 2000. *Total diet study statistics on element results, 1991–1998.* Rev. 1, 25 April. Washington, D.C.: FDA.

U.S. Navy. 2000a. *Overview of metals bioavailability: NFESC user's guide.* Part 1 of *Guide for incorporating bioavailability adjustments into human health and ecological risk assessments at U.S. Navy and Marine Corps facilities.* UG-2041-ENV. Washington, D.C.: U.S. Navy. Available at http://web.ead.anl.gov/ecorisk/methtool/dsp_bioavail.cfm (accessed 28 November 2003).

———. 2000b. *Technical background document for assessing metals bioavailability: NFESC user's guide.* Part 2 of *Guide for incorporating bioavailability adjustments into human health and ecological risk assessments at U.S. Navy and Marine Corps facilities.* UG-2041-ENV. Washington, D.C.: U.S. Navy. Available at http://web.ead.anl.gov/ecorisk/methtool/dsp_bioavail.cfm (accessed 28 November 2003).

U.S. Office of Technology Assessment. 1987. *Identifying and regulating carcinogens.* OTA-BP-FI-42. Washington, D.C.: GPO.

Welshons, K. A. T., B. M. Judy, J. A. Taylor, E. M. Curran, and F. S. vom Saal. 2003. Large effects from small exposures. Part 1, Mechanisms for endocrine-disrupting chemicals with estrogenic activity. *Environ. Health Perspect.* 111:994–1006.

WHO (World Health Organization). 1995. *Updating and revision of the air quality guide- lines for Europe: Report on a WHO working group on PCBs, PCDDs, PCDFs.* May. Maastricht, The Netherlands: WHO.

———. 1999. *Air quality guidelines.* Available at http://www.who.int/peh/air/Air qualitygd.htm (accessed 28 November 2003).

———. 2000. *WHO air quality guidelines.* 2d ed. Regional Office for Europe. Available at http://www.euro.who.int/air/Activities/20020620_1 (accessed 26 December 2003).

Wilson, J. D. 1996. *Thresholds for carcinogens: A review of the relevant science and its implications for regulatory policy.* Toronto: Canadian Institute for Environmental Law and Policy. Available at http://www.rff.org/rff/Documents/RFF-DP-96–21.pdf (accessed 28 November 2003).

Winfield, M. S., and P. Muldoon. 1999. *An environmental agenda for Ontario.* Draft chapter, Democracy. Toronto: Canadian Institute for Environmental Law and Policy. Available at http://www.cielap.org/infocent/research/democr.html (accessed 28 November 2003).

Yang, R. S. H., H. A. El-Masri, R. S. Thomas, A. A. Constan, and J. D. Tessari. 1995. The application of physiologically based pharmacokinetic pharmacodynamic (PBPK/PD) modeling for exploring risk assessment approaches of chemical mixtures. *Toxicol. Lett.* 79:193–200.

Acknowledgements

The authors wish to gratefully acknowledge peer review comments from Lynn Flowers, Ph.D., DABT, and Gene Hsu, Ph.D, National Center for Environmental Assessment, USEPA; Robert G. Liteplo, Ph.D., Existing Substances Division, Environmental Contaminants Bureau, Health Canada; and Cecile Willert, B.A.Sc., P.Eng.Manager–Risk Assessment/Management, Jacques Whitford Environment Ltd.

Notes

1. The more general term now in use to describe "threshold" is *nonlinear dose-response relationship.*

2. The more general term now in use to describe "non-threshold" is *linear dose-response relationship.*

3. The FDA was established in 1927 as the Food, Drug, and Insecticide Administration and was renamed the Food and Drug Administration in 1930.

4. Potency is a measure reflecting the dose or concentration necessary to induce a given effect. USEPA now reserves this term for instances where true potencies of various chemicals are compared (Flowers and Hsu 2004). We use the term *estimated potency* in this review to differentiate it from "true" potency.

5. "Risk" is unitless and is usually omitted in the description, but when doing unit conversions, it is useful to include risk in the equation.

6. *Aggregate risk assessment* considers exposure of a population to a chemical by multiple routes. For a more detailed description of aggregate risk assessment, refer to section 4.4.

7. *Point of departure* is defined as "the dose-response point that marks the beginning of a low-dose extrapolation. This point is most often the upper bound on an observed incidence or on an estimated incidence from a dose-response model" (USEPA 2003e).

8. Terminologies differ. We found the terminology used by the USEPA (2000a) clearest and this terminology is used throughout the discussion of the benchmark methodology.

9. Benchmark Dose$_{05}$ (Benchmark Concentration$_{05}$) is the dose (concentration) of a substance that induces a 5 percent increase in the incidence of a health effect in an exposed population.

10. Tumorigenic Dose$_{05}$ (Tumorigenic Concentration$_{05}$) is the dose (concentration) of a carcinogen that induces a 5 percent increase in the incidence of, or deaths due to, tumors in an exposed population.

11. See section 3.2.7 for a description of physiologically based pharmacokinetic modeling.

12. Bioaccessibility, bioavailability, and desorption are discussed in section 3.3.5.

13. Hazard quotient is defined only for assessment of chemicals assumed to have thresholds, while exposure ratio can also be used for assessment of non-threshold carcinogens, treating the dose corresponding to a specific risk level (e.g., one in a million) as the threshold (or reference exposure level).

14. See the "Overview" at http://ecb.jrc.it/existing-chemicals.

15. See IPCS 2000 for a more detailed description.

16. It should be noted that Jager and RIVM refer to an uncertainty assessment on the dose-response data. On the other hand, U.S. sources discuss uncertainty assessment as related to exposure data. USEPA does not recommend application of probabilistic risk assessment to dose-response data at this time.

17. WHO (1999) established a guideline for benzo[a]pyrene (B[a]P) and not for PAHs. However, the methodology used to develop this number is based on B[a]P actually serving as a surrogate for the entire PAH fraction of the mixture in a similar way to particles in diesel exhaust. Exhaust particles assume the potency of the entire diesel exhaust; the particles therefore serve as a surrogate for the entire diesel exhaust.

Chapter 2
Air Quality Monitoring and Associated Instrumentation

Ashok Kumar, Harish G. Rao,
Siva Sailaja Jampana, and Charanya Varadarajan

Air contaminants are substances added to the atmosphere that cause a deviation from the mean composition and are introduced by natural or anthropogenic sources. Contaminants of anthropogenic origin or which are present in sufficient quantities to have adverse effects on human health, environment, animal life, plant growth, and materials are labeled *air pollutants* (Kumar 2004). People who have respiratory difficulties can be highly affected by elevated levels of such air pollutants. Many urban areas throughout the globe are facing deteriorating air quality, with increases in the air pollutant concentrations of fine particulate matter, ozone, nitrogen oxides, carbon monoxide, sulfur dioxide, and lead.

Monitoring air quality helps us better understand the sources of contaminants, levels of different air pollutants, exposure to various substances in the air we breathe (Valicenti and Wenger 1997), and impact of air pollution control programs or policies. The data collected during a monitoring program assist us in assessing air quality and in improving or developing control programs to mitigate the adverse effect of air pollution (Terry and Kumar 1985; Terry, Chandhok, and Kumar 1987).

The most common purposes for conducting air quality monitoring are to:

- Determine air quality levels to identify spatial and temporal variations
- Assess population exposure and health impact
- Determine background air quality levels
- Inform and raise the awareness of the public regarding air quality
- Identify threats to natural ecosystems
- Determine compliance with national or international standards
- Provide objective inputs to air quality management, traffic, and land-use planning in a city
- Develop policy and prioritize management actions related to air pollution problems
- Develop or validate air management tools such as dispersion models (Kumar, Luo, and Bennett 1993) and Geographic Information Systems
- Assess point or area source impacts
- Quantify trends in air quality to identify future problems or progress against management and control targets
- Conduct source apportionment and identification
- Serve as an early warning system for workers during site remediation projects

In the United States, monitoring of air pollutants is carried out by the U.S. Environmental Protection Agency (USEPA) with the help of state and local agencies. The Federal Clean Air Act of 1970 required USEPA to assist states and localities in establishing ambient air quality monitoring networks to characterize the human health exposure and public welfare effects of criteria pollutants. Over the last three decades, USEPA has put in place several monitoring programs to respond to public concerns over air pollution from acid rain, toxic air pollutants, urban smog, and climate change. These monitoring stations continuously monitor and collect information on the presence and level of atmospheric contaminants as well as a number of meteorological indices.

The goal of air monitoring is not merely to collect data but also to provide the information necessary for engineers, scientists, policy makers, politicians, and planners to make informed decisions on managing and improving the air environment. Demerjian (2000) presented a review of the national monitoring networks in North America and noted that there are more than 4,300 monitoring sites operating as part of the air quality networks of three nations: Canada (the National Air Pollution Surveillance [NAPS] network); the United States (State and Local Air Monitoring Stations [SLAMS] and National Air Monitoring Stations [NAMS]); and Mexico (metropolitan networks). In his review, Demerjian described the current state of the national air quality monitoring networks in North America and assessed the effectiveness and adequacy of these networks.

Large industrial complexes have also developed their own air quality monitoring networks. For example, Ashok Kumar (1979) provided a description of the air monitoring network at Syncrude, the world's largest tar sands plant. He explained how the network helps in measuring wind direction, wind speed, sulfur dioxide, hydrogen sulfide, and other variables. He also described the static air quality network and various other monitoring methods implemented by Syncrude. The air quality monitoring at Syncrude included (1) baseline monitoring of meteorological conditions and (2) ambient air quality monitoring through the operation of a network of five continuous monitoring stations within 15 kilometers of the main stack, a static air quality monitoring network, sulfur dust fall stations, a biological air quality monitoring network, and source emissions monitoring.

On a global scale, many initiatives have been taken to monitor the levels of air pollutants and climate-related data by the United Nations, World Health Organization (WHO), and World Meteorological Organization. The use of satellites to track the movement of air pollutants and weather conditions is not uncommon.

A typical monitoring station (see figures 2.1, 2.2, 2.3, and 2.4) may include sophisticated gaseous pollutant analyzers, particle collectors, and weather sensors that are continuously maintained and operated. The formation and movement of air pollution are mainly affected by the weather conditions. Hence, a range of meteorological parameters, such as relative humidity, ultraviolet and solar radiation, barometric pressure, visibility, surface temperature, wind speed and direction, and precipitation are also measured at many monitoring stations. Figure 2.5 shows the location of air monitors in Lucas County, Ohio.

This chapter provides an overview of the design approaches for establishing air monitoring networks and a discussion of national air quality monitoring networks operating in the United States, Canada, and Mexico that are in the public domain. The efforts by WHO and the National Aeronautics and Space Administration (NASA) in this

Figure 2.1 Sulfur dioxide (SO$_2$) monitoring station operating in Toledo, Ohio
Source: Toledo Environmental Services

Figure 2.2 PM$_{10}$ monitor in Collins Park, Oregon
Source: Toledo Environmental Services

area are further discussed. A section is also devoted to source-monitoring programs. The final section contains a brief discussion of the air monitoring instrumentation needed to support air monitoring programs.

I   Design of Air Quality Monitoring Networks

The design of air monitoring networks requires knowledge from different disciplines. Technical considerations for design now routinely involve a survey of advances in computer technology, which can influence decisions on scope and design of the network. Availability of wireless transfer of data in recent years, for example, has given far more flexibility to develop new monitoring programs. Remote-sensing and satellite-based monitoring systems are also in use.

The design and efficiency of air monitoring networks has been a subject of many research papers. Many approaches have been developed since the 1970s. Early methods of locating monitoring networks were based on either field studies or simple modeling work to identify areas of high concentrations (Darby, Ossenbruggen, and Gregory 1974; Arbeloa, Caseiras, and Andres 1993). Such networks were generally designed for large industrial complexes or urban areas. In the 1990s, as the discipline advanced, rigorous mathematical techniques were developed and used to evaluate the effectiveness of the design of current monitoring networks for multipollutants and to locate monitors in complex flow conditions.

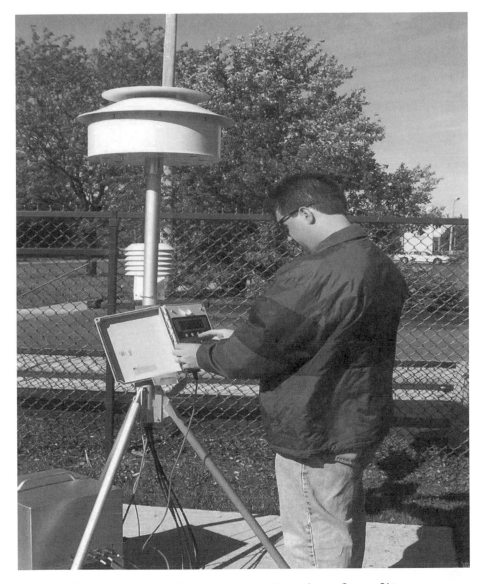

Figure 2.3 PM$_{2.5}$ speciation monitoring instrumentation in Lucas County, Ohio
Source: Toledo Environmental Services

Design methods continue to evolve. Chang and Tseng (1999a) documented many papers available on the design of monitoring networks to meet multiple criteria. They studied the multipollutant design principles and optimal searches for siting patterns of an air quality monitoring network.

Peterson (2000) focused on the design of monitoring networks in mountainous regions and stressed the need for paying special attention to the temporal and spatial scale of measurement while designing a network in such a region. Silva and Quiroz (2003) focused on the statistical evaluation of optimal cost modifications to the monitoring network at Santiago, Chile, as the city grew. Their work attempts to optimize the city's air monitoring networks by excluding the least informative stations with respect to a number of variables under study.

Figure 2.4 Meteorological station at Desert Ridge, Arizona
Source: College of Agriculture and Life Sciences, University of
Arizona (ag.arizona.edu/azmet/27gallery.htm)

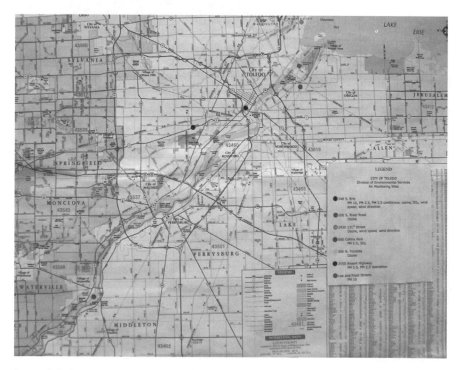

Figure 2.5 Location of air monitors in Lucas County, Ohio
Source: Toledo Environmental Services

Air monitoring networks for addressing health-related questions are also being designed. Baldauf and colleagues studied the design of an ambient air quality monitoring network for assessing the health impacts from exposures to airborne contaminants (Baldauf, Lane, and Marote 1999). Their study concluded that ambient air quality monitoring networks designed for the protection of public health or for epidemiological studies evaluating adverse health impacts from exposures to ambient air contaminants should account for both contaminant characteristics and human health parameters such as those used in human health risk analysis.

In their statistical study of redundant measurements of air quality monitoring networks, Hwang and Chan (1997) developed a statistical method to identify significant redundancy in monitoring sites for one-year measurements at two air monitoring networks in Taiwan, which could be used to downsize the monitoring networks. Because the operation and maintenance of air quality monitoring stations is very expensive, this method can be used to determine a reasonable number of stations needed to report daily pollutant standard index (PSI), average concentrations, and the number of exceedances of the national ambient air quality standard (NAAQS) in a monitoring area. It is desirable to use as few stations as possible to meet monitoring objectives. Such a study also helps monitoring officials to prioritize their site locations in an existing monitoring network and recheck their monitoring objectives.

Various issues related to air monitoring network design, such as creating new knowledge using monitored data (Rao, Terry, and Kumar 1988; Taha, Konopacki, and Akbari 1998; Kumar, Vedula, and Sud 2000), meeting air pollution levels set by regulations, and availability of detailed data on the Internet, are being debated in the current literature. Advances in monitoring network design will continue to evolve to meet the specific needs of agencies and organizations interested in monitoring exposures to air pollutants and protecting public health and the environment.

The fundamental design parameters—namely the number of monitoring stations, location of sites, and types of instruments—depend on the following factors:

- Objectives of the monitoring
- Extent of air pollution problem in the area
- Type or intent of monitoring
- Characteristics of pollutants
- Type of area (rural, urban, industrial, etc.)
- Sources of emission
- Emission density
- Averaging period
- Topography
- Population density
- Population likely to be exposed
- Expected concentrations of pollutants
- Traffic conditions
- Land use planning
- Applicable environmental laws
- Meteorological conditions

Figure 2.6 shows the air quality monitoring network operating in the Helsinki, Finland, metropolitan area (Hamekoski and Koskentalo 1998). Table 2.1 provides infor-

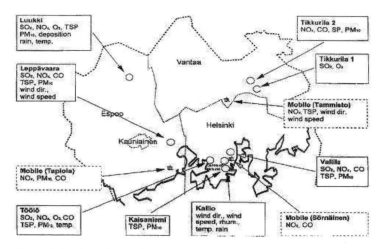

Figure 2.6 Air quality monitoring network operating in the Helsinki, Finland, metropolitan area (Hamekoski and Koskentalo 1998)

mation on population categories and the range in the density of stations required, based on the U.S. NAMS network design criteria.

The placement of probes used in monitors included in a network needs to be carefully considered and should:

•   Avoid direct influence of emissions

•   Avoid nearby buildings that interfere with air flow and pollutant concentrations

•   Avoid building surfaces that interfere with pollutant concentrations

•   Review the vertical distribution of pollutant concentrations for determining probe height

Modeling studies may be carried out in order to determine the number and location of monitors for the study area. The idea is to make sure that suitable representative samples are taken during monitoring. Agreeing on what constitutes a representative ambient air sample has been the subject of debate among scientists and the legal profession. This can be illustrated by the measurement of carbon monoxide (CO) to determine violations of air quality standards in urban areas. Business districts in a city generally have high CO concentrations. The monitor could be located in a parking garage, a block away from the parking garage, at a heavily trafficked intersection, or on the side of a busy street—all of which could produce very different readings that require careful interpretation.

Care should also be taken in drawing a representative air sample to the monitoring device. Unwanted material can easily lead to erroneously high readings. For example, a bug screen for air particulate measurements or steps that prevent unnecessary reactions of gases with solids captured could help prevent biased measurements.

Over the years, the following generally accepted guidelines for the design of air monitoring networks have emerged:

•   Ambient air monitors should be placed at a location where public has unrestricted access and where the pollution concentration is highest

Table 2.1 Design criteria for the U.S. National Air Monitoring Stations (NAMS)

Pollutant	Population category[b]	Range in station density[c]	Site selection
Carbon monoxide	>500,000	⩾2	Major traffic arteries and heavily traveled streets in downtown urban areas; Neighborhood areas
Nitrogen dioxide	>1,000,000	⩾2	Neighborhood areas with highest NO_x emission densities
Ozone	>200,000	⩾2	Urban areas considering peak downwind ozone transport and population exposure; Neighborhood areas on fringe of central business districts and considering peak downwind ozone transport and population exposure
Sulfur dioxide	>1,000,000 500,000–1,000,000 250,000–500,000 100,000–250,000	$H = 6{-}10, M = 4{-}8, L = 2{-}4$ $H = 4{-}8, M = 2{-}4, L = 1{-}2$ $H = 3{-}4, M = 1{-}2, L = 0{-}1$ $H = 1{-}2, M = 0{-}1, L = 0$	Urban and neighborhood areas that are impacted by one or more point sources which extend over a broad geographic scale
PM_{10}	>1,000,000 500,000–1,000,000 250,000–500,000 100,000–250,000	$H = 6{-}10, M = 4{-}8, L = 2{-}4$ $H = 4{-}8, M = 2{-}4, L = 1{-}2$ $H = 3{-}4, M = 1{-}2, L = 0{-}1$ $H = 1{-}2, M = 0{-}1, L = 0$	Urban and neighborhood areas that are impacted by motor vehicle diesel exhaust, industrial/combustion sources, and residential oil, coal or wood burning for space heat

Source: Demerjian, 2000

[a] The principle monitoring objective of NAMS is to measure pollutant concentration in areas which are expected to have the highest levels and population exposures in terms of NAAQS averaging times. The criteria identify two categories of stations: category (a) stations located in area(s) of expected maximum concentration and category (b) stations having the combined attributes of poor air quality in highly populated areas, but not necessarily in area(s) of expected maximum concentration.

[b] The selection of urban areas and the required number stations per area are jointly determined by USEPA and the State agency.

[c] The range in the approximate number of stations required per area is based on the expected levels of pollution in the case of SO_2 and PM_{10}: H = exceeding the NAAQS, M = exceeding 60% of the primary or 100% of the secondary NAAQS, and L = exceeding ⩽60% of the primary or 100% of the secondary NAAQS.

- The location should have power, shelter from rain and snow, easy access, protection from vandalism and animals, and a constant thermal environment, if possible

USEPA has developed detailed guidelines for the placement of air monitors as outlined in 40 CFR 58 (USEPA 2001a) and analytical methods for measuring air quality parameters as given in 40 CFR 50 (USEPA 2001b).

The design of air monitoring networks falls into three categories:

1. Air monitoring design using dispersion/deposition modeling
2. Air monitoring design using dispersion/deposition modeling and/or statistical/optimization techniques
3. Air monitoring network design using risk calculations

1.1   Dispersion/Deposition Modeling

An example of air monitoring design under the first category is illustrated using USEPA's latest AERMOD dispersion model for sulfur dioxide (SO_2) and nitrogen oxides (NO_x) emission sources within Lucas County in Ohio. Urban and flat terrain options were selected, along with 3-hour and 24-hour averaging times for SO_2 and 1-hour and 24-hour averaging times for NO_x. A detailed emission inventory of all stacks in Lucas County for 1990 was obtained from the Ohio Environmental Protection Agency. The emission inventory consisted of yearly emission rates of various sources, the source identification, its location in UTM (Universal Transverse Mercator) coordinates, and stack properties, including stack diameter, velocity, and exit temperature.

There were 123 stacks releasing SO_2 in Lucas County, and these were divided into three groups. The first group consisted of 16 stacks representing approximately 96 percent of the emissions, with each of the individual stacks emitting more than 210 tons annually. The second group consisted of 28 stacks representing 3.7 percent of the pollutant emissions; each of the individual stacks emitted between 5 and 210 tons annually. The third group, comprising the remaining 79 stacks, represented 0.24 percent of the pollutant emissions, with each of the individual stacks emitting less than 5 tons annually. The second and third groups were modeled as "superstacks" (see Kumar, Bellam, and Sud 1996).

There were 126 stacks releasing NO_x in Lucas County, which were again divided into three groups. The first group of 15 stacks, each emitting more than 210 tons annually, represented approximately 88 percent of the county's emissions. The second group (5–210 tons annually) consisted of 86 stacks representing 11.6 percent of the pollutant emissions. The final 25 stacks that made up the third group (less than 5 tons annually) represented just 0.1 percent of the pollutant emissions. As with the SO_2 modeling, the second and third groups of stacks were modeled as superstacks.

A total of 18 sources, shown in figure 2.7, were selected within the area to simulate SO_2 and NO_x emissions for this application, and a uniform Cartesian grid receptor network was laid over the Lucas County shapefile. A total of 441 receptors were defined for the selected grid receptor network. The final dimensions of the grid, as well as the spacing along the x- and y-axes, were automatically calculated in AERMOD and associated software.

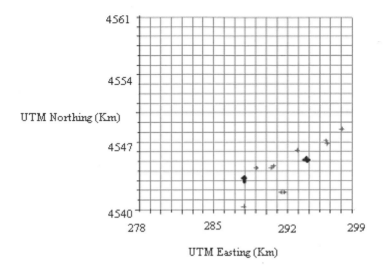

Figure 2.7 Sources and monitors used in dispersion modeling for Lucas County, Ohio
Note: Some stack locations overlap each other

AERMOD's meteorological pathway was used to specify the meteorological data file and other meteorological variables, including period to process from the meteorological file. The raw hourly surface observations and raw upper air data were obtained for the year 1990 and were prepared using AERMET, a meteorological preprocessor. The weather stations at Toledo Express Airport and Ann Arbor, Michigan, were the nearest available stations with all the required hourly surface data and upper air data. The surface data are hourly observations of surface-level parameters such as wind speed, temperature, and cloud cover that are used by AERMET to generate a surface file for use in AERMOD. The upper air data file provides information on the vertical structure of the atmosphere. This includes the altitude, pressure, dry bulb temperature, and relative humidity. Annual values of SO_2 concentration for 1990 were obtained at every receptor point within the Lucas County area.

Figure 2.8 shows the SO_2 and NO_x concentration contours for various averaging times and the location of existing monitors. An examination of these figures indicates that the monitoring locations are fairly close to the areas of high concentrations for certain averaging times and certain pollutants. This poses a challenge to environmental engineers who are trying to decide the optimum locations and number of monitors required for regulatory purposes, when dealing with multiple pollutants, each with different sampling times.

1.2 Statistical/Optimization Techniques

Another approach to modeling design is the accounting of uncertainties in growth and planning in order to expand existing networks optimally. Chang and Tseng (1999b) present a new approach, which they call *gray compromise programming*, for siting new air quality monitoring stations in a metropolitan region. They postulate, in particular, cost, effectiveness, and efficiency characteristics for expanding the existing air quality monitoring network in the multicriteria decision-making process. In order to test the impacts of the growth of population and emission sources, a series of technical set-

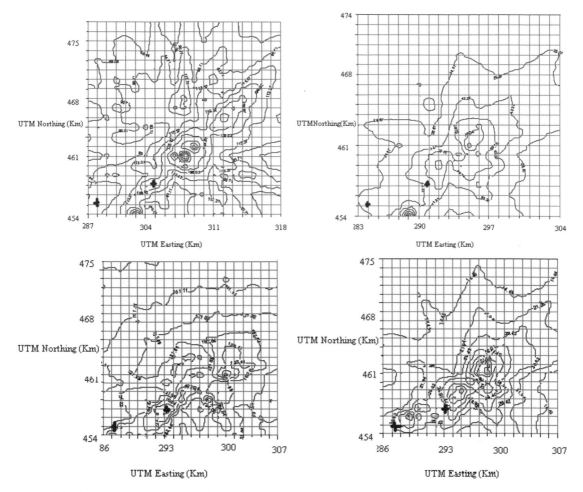

Figure 2.8 Concentration contours of SO$_2$ and NO$_x$ for various averaging times and monitor locations in Lucas County, Ohio
Note: Location of existing monitors is shown by the symbol +

tings in the gray compromise programming model were examined, and the uncertainties embedded in the planning procedure were recognized for assessing the optimal expansion alternatives in the air quality monitoring network. The practical implementation of this approach was demonstrated by a case study for the city of Kaohsiung in Taiwan. It appears that such an approach is useful for screening the expansion alternatives in the air quality monitoring network under uncertain environments. Further, this approach can help planners and decision makers find more-flexible solutions based on those multiobjective design principles.

1.3   Risk Calculations

The identification of high-risk areas using air toxics data can also be used in designing monitoring networks. For example, Joshi and Ashok Kumar (2002) suggested the following methodology to identify the high-cancer-risk areas in Ohio:

1. USEPA's Toxic Release Inventory (TRI) database was chosen because it is widely available and provides comprehensive annual data on hazardous environmental chemical emissions.

2. Carcinogenic toxicity potentials were used to convert the total air emissions in terms of benzene equivalents.

3. Zip code was chosen as the area unit for analysis.

4. The top 10 zip codes for 1999 were chosen for further analysis. A period of four years (1995–98) was considered for analyzing the increase/decrease in emissions in each zip code. The data from 1995 were truncated because the largest expansion in reporting of TRI data took place for the 1995 reporting year, when 286 chemicals were added, increasing the number on the list to 643 chemicals (USEPA).

5. The zip codes were ranked based on the equivalent benzene emissions.

6. Reference hazard values were calculated and aggregated for every zip code for the past five years.

7. Again the top 10 zip codes were chosen to carry out further analysis.

8. Socioeconomic factors were chosen from the 1990 Census, including total population, population density, percentage minority, percentage of high school non-graduates (above 25 years of age), percentage poverty, and per capita income.

9. GIS was used to project the facilities of the zip codes ranking in the top 10 on the demographic parameters (see figure 2.9).

10. Finally, the high-risk zip codes were identified in two lists: One list contained zip codes based on benzene equivalents and the other contained zip codes based on both benzene equivalents and socioeconomic factors.

The high-risk zip codes identified through this methodology provide information to locate monitors in such areas. The above approach can also be modified depending on monitoring needs.

2 Types of Air Quality Monitoring Networks

Since the 1970s, new monitoring networks have been developed as our understanding of air pollution problems has improved. Different types of networks have been designed both in response to public concerns and as a way to improve our understanding of the role of air pollution in daily life. Some requirements for developing new networks grew out of requirements in federal statutes and international agreements. For example, USEPA is required to assess the effectiveness of air pollution control efforts under Title IX of the Clean Air Act Amendments, the National Acid Precipitation Assessment Program, the Government Performance and Results Act, and the U.S.–Canada Air Quality Agreement.

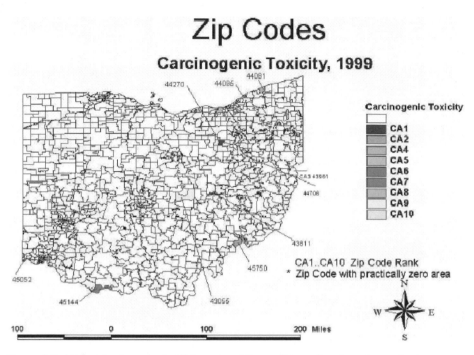

Figure 2.9 High-risk zip codes in Ohio as possible candidates for air toxic monitoring

Air quality trends are often obscured by the wide variability of measurements and climate because the changes in the atmosphere occur at a slow pace. Thus there is a need for continuous and consistent data to overcome the variability. The procedures for filling missing values in monitored data are available (Terry, Lee, and Kumar 1986). Therefore, long-term monitoring networks should be considered to characterize concentration and deposition levels that can be used to identify relationships among emissions, atmospheric loadings, and effects on human health and the environment. To observe significant changes in atmospheric composition over the long term, consistent procedures and quality-assured practices need to be implemented. The results from the air quality and deposition monitoring networks provide insights regarding the effectiveness of the ongoing emission control policies and new regulatory approaches for improving air quality and safeguarding the environment.

In general, air monitoring can be grouped into the following types, depending on the purpose and scope of the monitoring effort:

Emissions monitoring: This type of monitoring focuses on emissions coming out of natural and manmade sources. The idea is to characterize sources of emissions and the amount of contaminants released from these sources.

Ambient monitoring: The emphasis here is on ambient air concentrations of toxic as well as nontoxic contaminants. Chemical speciation is becoming an important component of new network designs.

Deposition monitoring: This type of monitoring measures the dry and wet deposition of atmospheric contaminants.

Visibility monitoring: The ability to see things (i.e., the visual and aesthetic component of air quality) is the primary focus of this type of monitoring.

Upper air monitoring: This type of monitoring focuses on ambient air concentrations in the upper atmosphere, with the help of satellites, airplanes, and so forth.

Health monitoring: This type of monitoring recognizes the importance of risk assessment and risk management in public health studies.

Biomonitoring: This type of monitoring measures the quality of the environment by using sensitive species as indicators to identify and assess changes in the environment caused by natural and manmade activities.

3   Descriptions of Air Quality Monitoring Networks

This section describes different types of air quality monitoring networks operating today in the world. They were developed successfully over time and are examples of networks that help us understand the concept of air quality monitoring. Because of the large number of monitoring networks operating around the globe, only selected examples were chosen from ambient air monitoring programs in the United States, Canada, and Mexico, along with emission monitoring programs at industrial plants, the health monitoring programs of WHO, satellite monitoring by NASA and USEPA, and biomonitoring.

3.1   The U.S. Ambient Air Monitoring Program

Ambient air is the air around us. Monitoring instruments are placed to test ambient air quality, which gives us an idea of the background concentrations and releases of pollutants from both human activity and natural sources. It also reflects the effects of factors such as temperature, sunlight, air pressure, humidity, wind, rain, and landscape. The type of readings obtained and time required for testing are dependent on the nature of the contaminant. For gaseous constituents, real-time monitors can generate raw data at the site and transmit it to a central computer. In some cases, however, such as volatile organic compounds, an air sample must be drawn into a stainless steel cylinder at the monitoring site and then transported to a laboratory for detailed analysis.

The United States has developed a comprehensive Ambient Air Monitoring Program to characterize and assess air quality throughout America. The USEPA's network has been designed to meet one of the following four basic monitoring objectives:

1.   To determine highest concentrations expected to occur in the area covered by the network

2.   To determine representative concentrations in areas of high population density

3.   To determine the impact of significant emission sources or source categories on ambient pollution levels

4.   To determine general background concentration levels

The United States has established a long-term national monitoring program to determine the status and trends of air pollutant emissions, ambient air quality, and pollutant deposition. Table 2.2 shows the changes in the number of criteria air pollutant monitors operating between 1970 and 1999 in the United States.

USEPA's ambient air quality monitoring program consists of seven major categories of monitoring stations:

1. *State and Local Air Monitoring Stations (SLAMS):* The SLAMS consist of approximately 4,000 monitoring stations (see figure 2.10). The size and distribution are largely determined by the needs of state and local air pollution control agencies to meet their respective State Implementation Plan (SIP) requirements.

2. *National Air Monitoring Stations (NAMS):* These 1,080 stations are key sites under the SLAMS network, with emphasis on areas of maximum concentrations and high population density, that is, multisource and urban areas (see figure 2.11). The design criteria for the NAMS are given in table 2.1.

3. *Special-Purpose Monitoring Stations (SPMS):* These stations are used to measure criteria pollutants and provide for special studies needed by the state and local agencies to support state

Table 2.2 Number of criteria air pollutant monitors in the United States, 1970–1999

Year	PM_{10}	O_3	NO_2	SO_2	CO	Total
1970	245	1	43	86	82	457
1975	1120	321	303	827	494	3065
1980	1135	546	375	1088	511	3655
1985	970	527	305	906	458	3166
1990	720	627	345	743	493	2928
1999	1214	1086	424	637	531	4184

Note: Compiled from the USEPA website

Figure 2.10 SLAMS air monitoring network in the United States
Source: USEPA

Figure 2.11 NAMS air monitoring network in the United States
Source: USEPA

implementation plans and other air program activities. The SPMS are not permanently established. These sites can be adjusted easily to accommodate changing needs and priorities. The SPMS are used to supplement the fixed monitoring network as circumstances require and resources permit. If the data from SPMS are used for SIP purposes, they must meet all quality assurance and methodology requirements for SLAMS monitoring.

4.   *Photochemical Assessment Monitoring Stations (PAMS):* The PAMS are designed to measure ozone, ozone precursors (approximately 60 volatile hydrocarbons, nitrogen oxides, and carbonyl compounds), and meteorological parameters required by the 1990 amendments to the Clean Air Act. A PAMS network is required in each ozone nonattainment area that is designated serious, severe, or extreme. The required networks will have from two to five sites, depending on the population of the area. There was a phase-in period of one site per year starting in 1994. Currently there are 73 active PAMS sites as shown in figure 2.12.

Figure 2.12 PAMS air monitoring network in the United States
Source: USEPA

5.   *Interagency Monitoring of Protected Visual Environments (IM-PROVE):* National parks and wilderness areas possess many stunning vistas and scenery. IMPROVE is a collaborative monitoring program, the purpose of which is to establish present visibility levels and trends and to identify sources of anthropogenic impairment. The National Park Service and USEPA are the lead funding agencies for this network. It started with 20 long-term monitoring sites in 1987 and currently encompasses 108 sites in parks and wilderness areas across the United States. Monitor locations of the IMPROVE network are shown in figure 2.13.

6.   *National Atmospheric Deposition Program (NADP)* and *National Trends Network (NTN):* The NADP was developed to monitor wet acid deposition. Rural monitoring sites of NADP collect data in sensitive ecosystems and provide insight into natural background levels of pollutants where urban influences are minimal. These data help in the study and evaluation of various environmental effects. The NADP was initiated in the late 1970s as a cooperative program between federal and state agencies, universities, electric utilities, and other industries to determine geographical patterns and trends in precipitation chemistry in the United States. In 1978, collection of weekly wet deposition samples began. The network size grew rapidly during the 1980s as the National Acid Precipitation Assessment Program called for characterization of acid deposition levels. The NADP had

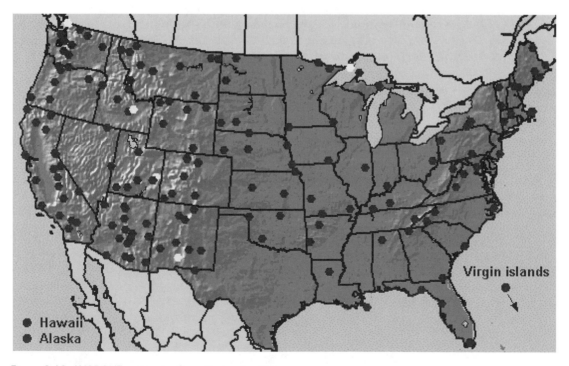

Figure 2.13 IMPROVE network of monitoring locations
Source: http://vista.cira.colostate.edu/improve/Overview/IMPROVENetworkExp.htm

grown to nearly 200 sites by the mid-1980s. Presently it stands as the longest-running national atmospheric deposition monitoring network. The NTN measures sulfates, nitrates, hydrogen ions (a measure of acidity), ammonia, chloride, and base cations (calcium, magnesium, potassium). To ensure comparability of results, the NADP's Central Analytical Laboratory at the Illinois State Water Survey conducts laboratory analyses for all samples. A new subnetwork of the NADP, the Mercury Deposition Network (MDN), measures mercury in precipitation.

7.   *Clean Air Status and Trends Network (CASTNET):* USEPA established the National Dry Deposition Network (NDDN) in 1986 in order to obtain field data on rural deposition patterns and trends at different locations throughout the United States. The network consisted of 50 monitoring sites deriving dry deposition based on measured air pollutant concentrations and modeling dry deposition velocities estimated from meteorology, land use, and site characteristic data. USEPA then created CASTNET from NDDN in 1987 in cooperation with the National Oceanic and Atmospheric Administration (NOAA) to meet the requirements of the amendments to the Clean Air Act in 1990. This network now comprises more than 70 monitoring stations across the United States. CASTNET was developed to monitor dry acid deposition, as well as rural ozone and the chemical constituents of $PM_{2.5}$. Like the NADP, rural CASTNET monitoring sites provide data on sensitive ecosystems and natural background levels of pollutants. Measurements from CASTNET are also important for understanding nonecological impacts of air pollution such as visibility impairment and damage to materials, particularly those of cultural and historical importance. CASTNET dry deposition stations primarily measure:

   •   Weekly average atmospheric concentrations of sulfate, nitrate, ammonium, sulfur dioxide, and nitric acid

   •   Hourly concentrations of ambient ozone levels

   •   Meteorological conditions required for calculating dry deposition rates

3.2   The U.S. Atmospheric Integrated Research Monitoring Network (AIRMoN)

The Atmospheric Integrated Research Monitoring Network, a program under NOAA, is designed to provide a research-based foundation for the routine operations of the nation's deposition monitoring networks, the NADP for wet deposition and CASTNET for dry deposition. AIRMoN is a research program aimed at developing and implementing improved dry and wet deposition monitoring methodologies. The results of emissions controls mandated by the Clean Air Act Amendments of 1990 are studied, and these results are quantified in terms of deposition to sensitive areas.

AIRMoN combines two previously existing deposition research networks—the MAP3S precipitation chemistry network and the CORE/satellite Dry Deposition Inferential Method network—resulting in a new monitoring activity that allows for online modeling and analysis. An air-sampling component of AIRMoN provides some unique information on changes in air quality, and the techniques are designed to quantify the extent to which changes in emissions affect air quality and deposition at selected locations. AIRMoN sites are chosen so that the probability of detecting the needed change is optimized and the needs of researchers studying the effects of emission controls are served. For example, Hicks and colleagues (2001) studied the climatological features of a region's surface air quality from the AIRMoN data in the United States.

3.3   Canadian Air Monitoring Networks

Canada has developed three air monitoring networks. The first was established in 1969 and is known as the National Air Pollution Surveillance (NAPS) network. It is primarily an urban network, with 239 air monitoring stations at more than 136 sites. The second network, known as the Canadian Air and Precipitation Monitoring Network (CAPMoN), is a rural network with 23 air monitoring stations in Canada and one in the United States.

The NAPS network (see figure 2.14) gathers data on ozone, particulate matter, SO_2, CO, NO_x, and volatile organic compounds. Air quality data collected by the NAPS network have been used to establish the relationships between air pollution and human health and also to evaluate air pollution control strategies, identify urban air quality trends, and look at emerging air pollution issues. Land-use planners, public transportation and urban planners, and many others who must take air quality into account

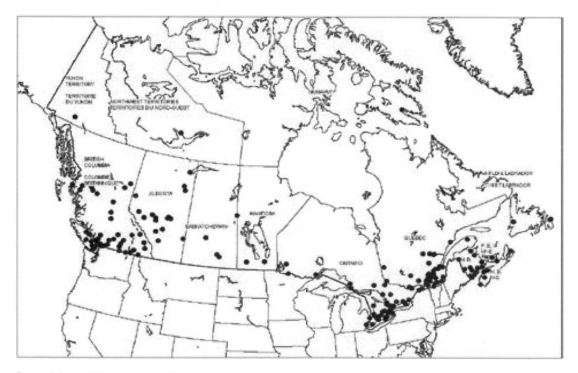

Figure 2.14 NAPS network in Canada
Source: National Air Pollution Surveillance (NAPS) Network Annual Data Summary for 2002 Report EPS 7/AP/ 35 Rapport SPE 7/AP/35, December 2003

in their decisions use information from the NAPS network. Current plans are to add 10 new monitoring stations by 2005.

The CAPMoN was established for acid rain and transboundary transport of pollutants studies and has been in operation for more than 20 years. At present, NO_x, PM, and ozone are also measured at some sites. The stations are located to ensure measurements are regionally representative and are not affected by local sources of air pollution.

The maps produced by the networks are available during the smog season—May through September—on Environment Canada's website, www.ec.gc.ca/air/ozone-maps_e.shtml.

3.4   The Canada–U.S. Deposition Network

The Integrated Atmospheric Deposition Network (IADN) was formed in 1990 through mandates of the Clean Air Act and Great Lakes Water Quality Agreement to study the deposition of persistent organic pollutants (POPs) from a variety of sources. The project is a joint venture of Environment Canada, the Ontario Ministry of the Environment, and USEPA's Great Lakes National Program Office. Master sampling stations in rural areas near each of the five Great Lakes and a series of satellite stations throughout the Great Lakes Basin (see figure 2.15) are used to collect vapor and particle phase air samples concurrently every 12 days for 24 hours. Precipitation samples are also taken

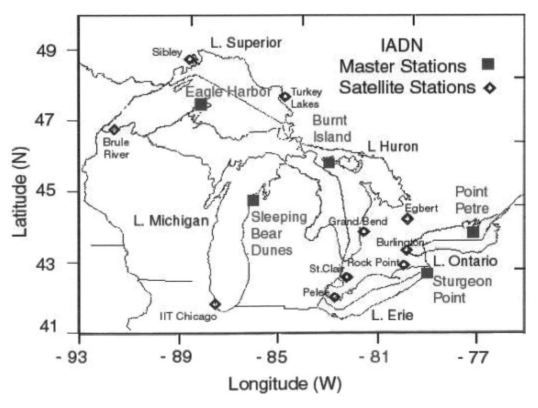

Figure 2.15 IADN monitoring network in Canada
Source: http://www.epa.gov/glnpo/glindicators/air/airb.html

and represent a composite over 28 days at U.S. sites and 14 days at Canadian sites. Samples are also analyzed for 56 polychlorinated biphenyl (PCB) congeners or congener groups, 18 organochlorine pesticides (both banned and in use), and 16 polycyclic aromatic hydrocarbons (PAHs) in each of the phases.

The long-term data from the IADN monitoring network provides trends in POPs in the Great Lakes atmosphere. The most pronounced spatial trend has been in the concentration gradient of PCBs and PAHs between urban and rural sites. PCBs were banned in the late 1970s, while PAHs continue to be released as combustion by-products, but both compound classes are strongly associated with urban areas. In fact, concentrations for these compounds measured at Chicago are an order of magnitude higher than at any other sampling site. Analysis of IADN data has also shown that temporal trends in vapor phase concentrations of banned POPs (such as hexachlorocyclohexane, an insecticide banned in the United States and Canada in 1978) indicate an overall decline in atmospheric concentrations. These long-term data trends from monitoring provide a means for demonstrating that bans on these POPs have worked and provide an essential link between science and policy.

3.5   The Mexican Network

Air quality monitoring activities in Mexico began in late 1992. SIMA (Sistema Integral de Monitoreo Ambiental) was born that year, consisting of five automatic monitoring stations in the Monterrey area of Mexico. IMECA (Indice Metropolitano de la Calidad del Aire) is the air quality or pollutant level index used in Mexico, through which air quality is reported to the public. For example, the Monterrey network is divided into five zones (Northwest, Northeast, Southeast, Southwest, and Central). In each zone, a fixed monitoring station is located at a strategically important point of the metropolitan area, so as to be representative of the air quality in that zone. The exact location of the monitoring stations depends on a number of factors, such as the size of the monitoring area, local meteorology, population density, topography of the zone, dispersion of pollutants, and the representative considerations of scales for the specific objectives of the network. The IMECA air quality index for NO_x, SO_2, ozone, CO, PM_{10}, and ultraviolet radiation is reported for each zone on an hourly basis. Studies reveal that PM_{10} and ozone air pollution problems are severe in the Mexico Metropolitan Area. SIMA reports for 2001 and 2002 also reveal a great increase in these two pollutants.

CICA is the U.S.–Mexico Information Center on Air Pollution, which provides technical support and assistance in evaluating air pollution problems along the border. CICA offers technical assistance in a wide range of areas, ambient monitoring being one. There are many U.S.–Mexican joint programs aimed at improving air quality. For example, the Texas Commission on Environmental Quality (TCEQ) has reached a major milestone in characterizing and improving air quality in the El Paso region whereby real-time air quality data from Ciudad Juarez, Mexico, are being transmitted via a two-way radio system to the TCEQ.

A highly successful project was completed during 1990–93 by Los Alamos National Laboratory and Mexico. This resulted in the development of a set of analytical tools for making decisions on air quality management by Mexico City Air Quality Research Initiative, known as MARI. The Ambient Air Monitoring Network of Mexico City consists of 32 remote sites with 108 analyzers. There are 19 remote sites for the manual network for suspended particles and 16 for the atmospheric deposition network.

3.6   Emission Monitoring at Industrial Plants

Monitoring of emissions from individual sources plays an important role in understanding the nature of air emission sources and interpreting the data from ambient air monitoring. Facilities releasing pollutants into the air as well as regulatory agencies use this type of monitoring to ensure that emission sources stay within the limits established by regulatory standards, permits, and guidelines. All industrial plants, unless exempt, must obtain an air quality approval or permit that allows the facility to install and operate air emission sources within established limits for regulated pollutants. In many instances, these facilities find it useful to perform continuous monitoring. This helps in rapid detection and recognition of irregular conditions at the plant and also allows the operating staff to run the plant under optimum conditions and return to standard operating conditions as quickly as possible following a malfunction or upset. However, for many other plants, emission monitoring at regular intervals may be sufficient to monitor the performance of equipment and to estimate the emission levels.

In general, the frequency of monitoring depends on the type of process and the process equipment installed, the stability of the process, the reliability of the analytical methods, and the regulatory requirements. The frequency of monitoring is generally selected by balancing the cost of obtaining additional samples against the information required to meet the intent of the regulations or permit requirements.

The USEPA website defines a *continuous emission monitoring system* (CEMS) as the total equipment necessary for the determination of a gas or particulate matter concentration or emission rate using pollutant analyzer measurements and a conversion equation, graph, or computer program to produce results in units of the applicable emission limitation or standard. Some of the USEPA regulations require the use of a CEMS for either continual compliance determinations or determination of exceedances of the standards. The individual subparts of the USEPA rules specify the reference methods that are used to substantiate the accuracy and precision of the CEMS. *Performance specifications* are used for evaluating the acceptability of the CEMS at the time of or soon after installation and whenever specified in the regulations. Quality assurance procedures in appendix F to 40 CFR 60 (USEPA 2001b) are used to evaluate the effectiveness of quality control and quality assurance procedures and the quality of data produced by any CEMS that is used for determining compliance with the emission standards on a continuous basis as specified in the applicable regulation. A *predictive emission monitoring system* (PEMS) may be permitted for the determination of a gas concentration or emission rate.

A typical CEMS includes:

*   A monitor to measure concentrations of gaseous compounds
*   A monitor to measure particulate matter
*   A volumetric flow monitor
*   An opacity monitor
*   A diluent gas (O_2 or CO_2) monitor
*   A computer-based data acquisition and handling system (DAHS) for recording and performing calculations with the data
*   A sampling system

CEMS for gases are divided into three types: extractive, in situ, and Fourier Transform Infrared (FTIR) spectroscopy systems. *Extractive* CEMS extract a gas sample from the exhaust at a measurement site and transport the sample through a conditioning system and into separate analyzers. Each analyzer measures its designated pollutant gas concentration. *In-situ* CEMS allow the effluent gas to enter a measurement cell inserted in the stack or duct. The concentration of the pollutant in the effluent is then measured by a variety of techniques. Many in-situ instruments are capable of measuring more than one pollutant. The *FTIR* CEMS is basically the same design as the extractive CEMS except that all pollutant gas concentrations are measured with a single instrument.

Particulate emissions that are measured in stacks or ducts before they are exhausted into air need to be sampled isokinetically. This may be done to provide a routine baseline manual check for any continuous particulate monitoring or as a routine for control purposes where continuous monitoring methods do not exist. It may be possible in some situations to adapt the sample collection system to provide for continuous monitoring. Isokinetic sampling subjected to a variety of national standards and appropriate methods will generally need to be agreed upon with the regulatory authorities. Typically, it consists of combined airflow measurement and extraction sampling equipment that can be controlled to maintain the same velocity in the sampling nozzle as is present in the duct. These can be combined to give mass emissions.

The opacity of particulate matter in stack emissions is measured based upon the principle of transmissometry. A CEMS for opacity is also called a *continuous opacity monitoring system* (COMS). A light source having specific spectral characteristics is projected through the effluent in the stack or duct, and the intensity of the projected light is measured by a sensor. The projected light is attenuated because of absorption and scatter by the particulate matter in the effluent; the percentage of visible light energy attenuated is defined as the opacity of the emission. The opacity monitor used to determine the opacity of the effluent consists of a transmissometer, from which the in-situ opacity is determined, and all other interface and peripheral equipment necessary for continuous operation. The opacity monitor includes sample interface equipment such as filters and purge air blowers (shutters or other devices may be included to provide protection during power outages or failure of the sample interface) and a remote control unit to facilitate monitoring the output of the instrument, initiation of zero and upscale calibration checks, or control of other monitor functions.

3.7 WHO Health Monitoring Programs

Ambient air quality in urban areas is an important issue in the major cities of the world. Health-related environmental monitoring programs were implemented by WHO after considering the importance of public health studies with respect to risk assessment and risk management. WHO's health-related environmental monitoring programs include the implementation of the following five separate global projects:

- Air quality monitoring (GEMS/Air)
- Water quality monitoring (GEMS/Water)
- Food contamination monitoring (GEMS/Food)
- Environmental radiation monitoring network (GERMON)
- Human exposure monitoring (GEMS/Heal)

The above projects are being implemented in cooperation with United Nations Environment Program (UNEP) and with the support of several other agencies. In this chapter we will discuss the Global Environmental Monitoring System/Air Pollution (GEMS/Air) program, which was operated from 1973 to 1995 to assess ambient air pollution trends globally, and its successor program, the Air Management Information System (AMIS).

During its operation, GEMS/Air was the only global program providing long-term air pollution monitoring data for cities in developing countries. Some of the program benefits of GEMS/Air were:

* Megacities Reports
* GEMS/Air Methodology Review Handbook series
* Series of reports entitled "GEMS/Air City Air Quality Trends"
* Reports on emission inventories and control measures
* Cost-benefit analyses

These reports and studies have been used to assess the risks of increase in total mortality from air pollution.

AMIS is the successor to the GEMS/Air. In 1996, WHO developed AMIS as a new program under the umbrella of WHO's Healthy Cities Program. This program is an information exchange system within the framework of the Global Air Quality Partnership providing information on all issues of air quality management among its many participants, including municipalities, national environmental protection agencies, international organizations, the World Bank and international development banks, and nongovernmental organizations.

The various objectives of the AMIS program are to:

* Conduct global assessments of air quality
* Act as a global data and information exchange medium for air quality management issues in the context of the Global Air Quality Partnership
* Facilitate review and validation of assessments to establish codes of best practice
* Identify and establish AMIS Regional Collaborating Centers and experts for coordinating and supporting activities relevant to the needs of the regions
* Produce technical documents in support of all aspects of air quality management
* Build and maintain a global database with validated data from an expanded number of cities
* Conduct annual reviews and distribute the results widely

Under the AMIS program, information is collected on all issues of air quality management from its participants. This information is distributed from a centralized WHO information center. Several databases have already been developed. The AMIS core database of ambient air pollutant concentrations consists of summary data, including annual means, 95th percentiles, and the number of days on which WHO Air Quality Guidelines are exceeded, from more than 100 cities around the world. The updated core database has been made available on a CD-ROM since 1998. Table 2.3 shows the participating countries, cities, and monitored contaminants of the AMIS database.

Table 2.3 Countries and cities participating in WHO's Air Management Information System and compounds monitored

Country	City	Compounds
Argentina	Cordoba City	NO_2, SPM
	Mendoza	SO_2, NO_2, black smoke, lead
	Santa Fe	Black smoke
Australia	Melbourne	SO_2, NO_2, O_3, CO, SPM, PM_{10}, lead
	Perth	SO_2, NO_2, O_3, CO, SPM, PM_{10}, lead
	Sydney	SO_2, NO_2, O_3, CO, SPM, PM_{10}, lead
Austria	Vienna	SO_2, NO_2, O_3, CO, PM_{10}
Belgium	Brussels	SO_2, NO_2, O_3, SPM
Brazil	Sao Paulo	SO_2, NO_2, O_3, CO, SPM, PM_{10}
Bulgaria	Sofia	SO_2, NO_2, SPM, lead
	Plovdiv	SO_2, NO_2, SPM, lead
Canada	Hamilton	SO_2, NO_2, O_3, CO, SPM, lead
	Montreal	SO_2, NO_2, O_3, SPM, lead
	Toronto	SO_2, NO_2, O_3, CO, SPM, lead
	Vancouver	SO_2, NO_2, O_3, CO, SPM, lead
Chile	Santiago	SO_2, NO_2, O_3, CO, PM_{10}, $PM_{2.5}$
China	Beijing	SO_2, SPM
	Chongqing	SO_2, NO_2, CO, SPM
	Guangzhou	SO_2, SPM
	Shanghai	SO_2, SPM
	Shenyang	SO_2, SPM
	Xi'an	SO_2, SPM
Croatia	Zagreb	SO_2, SPM, black smoke, lead, cadmium
Cuba	Havana	SO_2, NO_2, ammonia
Denmark	Copenhagen	SO_2, NO_2, O_3, CO, SPM, black smoke, lead
Ecuador	Ambato	SO_2, black smoke
	Cuenca	SO_2, black smoke
	Esmeraldas	SO_2, black smoke
	Guayaquil	SO_2, SPM, black smoke
	Quito	SO_2, SPM, PM_{10}, black smoke, lead
Finland	Helsinki	SO_2, NO_2, O_3, CO, SPM, PM_{10}, lead
Germany	Frankfurt	SO_2, NO_2, O_3, CO, SPM, PM_{10}, lead
	Munich	SO_2, NO_2, O_3, CO, SPM
Ghana	Accra	SO_2, SPM
Greece	Athens	SO_2, NO_2, O_3, CO, black smoke
India	Mumbai	SO_2, NO_2, SPM, PM_{10}, lead
	Calcutta	SO_2, NO_2, SPM, PM_{10}, lead
	New Delhi	SO_2, NO_2, SPM, PM_{10}, lead
Japan	Osaka	SO_2, NO_2, O_3, CO, PM_{10}, lead
	Tokyo	SO_2, NO_2, O_3, CO, PM_{10}, lead
Mexico	Mexico City	SO_2, NO_2, O_3, SPM
New Zealand	Auckland	SO_2, NO_2, SPM, black smoke, lead
	Christchurch	SO_2, NO_2, CO, SPM, lead
Portugal	Lisbon	SO_2, NO_2, SPM, PM_{10}, black smoke, lead

Romania	Bucharest	SO_2, NO_2, SPM, lead
South Africa	Johannesburg	SO_2, NO_2, O_3, CO, black smoke, lead
	Capetown	SO_2, NO_2, PM_{10}
	Durban	SO_2, lead
Spain	Madrid	SO_2, NO_2, black smoke
Switzerland	Geneva	SO_2, NO_2, SPM, lead
Thailand	Bangkok	SO_2, NO_2, O_3, CO, SPM, lead
United Kingdom	Birmingham	SO_2, NO_2, O_3, CO, PM_{10}, black smoke
	Edinburgh	SO_2, NO_2, O_3, CO, PM_{10}, black smoke
	Glasgow	SO_2, NO_2, O_3, CO, PM_{10}, black smoke
	Liverpool	SO_2, NO_2, O_3, CO, PM_{10}, black smoke
	London	SO_2, NO_2, O_3, CO, PM_{10}, black smoke
	Manchester	SO_2, NO_2, CO, black smoke, lead
	Sheffield	SO_2, NO_2, CO, black smoke, lead
United States	Los Angeles	SO_2, NO_2, O_3, CO, PM_{10}, lead
Venezuela	Caracas	SO_2, NO_2, SPM, lead
Philippines	Metropolitan Manila	SO_2, SPM

Source: WHO
NO_2 = nitrogen dioxide; SPM = suspended particular matter; SO_2 = sulphur dioxide; O_3 = ozone; CO = carbon monoxide; PM_{10} = particular matter 10 μm or less in diameter; $PM_{2.5}$ = particular matter 2.5 μm or less in diameter.

The data collected under the AMIS program can be used in assessing the global risk of exposure to air pollution with respect to mortality and morbidity in human populations of the various regions of the world. A database on air quality guidelines and air quality standards contains data from about 60 countries, while that on pollution management capabilities contains data from 70 cities. These databases help AMIS participants in different countries to communicate with each other. Future databases that are planned under the AMIS program include:

- Air quality guidelines and standards
- Monitoring and management capabilities of cities participating in AMIS
- Indoor air pollutants and fuel use
- Noise levels in urban areas
- Dispersion models with information on their use and where to access them
- Human exposure estimates
- Control measures and the magnitude of their costs
- Health effects and the magnitude of their costs as assessed in national studies

WHO also has plans to use AMIS to provide electronic access to the health data that serve as the basis for WHO's air quality guidelines. It is also being modified to provide essential ingredients for estimating the disease burden after exposure to air pollution in urban areas. It is likely that in the future the program will evolve, as a component of the Global Air Quality Partnership, to have all participants provide and have access to the information contained in AMIS.

3.8 Satellite Monitoring by NASA and USEPA

Large ground-level monitoring networks to assess air quality impacts are impractical to implement on a global basis. The need for more frequent air quality assessments to help combat global air pollution will increase as the population grows, the world moves toward a global economy, and air pollution continues to cross geopolitical boundaries.

In 1974, Rowland and Molina made the seminal discovery that the release of chlorofluorocarbons (CFCs), manmade chlorine compounds, can cause dramatic decreases in the stratospheric ozone layer, especially over Earth's polar regions. Satellite data measured by NASA's Total Ozone Mapping Spectrometer (TOMS) since 1979 confirmed that the ozone layer was being depleted. This detection led to regulations and the phaseout of CFC production under the 1986 Montreal Protocol and its successor amendments.

Satellites can provide measurements of trace gases and aerosols such as ozone (O_3), NO_2, SO_2, CO, and particulate matter/aerosols. The cost of collecting data over the entire globe using satellites is far less than what a network of ground-based systems could achieve over a fraction of the area.

Research is under way to explore the capabilities of satellites to identify the sources of air pollution and where polluted air travels. For example, the Earth Observing System (EOS) Aura satellite (see figures 2.16 and 2.17), which was launched by NASA in July 2004, is designed to study Earth's ozone layer and changes in air quality and climate (NASA 2004). This is the third in a series of major EOS satellites to study the environment and climate change and is part of NASA's Earth Science Enterprise. The Aura satellite carries a payload of four instruments, each of which will measure ozone or its precursors in different and overlapping ways. Combining this data together, researchers will be able to extract information on the horizontal and vertical distribution of key atmospheric pollutants and greenhouse gases.

USEPA's air program is considering OMI (Ozone Monitoring Instrument) and CALIPSO (Cloud-Aerosol Lidar and Infrared Pathfinder Satellite) as possible instruments to be used on satellite trips.

NASA has developed a technology for measurement of gases such as CO, methane, and NO_x in Earth's atmosphere from aircraft and satellite platforms. This technology is being improved now to help U.S. industries reduce smokestack pollution. The gas filter correlation radiometer (GFCR) is a device that has many distinct advantages over other conventional gas sensors. It has advanced features, capabilities for remote sens-

Figure 2.16 Artist's rendition of the Aura satellite
Source: NASA

Figure 2.17 Aura satellite under construction at
Northrop Grumman in Redondo Beach, California
Source: NASA

ing and area source monitoring, higher reliability, faster response, and a more compact design.

3.9 Biomonitoring

Biomonitoring, short for biological monitoring, is a monitoring technique to measure the quality of the environment by using sensitive species as indicators to identify and assess changes in the environment caused by natural phenomena, human activity, or a combination of the two. The main purpose of biomonitoring is to improve awareness of environmental issues and identify and prioritize pollution-related problems.

Lichens, bryophytes, fungi, algae, and higher plants have been used in biomonitoring of air pollution. Lichens and bryophytes are the most widely used plant groups in air pollution monitoring. The amount of degeneration of the plant after exposure to pollutants is a measure of the kind and quantity of emissions to which the plant is exposed. The absence of lichen and moss are indicators of the presence of SO_2 and carbon dust discharged from power plants. The presence of SO_2 can also be confirmed from the damaged leaves of beeches, blackberries, and pines. Fluorhydrogene damages the leaves of willows, pears, and black and red currents, and lead damages cabbage. Table 2.4 discusses some air pollutants and the bioindicators used to monitor them.

Several air quality biomonitoring methods are available (Mulgrew and Williams 2000). Monitoring can be qualitative or quantitative, using changes in single indicator species or community changes. Other methods include physiological/biochemical plant responses or visible injury as indicators of air pollution. By analyzing plant tissue samples at different distances from a pollution source, the type of pollution and the size of the fallout zone can be determined.

Either passive or active air quality biomonitoring programs can be developed using plants. Passive monitoring is generally quicker and easier and may be used to assess long-term pollution exposure. In contrast, active methods require transplantation and the use of control plants for comparison and maintenance of a pollution regime over

Table 2.4 Typical instrumentation/measurement methods used in North American monitoring networks

Pollutant	Detection principle	Lowest detectable limit/noise[a,b]	Quality assurance requirements	Concentration ranges for audit levels (ppm)[a]
Carbon monoxide	Non-dispersive IR spectrometry	1.0 ppm/0.50 ppm	Routine zero and span checks Routine precision checks Annual audits	1. 3–8 2. 15–20 3. 35–45 4. 80–90
Nitric oxide and Nitrogen dioxide (NO_2-NO)	Chemiluminescence	10 ppb/5 ppb	Routine zero and span checks Routine precision checks Annual audits	1. 0.03–0.08 2. 0.15–0.20 3. 0.35–0.45
Ozone	1. Chemiluminescence 2. UV Photometry	10 ppb/5 ppb	Routine zero and span checks Routine precision checks Annual audits	1. 0.03–0.08 2. 0.15–0.20 3. 0.35–0.45 4. 0.80–0.90
Sulfur dioxide	1. Coulometry 2. UV fluorescence	10 ppb/5 ppb	Routine zero and span checks Routine precision checks Annual audits	1. 0.03–0.08 2. 0.15–0.20 3. 0.35–0.45 4. 0.80–0.90
High volume sampler				
Total suspended particulate	Gravimetric	1 $\mu g\ m^{-3}$		
Particulate lead	X-ray fluorescence	0.01 $\mu g\ m^{-3}$		
Sulfate	Colorimetry	0.1 $\mu g\ m^{-3}$		
Dichotomous/PM_{10} samplers				
Total suspended particulate	Gravimetric	1 $\mu g\ m^{-3}$		
Particulate lead	X-ray fluorescence	0.01 $\mu g\ m^{-3}$		
Sulfate	Colorimetry	0.1 $\mu g\ m^{-3}$		

Source: Demerjian, 2000
[a] Performance as specified in Subpart B—Procedures for testing performance characteristics of automated methods SO_2, CO, O_3, and NO_2; 40 FR 7049, 18 Feb. 1975 as amended at 40 FR 18168, 25 Apr. 1975; 41 FR 52694, 1 Dec. 1976 as required in US Environmental Protection Agency SLAMS/NAMS networks.
[b] Performance specifications in NAPS (Environment Canada, 1997) are a factor of two more sensitive in LDL and noise for gas measurements.

a prescribed period of time, but they can provide quantitative assessments such as deposition rates. In addition, if the genetic state and physiology of the plant is known, the results can be more reliably related to air pollution. The choice of method depends, among other factors, on the purpose of the survey, the size of study area, the resources available, and the desired outcome of the study.

A number of biomonitoring networks have been established to study the effect of atmospheric pollutants on sensitive species. The Ohio Department of Forestry studied lichens in areas of eastern Ohio where air quality was historically poor during the 1970s. Lichen populations were degenerating and key indicator species could not be found. A 1997 lichen survey of the same area showed that lichen species richness was lower in that region than other areas of the state, but key indicator species were returning and lichens were recovering. This indicated that the air quality was improving.

The objective of using lichens as bioindicators was to qualitatively document the historical trends in SO_2 exposure. Figure 2.18 shows two lichens that serve as good indicators of air quality because they survive only where airborne SO_2 levels are low.

The Ozone Monitoring Network of the Wisconsin Department of Natural Resources has used plants as bioindicators to monitor ozone. The milkweed monitoring sites use the common milkweed, which is very sensitive to ozone. Milkweed biomonitoring sites have been established along the Lake Michigan shoreline where ozone concentrations are typically high. The Forest Health Monitoring Bio Indicator Network is a component of the national effort to monitor ozone injury on a variety of native plants and trees, including milkweed, dogbane, blackberry, black cherry, white ash, sassafras, sweet gum, pin cherry, and big leaf aster. A minimum of 10 and a maximum of 30 plants of each species are evaluated in mid-August, and visible symptoms of ozone-induced injury on plants are evaluated.

A European network for the assessment of air quality called the Euro Bionet was founded in 1999 (Euro Bionet 2004) and is established in a number of cities in Europe with more than 80 monitoring sites. Bioindicator plants such as tobacco (ozone), poplar, spiderwort, Italian rye grass (sulfur), and curly kale are studied for visible injuries, and growth effects and pollutant accumulation in their leaves are investigated to determine the quality of the ambient air.

Figure 2.18 Lichens: Flavoparmelia caperata *(top) and* Punctelia rudecta
Source: http://dnr.state.oh.us/forestry/health/lichen/ lichenstudy.htm

Biomonitoring using plants can be a simple and inexpensive process that lends itself as an adaptable method of assessing air quality in developing countries. Biological monitoring becomes highly applicable in remote areas where continuous, direct air sampling is expensive and impractical. Biomonitoring is a valuable assessment tool that is receiving increased use in air quality monitoring programs. Bioindicators can provide complete information concerning the smallest changes in the environmental conditions in specific environments.

4 Air Monitoring Instrumentation

Air pollution monitoring instruments are available for the measurement of indoor and outdoor air pollution. The available instruments can be grouped into the following major categories:

> *Concentration measurement instruments:* This group includes the instruments available for sampling and measuring gaseous and particulate concentrations in air.

> *Continuous emission monitoring systems (CEMS):* Real-time monitoring of stack gases is the basic thrust behind these systems.

> *Air measuring devices:* This category includes volume meters, rate meters, and velocimeters.

> *Meteorological instruments:* Basic devices used for measuring atmospheric variables are included in this category.

A detailed description of these instruments is given in many books on instrumentation (e.g., USEPA 1983; World Meteorological Organization 1996; Randerson 1984; ACGIH 1983).

4.1 Concentration Measurement Instruments

An important aspect of maintaining outdoor air quality remains the measurement of pollutant concentration. There are more than 15,000 air sampling and analyzing devices on the market. The companies that deal with concentration measurement instruments can be easily searched on the Internet. Wark and colleagues (1998, 573) provide a summary of various methods used for continuous monitoring of SO_2, NO_x, CO, hydrocarbons, and photochemical oxidants. Table 2.5 lists commonly used methods for measuring particulate (PM_{10}, $PM_{2.5}$, and TSP), CO, NO_x, O_3, SO_2, lead, and sulfate in U.S. monitoring networks.

4.2 Continuous Emission Monitoring Systems

A comprehensive treatment of CEMS is provided by Jahnke (2000, 416), covering the technical and regulatory issues associated with this monitoring method, which includes the installation of a variety of instruments and procedures for data acquisition and quality assurance to continuously obtain direct emission measurements. Increasingly, CEMS are being considered and required for use in a variety of regulatory monitoring and compliance programs. Devices used in CEMS measure emissions directly for comparison to an applicable standard and are widely understood. Plants using

Table 2.5 Bioindicators for different pollutants.

Pollutant	Biomonitors Plant Group	Use
Heavy metals	1) Lichens and moss	1) Very widely used as effective biomonitors due to their bioaccumulaitve property
	2) Fungi	2) Limited use but some groups are more effective accumulators than others
	3) Higher plants	3) Used in highly polluted areas where lichens and moss are absent
Atmospheric gases	1) Lichens and mosses	1) Lichens are widely used plant group for air pollution monitoring. Bryophytes (e.g., Moss) are used in regional and urban SO_2 studies but bryophytes are used mostly for heavy metal indicators.
	2) Fungi	2) Leafyeats proposed as a SO_2 bioindicator in rural and urban areas
	3) Algae	3) Used as a nitrogen indicator
	4) Higher plants	4) Considered as promising indicators of Ozone
Organic compounds	1) Lichens and mosses	1) Widely used and effective due to the bioaccumulative property
	2) Fungi	2) No Literature Cited
	3) Higher plants	3) Can be used where lichens and mosses are absent. Tree leaves and needles can act as bioaccumulative indicators but interpretation of results complicated due to root uptake of chemicals

Source: Compiled from (*http://www.umweltbundesamt.de/whocc/AHR10/content2.htm*)
WHO Collaborating Centre for Monitoring And Assessment at MARC—Monitoring and Assessment Research Centre, London, UK

CEMS make adjustments to operating parameters to maximize production and minimize operating cost. The cost of the systems has dropped by 50 percent in the past decade. Information on the instruments and manufacturers is widely available on the Internet.

4.3 Air Measuring Devices

A number of instruments or devices are used to measure the total volume of air, flow rate, or velocity. The devices are broadly classified as volume meters, rate meters, and velocimeters. Volume meters measure the volume of gas and (with a timing device) flow rate passing through the meter. There are various types of volume meters, such as the wet test meter and dry test meter. Rate meters are mainly used to measure the flow rate. Velocimeters are used to measure the linear velocity of a gas. Various types of velocimeters, such as pitot tubes, anemometers, and mass flow meters, are available in the market. Air measuring devices are an integral part of air monitoring. Kumar and his colleagues (2004) provide a guide to gas flow measuring devices.

4.4   Meteorological Instruments

Meteorological instruments are used in the study of weather and to measure a variety of atmospheric dispersion–related parameters. The selection and use of these instruments mainly depends on the purpose of the investigation. They are classified as primary meteorological instruments, which are used to measure variables such as wind direction and temperature, and secondary meteorological instruments, which deal with solar radiation, visibility, humidity, and atmospheric pressure.

A basic meteorological (met) station provides sensors that measure five basic environmental variables: air temperature, wind speed, wind direction, relative humidity, and barometric pressure. In addition, other types of sensors that may be part of a met station measure parameters such as precipitation, solar irradiance (longwave), solar irradiance (shortwave), and maximum wind gust. Met stations located on buoys or at the coast may also provide sea surface measurements such as sea surface temperature, wave period, and wave height.

The Jet Propulsion Laboratory and the University of California at Davis maintain a full meteorological station (wind speed, wind direction, air temperature, and relative humidity), a full radiation station (long- and shortwave radiation up and down), a shadow band radiometer, and an all-sky camera at the U.S. Coast Guard Station at Lake Tahoe, California (see figure 2.19). The shadow band radiometer provides information on total water vapor and aerosol optical depth.

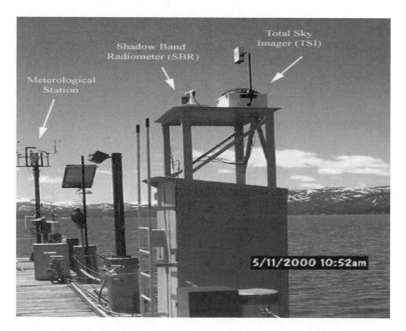

Figure 2.19 Meteorological station at the U.S. Coast Guard Station, Lake Tahoe, California
Source: http://trg.ucdavis.edu/research/annualreport/contents/lake/article9.html

Figure 2.20 shows the Wolverine Glacier and Golkana Glacier met stations in Alaska. Their meteorological equipment has remained essentially the same since the station began operation in 1967. In September 1995, two digital temperature sensors, a digital wind speed sensor, digital data logger, and satellite telemetry were added to the station.

Figure 2.21 shows the meteorological station at the North Pole, which measures wind speed and direction, air temperature, and air pressure. Data are transmitted via the NOAA Argos satellite system.

5 References and Recommended Resources

ACGIH. 1983. Air sampling instruments for evaluation of atmospheric contaminants. 6th ed.

Arbeola, F. J., C. P. Caseiras, and P. M. Andres. 1993. Air quality monitoring: Optimization of a network around a hypothetical potash plant in open countryside. *Atmospheric Environment* 27 A (5): 729–38.

Baldauf, R. W., D. D. Lane, and G. A. Marote. 2001. Ambient air quality monitoring network design for assessing human health impacts from exposures to airborne contaminants. *Environmental Monitoring and Assessment* 66 (1): 63–76.

Chang, N., and C. C. Tseng. 1999a. Optimal design of a multi-pollutant air quality monitoring network in a metropolitan region using Kaohsiung, Taiwan, as an example. *Environmental Monitoring and Assessment* 57 (2): 121–48.

———. 1999b. Optimal evaluation of expansion alternatives for existing air quality monitoring network by grey compromise programming. *Journal of Environmental Management* 56:61–77.

Figure 2.20 Meteorological stations at the Wolverine Glacier region in Alaska
Source: http://ak.water.usgs.gov/glaciology/wolverine/met/

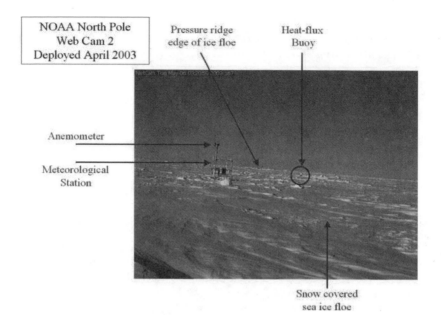

Figure 2.21 Meteorological station at the North Pole
Source: http://www.arctic.noaa.gov/gallery_np_instruments.html

Darby, W. P., P. J. Ossenbruggen, and C. J. Gregory. 1974. Optimization of urban air monitoring networks. *Journal of the Environmental Engineering Division of the American Society of Civil Engineers* 100 (EE3): 577–91.

Demerjian, K. L. 2000. A review of national monitoring networks in North America, 2000. *Atmospheric Environment* 34 (12:14): 1861–84.

Euro Bionet. 2004. www.uni-hohenheim.de/eurobionet/Acrobat/Summary-Leuven-Poster.pdf.

Hicks, B. B., T. P. Meyers, R. P. Hosker, Jr., and R. S. Artz. 2001. Climatological features of regional surface air quality from the Atmospheric Integrated Research Monitoring Network (AIRMoN) in the USA. *Atmospheric Environment* 35:1053–68.

Hwang, J. S., and C. C. Chan. 1997. Redundant measurements of urban air monitoring networks in air quality reporting. *Journal of the Air and Waste Management Association* 47 (5): 614–19.

Jahnke, J. A. 2000. Continuous emission monitoring. 2nd ed. John Wiley & Sons.

Joshi, A., and A. Kumar. 2002. Identification of high cancer risk zip codes in Ohio. In *Proceedings of the Ohio Air Pollution Research Symposium.* Available at http://p2tools.utoledo.edu/eiandtri/1999tridataanalysis.pdf.

Kumar, A. 1979. Air quality monitoring network at the Syncrude plant. *Environmental Science and Technology* 13 (6): 650–54.

———. 2004. *Introduction to Air Pollution.* Toledo: University of Toledo. Available at http://www.eng.utoledo.edu/~akumar/IAP1/mainpage.htm.

Kumar, A., N. K. Bellam, and A. Sud. 1996. Comparison, evaluation and use of industrial source complex models for estimating long term and short term ambient air concentrations of sulfur dioxide, nitrogen dioxide and particulate matter. *Environmental Progress* 18:93–100.

Kumar, A., J. Luo, and G. Bennett. 1993. Statistical evaluation of lower flammability distance (LFD) using four hazardous release models. *Process Safety Progress* 12 (1): 1–11.

Kumar, A., J. S. Sailaja, and H. G. Rao. 2004. Gas flow measurement, environmental instrument handbook. John Wiley & Sons.

Kumar, A., S. Vedula, and A. Sud. 2000. Development of an ozone forecasting model for non-attainment areas in the state of Ohio. *Environmental Monitoring and Assessment* 62:91–111.

Missouri Department of Natural Resources. 2004. National Ambient Air Quality Standards. Available at http://www.dnr.state.mo.us/alpd/esp/aqm/standard.htm.

Mulgrew, A., and P. Williams. 2000. Biomonitoring of air quality using plants. Air Hygiene Report no. 10. February. Berlin: WHO Collaborating Centre for Air Quality Management and Air Pollution Control.

NASA (National Aeronautics and Space Administration). 2004. *Aura: Up close.* Available at http://www.nasa.gov/vision/earth/technologies/AuraWebcastI.html.

Peterson, D. L. 2000. Monitoring air quality in mountains: Designing an effective network. *Environmental Monitoring and Assessment* 64 (1): 81–91.

Randerson, D. 1984. Atmospheric Science and Power Production. Washington, D.C.: U.S. Department of Energy, Office of Energy Research and Office of Health and Environmental Research.

Rao, H. G., W. R. Terry, and A. Kumar. 1988. Modeling non-homogenous variance time series: An application to sulfur dioxide data. *Environmental Monitoring and Assessment* 10 (2): 123–31.

Silva, C., and A. Quiroz. 2003. Optimization of the atmospheric pollution monitoring network at Santiago de Chile. *Atmospheric Environment* 37:2337–45.

Taha, H., S. Konopacki, and H. Akbari. 1998. Impacts of lowered urban air temperatures on precursor emission and ozone air quality. *Journal of the Air and Waste Management Association* 48:860–65.

Terry, W. R., and A. Kumar. 1985. Inferring causal models from operating data: An application in air quality. *Journal of Computers and Industrial Engineering* 9 (1): 231–35.

Terry, W. R., J. B. Lee, and A. Kumar. 1986. Time series analysis in acid rain modeling: Evaluation of filling missing values by linear interpolation. *Atmospheric Environment* 20:1941–45.

Terry, W. R., P. K. Chandhok, and A. Kumar. 1987. Modeling non-linear time series: An application to acid rain data. In *Meteorology of Acid Deposition*, 2:226–36. APCA Transactions TR-8. Air Pollution Control Association.

United Nations Environmental Program (UNEP). 2004. United Nations System Wide Earth Watch. Available at http://earthwatch.unep.net/about/docs/Pdwho.htm.

United Nations Environment Program (UNEP) and World Health Organization. 1994. *Quality Assurance in Urban Air Quality Monitoring.* Vol. 1 of the GEMS/Air Methodology Review Handbook Series. WHO/E0S/94.1, UNEPI GEMS/94.A.2. Nairobi: UNEP Nairobi.

———. 1995. Report on the GEMS/AIR Regional Training Course on Air Quality Monitoring, 23–27 October 1994, Amman, Jordan. Nairobi: UNEP/WHO.

U.S. Environmental Protection Agency (USEPA). 1983. *Atmospheric Sampling.* APTI Course 435. EPA 450/2–80–004. Washington, D.C.: USEPA.

———. 1995. http://www.epa.gov/tri/chemical/hazard_cx.htm.

———. 2001a. *Probe and monitoring path siting criteria for ambient air quality monitoring.* Code of Federal Register, 40 CFR 58, appendix E. Working copy, July. Available at http://www.epa.gov/ttn/amtic/files/cfr/pt58/40cfr58a.pdf.

———. 2001b. *National primary and secondary ambient air quality standards.* Code of Federal Register, 40 CFR 50, appendices A–H. Available at http://www.access.gpo.gov/nara/cfr/cfrhtml_00/Title_40/40cfr50_00.html.

———. 2004a. *Clean air markets: Programs and regulations.* Available at http://www.epa.gov/airmarkets/monitoring/factsheet.html.

———. 2004b. Clean Air Status and Trends Network. Available at http://www.epa.gov/castnet/.

———. 2004c. Laboratory and Field Operations. PM 2.5. Available at http://www.epa.gov/region4/sesd/pm25/p2.htm#1.

———. 2004d. *Visibility monitoring guidance.* Available at http://www.epa.gov/ttn/amtic/files/ambient/visible/r-99–003.pdf.

Valicenti, A., and J. Wenger. 1997. Air quality monitoring during construction and initial occupation of a new building. *Journal of the Air and Waste Management Association* 47:890–97.

Wark, K., C. F. Warner, and W. T. Davis. 1998. *Air pollution: Its origin and control.* Addison Wesley Longman.

Wisconsin Department of Natural Resources. 2004. *Air quality biomonitoring.* Available at http://www.dnr.state.wi.us/org/aw/air/MONITOR/bioweb/biohome.html.

World Health Organization (WHO). 1987. *Air quality guidelines for Europe.* World Health Organization Regional Publications, European Series no. 23. Copenhagen: World Health Organization.

World Meteorological Organization (WMO). 1996. *Guide to meteorological instruments and methods of observation.* 6th ed. WMO Publication no. 8. Geneva: WMO.

Chapter 3
Selection Criteria for Industrial Air Pollution Control Technologies

Nicholas P. Cheremisinoff

This chapter provides selection criteria for conventional technologies of industrial air pollution control. Extensive design, operational performance data, and theories of operations exist in printed form and on the World Wide Web. The reader is provided with a list of recommended printed publications in the reference section of this chapter.

The basis for selecting technologies is provided in three tables developed from available information in the public domain:

- Table 3.1 provides a listing of technologies and the most common applications to which each has been successfully applied.

- Table 3.2 provides data and information on the operational characteristics of technologies, including collection efficiency and emission stream characteristics such as airflow requirements, temperature, pollutant loading, and other considerations.

- Table 3.3 provides cost data, including ranges for capital cost, O&M (operational and maintenance), annualized cost, and cost-effectiveness (annualized cost per ton per year of pollutant controlled). Cost data are normalized to 1995 U.S. dollars.

Not all control technologies are covered. In some instances data or sufficient descriptive information could not be found. The reader should consult the list of references for further information and seek guidance from equipment suppliers.

Table 3.1 Control technologies and common applications

Pollutant Type[a]	Control Technology	Common Applications
Particulate matter	Gravity settling chambers	The metals-refining industries have used settling chambers to collect large particles, such as arsenic trioxide from the smelting of copper ore. Power and heating plants have used settling chambers to collect large unburned carbon particles for reinjection into the boiler. They are particularly useful for industries that also need to cool the gas stream prior to treatment in a fabric filter (Mycock, McKenna, and Theodore 1995). Settling chambers have been used to prevent excessive abrasion and dust loading in primary collection devices by removing large particles from the gas stream, such as either very high dust loadings or extremely coarse particles that might damage a downstream collector in series with the settling chamber. The upstream use of settling chambers has declined with improvements in acceptable loading of other, more efficient control devices and increasing space restrictions at facilities. In cases where sparks or heated material are present in the waste gas, settling chambers are still used to serve as "spark traps" to prevent a downstream baghouse or filter from catching fire (Wark and Warner 1981; USEPA 1998; Josephs 1999; Davis 1999). These devices are generally constructed for a specific application from duct materials, although almost any material can be used. Settling chambers have been replaced for most applications by cyclones, primarily due to the lower space requirements and the higher collection efficiency of cyclones. Multiple-tray settling chambers have never been widely used because of the difficulty in removing the settled dust from the horizontal trays (Mycock, McKenna, and Theodore 1995; Josephs 1999).
	Cyclone separators	Cyclones are designed for many applications. Cyclones themselves are generally not adequate to meet stringent air pollution regulations, but they serve an important purpose as precleaners for more expensive final control devices such as fabric filters or electrostatic precipitators (ESPs). In addition to use for pollution control work, cyclones are used in many process applications; for example, they are used for recovering and recycling food products and process materials such as catalysts (Cooper and Alley 1994). Cyclones are used extensively after spray-drying operations in the food and chemical industries and after crushing, grinding, and calcining operations in the mineral and chemical industries to collect salable or useful material. In the ferrous and nonferrous metallurgical industries, cyclones are often used as a first stage in the control of particulate matter (PM) emissions from sinter plants, roasters, kilns, and furnaces. PM from the fluid-cracking process is removed by cyclones to facilitate catalyst recycling. Fossil fuel- and wood-waste-fired industrial and commercial fuel combustion units commonly use multiple cyclones (generally upstream of a wet

scrubber, ESP, or fabric filter), which collect fine (less than 2.5-m) PM with greater efficiency than a single cyclone. In some cases, collected flyash is reinjected into the combustion unit to improve PM control efficiency (Air and Waste Management Association 1992; Avallone and Baumeister 1996; STAPPA/ALAPCO 1996; USEPA 1998).

Fabric filter, pulse-jet cleaned type

Fabric filters have had a long history in performing effectively in a wide variety of applications. Common applications using pulse-jet cleaning systems include:
- utility boilers (coal)
- industrial boilers (coal, wood)
- commercial/institutional boilers (coal, wood)
- ferrous metal processing (e.g., iron and steel production and steel foundries)
- mineral products (e.g., cement manufacturing, coal cleaning, and stone quarrying and processing)
- asphalt manufacturing
- grain milling

These applications are typical, but not at all definitive. Fabric filters can be used in almost any process where dust is generated and can be collected and ducted to a centralized location. In general, fabric filters come in many different sizes and configurations. In older plants, one can find makeshift operations.

Extended media cartridge filters

Cartridge collectors perform very effectively in many different applications. Common applications of cartridge filter systems with pulse-jet cleaning are comparable to the baghouses described earlier. In addition to these applications, cartridge collectors can be used in any process where dust is generated and can be collected and ducted to a central location.

Dry electrostatic precipitator: Wire-pipe type

Many older ESPs are of the wire-pipe design, consisting of a single tube placed on top of a smokestack. Dry pipe-type ESPs are occasionally used by the textile industry; pulp and paper facilities; the metallurgical industry, including coke ovens; hazardous waste incinerators; and sulfuric acid manufacturing plants, among others, although other ESP types are employed as well. Wet wire-pipe ESPs are used much more frequently than dry wire-pipe ESPs, which are used only in cases in which wet cleaning is undesirable, such as high temperature streams or wastewater restrictions. Typical applications by industry classification are as follows:

Utilities and industrial power plant fuel-fired boilers:
- Coal, pulverized
- Coal, cyclone
- Coal, stoker
- Oil
- Wood, bark
- Bagasse
- Fluid coke

(continues)

Table 3.1 Continued

Pollutant Type[a]	Control Technology	Common Applications
		Pulp and paper: • Kraft recovery boiler • Soda recovery boiler • Lime kiln *Rock products, kiln:* • Cement, dry • Cement, wet • Gypsum • Alumina • Lime • Bauxite • Magnesium oxide *Steel:* • Basic oxygen furnace • Open hearth • Electric furnace • Ore roasters • Cupola • Pyrites roaster • Taconite roaster • Hot scarfing *Mining and metallurgical:* • Zinc roaster • Zinc smelter • Copper roaster • Copper reverberatory furnace • Copper converter • Aluminum, Hall process • Aluminum, Soderberg process • Ilmenite dryer • Titanium dioxide process • Molybdenum roaster • Ore beneficiation *Miscellaneous:* • Refinery catalyst regenerator • Municipal incinerators • Apartment incinerators • Spray drying • Precious metal refining
	Wet electrostatic precipitator: Wire-pipe type and others	Wet ESPs are used in situations for which dry ESPs are not suited, such as when the material to be collected is wet, sticky, flammable, or explosive or has a high resistivity. Also, as higher collection efficiencies have become more desirable, wet ESP applications have been increasing. Many older ESPs are of the wire-pipe design,

consisting of a single tube placed on top of a smokestack. Wet pipe-type ESPs are commonly used by the textile industry; pulp and paper facilities; the metallurgical industry, including coke ovens; hazardous waste incinerators; and sulfuric acid manufacturing plants, among others, although other ESP types are employed as well (USEPA 1998; Flynn 1999).

Venturi scrubbers	Venturi scrubbers have been applied to control PM emissions from utility, industrial, commercial, and institutional boilers fired with coal, oil, wood, and liquid waste. They have also been applied to control emission sources in the chemical, mineral products, wood, pulp and paper, rock products, and asphalt manufacturing industries; lead, aluminum, iron and steel, and gray iron production industries; and municipal solid waste incinerators. Typically, venturi scrubbers are applied where it is necessary to obtain high collection efficiencies for fine PM. Thus, they are applicable to controlling emission sources with high concentrations of submicron PM.
Orifice scrubbers	Orifice scrubbers are used in industrial applications such as food processing and packaging (cereal, flour, rice, salt, sugar, etc.), pharmaceutical processing and packaging, and the manufacture of chemicals, rubber and plastics, ceramics, and fertilizer. Processes controlled include dryers, cookers, crushing and grinding operations, spraying (pill coating, ceramic glazing), ventilation (bin vents, dumping operations), and material handling (transfer stations, mixing, dumping, packaging). Orifice scrubbers can be built as high-energy units, but most devices are designed for low-energy service.
Condensation scrubbers	Condensation scrubbers are intended for use in controlling fine PM-containing waste-gas streams and are designed specifically to capture fine PM that has escaped a primary PM control device. The technology is suitable for both new and retrofit installations. Condensation scrubbing systems are a relatively new technology and are not yet generally commercially available. It may be argued that this is a pollution prevention type of technology since it replaces other approaches to controlling very fine PM, although the primary role is end-of-pipe treatment.
Gases and vapors	Packed tower wet scrubbers	The suitability of gas absorption as a pollution control method is generally dependent on the following factors:

The suitability of gas absorption as a pollution control method is generally dependent on the following factors:

1. availability of suitable solvent
2. required removal efficiency
3. pollutant concentration in the inlet vapor
4. capacity required for handling waste gas
5. recovery value of the pollutant(s) or the disposal cost of the unrecoverable solvent

Packed-bed scrubbers are typically used in the chemical, aluminum, coke and ferro-alloy, food and agriculture, and chromium electroplating industries. These scrubbers have had limited use as

(continues)

Table 3.1   Continued

Pollutant Type[a]	Control Technology	Common Applications
		part of flue gas desulfurization (FGD) systems, but the scrubbing solution flow rate must be carefully controlled to avoid flooding. When absorption is used for volatile organic compound (VOC) control, packed towers are usually more cost-effective than impingement plate towers (discussed below). However, in certain cases, the impingement plate design is preferred over packed-tower columns either when internal cooling is desired or where low liquid flow rates would inadequately wet the packing.
	Impingement plate/tray tower scrubbers	The suitability of gas absorption as a pollution control method is generally dependent on the following factors: 1. availability of suitable solvent 2. required removal efficiency 3. pollutant concentration in the inlet vapor 4. capacity required for handling waste gas 5. recovery value of the pollutant(s) or the disposal cost of the unrecoverable solvent Impingement plate scrubbers are typically used in the food and agriculture industry and at gray iron foundries. FGD is used to control SO_2 emissions from coal and oil combustion from electric utilities and industrial sources. Impingement scrubbers are one wet scrubber configuration used to bring exhaust gases into contact with a sorbent designed to remove the SO_2. On occasion, wet scrubbers have been applied to SO_2 emissions from processes in the primary nonferrous metals industries (e.g., copper, lead, and aluminum), but sulfuric acid or elemental sulfur plants are more popular control devices for controlling the high SO_2 concentrations associated with these processes. When absorption is used for VOC control, packed towers are usually more cost-effective than impingement plate towers. However, in certain cases, the impingement plate design is preferred over packed-tower columns either when internal cooling is desired or where low liquid flow rates would inadequately wet the packing.
	Fiber-bed scrubbers	Fiber-bed scrubbers are used to control aerosol emissions from chemical, plastics, asphalt, sulfuric acid, and surface-coating industries. They are also used to control lubricant mist emission from rotating machinery and mists from storage tanks. Fiber-bed scrubbers are also applied downstream of other control devices to eliminate a visible plume. Despite their potential for high collection efficiency, fiber-bed scrubbers have had only limited commercial acceptance for dust collection because of their tendency to become plugged.

^a Before selecting or sizing a specific control device, a careful evaluation of all aspects of the process and contaminants must be made. Improper terminology can lead to poor design and/or operation of any type of device. A list of contaminant definitions in accordance with the USA Standards Institute includes the following:

Liquid in Gas:

- *Mists and sprays.* There are numerous industrial chemical operations that involve liquid-in-gas dispersions. These operations generate mists and sprays that consist of particles in diameter ranges of 0.1 μm to 5 mm. Engineers most commonly encounter spray droplets that are particles often formed unintentionally in chemical plant operations. For example, vapors or fumes may condense onto piping, ducts, or stack walls. Under such conditions, liquid films form.

Solid in Gas:

- *Dusts* are fine solid particles often formed in such operations as pulverizing, crushing, grinding, drilling, detonation, and polishing. Other industrial sources are conveying operations and screening. Particle diameters generally are in the range of 1 μm to 1 mm. Dusts generally do not diffuse in air, but settle out by gravity.
- *Flyashes* are finely divided matter generally entrained in flue gases that arise from combustion. Particles range from 1 mm in size and smaller. This is not within the operational range of gravity settling chambers. Wet scrubbers are generally employed in flyash control. In some applications, high-efficiency electrostatic precipitators, baghouses, or cyclones are utilized.
- *Fumes* are finely divided solid particles that are generated by the condensation of vapors. Fumes are generally the by-products of sublimation, distillation, and molten metal processes. Particle diameters are generally in the range of 0.1 to 1 μm.
- *Smoke* constitutes fine, solid, gasborne matters that are products of incomplete combustion of organics (wood, coal, tobacco). Smoke particles are extremely small, ranging in size from less than 0.01 μm to 1 μm.
- *Smog* refers to a mixture of natural fog and industrial smoke.
- *Aerosols* are an assemblage of small particles, either solid or liquid, suspended in gas. Particle sizes range from 0.01 to 100 μm. There are several classes of aerosols. *Dispersion aerosols* are a common class, formed from processes such as grinding, solid and liquid atomization, and conveying powders in suspended state by vibration. Dispersion aerosols are usually composed of individual or slightly aggregated particles irregularly formed. *Condensation aerosols* are formed when supersaturated vapors condense or when gases react chemically to form a nonvolatile product. This latter class is usually less than 1 μm in size. Dispersion aerosols are considerably more coarse and contain a wide variety of particle sizes. Condensed aerosols usually consist of solid particles that are loose aggregates of a large number of primary particles of crystalline or spherical shape.
- *Grit.* The term *grit* is used to classify coarse particles that are unable to pass through 200-mesh screen. These particles are normally greater than 43 μm in diameter and are within the operating efficiency of gravity settling chambers.

Table 3.2 Operational and design characteristics for control technologies

Control Technology	Collection Efficiency	Emission Stream Characteristics			Other Considerations
		Airflow	Temperature	Pollutant Loading	
Gravity settling chambers	Varies as a function of particle size and design. Settling chambers are most effective for large and/or dense particles. Gravitational force may be employed to remove particles where the settling velocity is greater than about 13 ccm/sec (25 ft/min). In general, this applies to particles larger than 50 μm if the particle density is low, or down to 10 μm if the material density is reasonably high. Particles smaller than this would require excessive horizontal flow distances, which would lead to excessive chamber volumes. The collection efficiency for particulate matter (PM) less than or equal to 10 μm in aerodynamic diameter (PM_{10}) is typically less than 10%. Multiple-tray chambers have lower volume requirements for the collection of particles as small as 15 μm.	Size is usually restricted to a 4.25-m (14-ft) square shipping size. Units restricted by this shipping constraint will generally have flow rates that range up to 50 sm³/sec[a] (106,000 scfm[b]). Typical settling chamber waste gas flow capacity is 0.25–0.5 sm³/sec per cubic meter of chamber volume (15–30 scfm per cubic foot of chamber volume) (Wark and Warner 1981; Andriola 1999).	Inlet gas temperatures are limited only by the materials of construction of the settling chamber and have been operated at temperatures as high as 540°C (1,000°F) (Wark and Warner 1981; Perry and Green 1984).	Waste gas pollutant loadings can range from 20 to 4,500 g/sm³[c] (9–1,970 gr/scf[d]). Multiple-tray settling chambers can handle inlet dust concentrations of only less than approximately 2.3 g/sm³ (1.0 gr/scf) (Mycock, McKenna, and Theodore 1995; Parsons 1999; Steinbach 1999; Josephs 1999).	Leakage of cold air into a settling chamber can cause local gas quenching and condensation. Condensation can cause corrosion, dust buildup, and plugging of the hopper or dust removal system. The use of thermal insulation can reduce heat loss and prevent condensation by maintaining the internal device temperature above the dew point (USEPA 1982). No pretreatment is necessary for settling chambers.
Cyclone separators	Collection efficiency varies as a function of particle size and design, generally increasing with particle size and/or density, inlet duct velocity, body length, number of gas revolutions in	Typical gas flow rates for a single cyclone unit are 0.5–12 sm³/sec (1,060–25,400 scfm). Flows at the high end of this range and higher (up to approxi-	Inlet gas temperatures are limited only by the materials of construction of the cyclone and have been operated at temperatures as high as 540°C	Waste gas pollutant loadings typically range from 2.3 g/sm³ to 230 g/sm³ (1.0–100 gr/scf) (Wark and Warner 1981). For specialized applications,	Cyclones perform more efficiently with higher pollutant loadings, provided that the device does not become choked. Higher pollutant

the cyclone, ratio of cyclone body diameter to gas exit diameter, dust loading, and smoothness of the cyclone inner wall. Efficiency will decrease with increases in gas viscosity, body diameter, gas exit diameter, gas inlet duct area, and gas density. A common factor contributing to decreased control efficiencies in cyclones is leakage of air into the dust outlet. Control efficiency ranges for single cyclones are based on three classifications of cyclone: conventional, high-efficiency, and high-throughput. The control efficiency range for conventional single cyclones is 70–90% for PM, 30–90% for PM_{10}, and 0–40% for $PM_{2.5}$. High-efficiency single cyclones are designed to achieve higher control of smaller particles than conventional cyclones; high-efficiency single cyclones can remove 5-μm particles at up to 90% efficiency, with higher efficiencies achievable for larger particles. The control efficiency ranges for high-efficiency single cyclones are 80–99% for PM, 60–95% for PM_{10}, and 20–70% for $PM_{2.5}$. Higher efficiency cyclones come with higher pressure drops, which require higher energy costs to

mately 50 sm^3/sec or 106,000 scfm) use multiple cyclones in parallel (Cooper and Alley 1994). There are single cyclone units employed for specialized applications that have flow rates of up to approximately 30 sm^3/sec (63,500 scfm) and as low as 0.0005 sm^3/sec (1.1 scfm) (Wark and Warner 1981; Andriola 1999).

(1,000°F) (Wark and Warner 1981; Perry and Green 1994).

loadings can be as high as 16,000 g/sm^3 (7,000 gr/scf) or as low as 1 g/sm^3 (0.44 gr/scf) (Avallone and Baumeister 1996; Andriola 1999).

loadings are generally associated with higher flow designs (Andriola 1999).

(continues)

Table 3.2 *Continued*

Control Technology	Collection Efficiency	Emission Stream Characteristics			
		Airflow	Temperature	Pollutant Loading	Other Considerations
	move the waste gas through the cyclone. Cyclone design is generally driven by a specified pressure-drop limitation, rather than by meeting a specified control efficiency (see Andriola 1999; Perry and Green 1994). High-throughput cyclones are guaranteed to remove only particles greater than 20 μm, although collection of smaller particles does occur to some extent. The control efficiency ranges for high-throughput cyclones are 80–99% for PM, 10–40% for PM$_{10}$, and 0–10% for PM$_{2.5}$. Multicyclones are reported to achieve from 80 to 95% collection efficiency for 5-μm particles (USEPA 1998).				
Fabric filter, pulse-jet cleaned type	This equipment is used for the capture of PM, including particulate matter less than or equal to 10 μm in aerodynamic diameter (PM$_{10}$), particulate matter less than or equal to 2.5 μm in aerodynamic diameter (PM$_{2.5}$), and hazardous air pollutants (HAPs) that are in particulate form, such as most metals (mercury is the notable	Baghouses are separated into two groups—standard and custom—which are further subdivided into low, medium, and high capacity. Standard baghouses are factory-built, off-the-shelf units. They may handle from less than 0.10 sm^3/sec to more than 50 sm^3/sec ("hundreds" to	Typically, gas temperatures up to about 260°C (500°F), with surges to about 290°C (550°F), can be accommodated routinely with the appropriate fabric material. Spray coolers or dilution air can be used to lower the temperature of the pollutant stream and pre-	Typical inlet concentrations to baghouses are 1–23 g/sm^3 (0.5–10 gr/scf), but in extreme cases, inlet conditions may vary between 0.1 g/sm^3 and more than 230 g/sm^3 (0.05 gr/scf to more than 100 gr/scf).	Moisture and corrosives content are the major gas stream characteristics requiring design consideration. Standard fabric filters can be used in pressure or vacuum service, but only within the range of about ± 640 mm (25 inches) of the water column. Well-

exception, as a significant portion of its emissions are in the form of elemental vapor). Typical new equipment design efficiencies are between 99% and 99.9%; older existing equipment has a range of actual operating efficiencies of 95–99.9%. Several factors determine fabric filter collection efficiency, including gas filtration velocity, particle characteristics, fabric characteristics, and cleaning mechanism. In general, collection efficiency increases with increasing filtration velocity and particle size. For a given combination of filter design and dust, the effluent particle concentration from a fabric filter is nearly constant, whereas the overall efficiency is more likely to vary with particulate loading. For this reason, fabric filters can be considered to be constant outlet devices rather than constant efficiency devices. Constant effluent concentration is achieved because at any given time, part of the fabric filter is being cleaned. As a result of the cleaning mechanisms used in fabric filters, the collection efficiency is constantly changing. Each cleaning cycle removes at

more than 100,000 scfm). Custom baghouses are designed for specific applications and are built to the specifications prescribed by the customer. These units are generally much larger than standard units, i.e., from 50 sm³/sec to more than 500 sm³/sec (100,000 scfm to more than 1,000,000 scfm).

vent the temperature limits of the fabric from being exceeded. Lowering the temperature, however, increases the humidity of the pollutant stream. Therefore, the minimum temperature of the pollutant stream must remain above the dew point of any condensable in the stream. The baghouse and associated ductwork should be insulated and possibly heated if condensation may occur.

designed and -operated baghouses have been shown to be capable of reducing overall particulate emissions to less than 0.05 g/sm³ (0.010 gr/scf), and in a number of cases, to as low as 0.002–0.011 g/sm³ (0.001–0.0115 gr/scf).

(continues)

Table 3.2 Continued

Control Technology	Collection Efficiency	Emission Stream Characteristics			Other Considerations
		Airflow	Temperature	Pollutant Loading	
	least some of the filter cake and loosens particles which remain on the filter. When filtration resumes, the filtering capability has been reduced because of the lost filter cake, and loose particles are pushed through the filter by the flow of gas. As particles are captured, the efficiency increases until the next cleaning cycle. Average collection efficiencies for fabric filters are usually determined from tests that cover a number of cleaning cycles at a constant inlet loading.				
Extended media cartridge filters	Older existing cartridge collector types have a range of actual operating efficiencies of 99–99.9% for PM_{10} and $PM_{2.5}$. Typical new equipment design efficiencies are between 99.99% and in excess of 99.999%. In addition, commercially available designs are able to control submicron PM (0.8 m in diameter or greater) with a removal efficiency of better than 99.999%. Several factors determine cartridge filter col-	Cartridge collectors are currently limited to low airflow–capacity applications. Standard cartridge collectors are factory-built, off-the-shelf units. They may handle airflow rates from less than 0.10 sm³/sec to more than 5 sm³/sec ("hundreds" to more than 10,000 scfm).	Temperatures are limited by the type of filter media and sealant used in the cartridges. Standard cartridges utilizing paper filter media can accommodate gas temperatures up to about 95°C (200°F). Cartridge filters utilizing synthetic, nonwoven media such as needle-punched felts fabricated of polyester or Nomex can withstand	Typical inlet concentrations to cartridge collectors are 1–23 g/sm³ (0.5–10 gr/scf). Cartridge filters that utilize synthetic, nonwoven media such as needle punched felts fabricated of polyester or Nomex are able to handle inlet concentrations up to 57 g/sm³ (25 gr/scf).	Moisture and corrosive content in the gas streams are the major design considerations. Standard cartridge filters can be used in pressure or vacuum service, but only within the range of about ±640 mm (25 inches) of the water column. Baghouses have been shown to be capable of reducing overall particulate emissions to

lection efficiency, including gas filtration velocity, particle characteristics, filter media characteristics, and cleaning mechanism. In general, collection efficiency increases with increasing filtration velocity and particle size. For a given combination of filter design and dust, the effluent particle concentration from a cartridge collector is nearly constant, whereas the overall efficiency is more likely to vary with particulate loading. For this reason, cartridge collectors can be considered to be constant outlet devices rather than constant efficiency devices. Constant effluent concentration is achieved because at any given time, part of the filter media is being cleaned. As a result of the cleaning mechanisms used in cartridge collectors, the collection efficiency is constantly changing. Each cleaning cycle removes at least some of the filter cake and loosens particles that remain on the filter. When filtration resumes, the filtering capability has been reduced because the lost filter cake and loose particles are pushed

temperatures of up to 200°C (400°F) with the appropriate sealant material. Spray coolers or dilution air can be used to lower the temperature of the pollutant stream and prevent the temperature limits of the filter media from being exceeded. Lowering the temperature can result in higher humidity of the pollutant stream. Therefore, the minimum temperature of the pollutant stream must remain above the dew point of any condensable in the stream. The cartridge collector and associated ductwork should be insulated and possibly heated if condensation may occur.

less than 0.05 g/sm^3 (0.010 gr/scf). Penetration of PM in cartridge collectors is generally many times less than in traditional baghouse designs.

(continues)

Table 3.2 *Continued*

Control Technology	Collection Efficiency	Emission Stream Characteristics			Other Considerations
		Airflow	Temperature	Pollutant Loading	
	through the filter by the flow of gas. As particles are captured, the efficiency increases until the next cleaning cycle. Average collection efficiencies for cartridge collectors are usually determined from tests that cover a number of cleaning cycles at a constant inlet loading.				
Dry electrostatic precipitator: Wire-pipe type	Typical new equipment design efficiencies are between 99% and 99.9%. Older existing equipment has a range of actual operating efficiencies of 90–99.9%. While several factors determine ESP collection efficiency, ESP size is most important. Size determines treatment time: the longer a particle spends in the ESP, the greater its chance of being collected. Maximizing electric field strength will maximize ESP collection efficiency. Collection efficiency is also affected by dust resistivity, gas temperature, chemical composition (of the dust and the gas), and particle size distribution.	Typical gas flow rates for dry wire-pipe ESPs are 0.5–50 sm³/sec (1,000–100,000 scfm).	Dry wire-pipe ESPs can operate at very high temperatures, up to 700°C (1,300°F). Operating gas temperature and chemical composition of the dust are key factors influencing dust resistivity and must be carefully considered in the design of an ESP.	Typical inlet concentrations to a wire-pipe ESP are 1–10 g/sm³ (0.5–5 gr/scf). It is common to pretreat a waste stream, usually with a wet spray or scrubber, to bring the stream temperature and pollutant loading into a manageable range. Highly toxic flows with concentrations well below 1 g/sm³ (0.5 gr/scf) are also sometimes controlled with ESPs.	Dry ESPs operate most efficiently with dust resistivities between 5×10^3 and 2×10^{10} ohm-cm. In general, the most difficult particles to collect are those with aerodynamic diameters between 0.1 µm and 1.0 µm. Particles between 0.2 µm and 0.4 µm usually show the most penetration. This is most likely a result of the transition region between field and diffusion charging.

Wet electrostatic precipitator: Wire-pipe type and others	Typical new equipment design efficiencies are between 99% and 99.9%. Older existing equipment has a range of actual operating efficiencies of 90–99.9%. While several factors determine ESP collection efficiency, ESP size is most important. Size determines treatment time: the longer a particle spends in the ESP, the greater its chance of being collected. Maximizing electric field strength will maximize ESP collection efficiency. Collection efficiency is also affected to some extent by dust resistivity, gas temperature, chemical composition (of the dust and the gas), and particle size distribution.	Typical gas flow rates for wet wire-pipe ESPs are 0.5–50 sm³/sec (1,000–100,000 scfm) (Flynn 1999).	Wet wire-pipe ESPs are limited to operating at temperatures lower than 80–90C (170–190F) (USEPA 1998; Flynn 1999).	Typical inlet concentrations to a wire-pipe ESP are 1–10 g/sm³ (0.5–5 gr/scf). It is common to pretreat a waste stream, usually with a wet spray or scrubber, to bring the stream temperature and pollutant loading into a manageable range. Highly toxic flows with concentrations well below 1 g/sm³ (0.5 gr/scf) are also sometimes controlled with ESPs (Flynn 1999).	Dust resistivity is not a factor for wet ESPs because of the high-humidity atmosphere, which lowers the resistivity of most materials. Particle size is much less of a factor for wet ESPs compared to dry ESPs. Much smaller particles can be efficiently collected by wet ESPs due to the lack of resistivity concerns and the reduced reentrainment (Flynn 1999).
Venturi scrubbers	Venturi scrubbers' PM collection efficiencies range from 70% to greater than 99%, depending upon the application. Collection efficiencies are generally higher for PM with aerodynamic diameters of 0.5–5 m. Some venturi scrubbers are designed with an adjustable throat to control the velocity of the gas stream and the pressure drop. Increasing the venturi scrubber efficiency requires increasing the pressure drop, which in turn increases the energy consumption.	Typical gas flow rates for a single-throat venturi scrubber unit are 0.2–28 sm³/sec (500–60,000 scfm). Flows higher than this range use either multiple venturi scrubbers in parallel or a multiple-throated venturi.	Inlet gas temperatures are usually in the range of 4°–370°C (40°–700°F) (see Avallone and Baumeister 1996).	Waste gas pollutant loadings can range from 1 g/sm³ to 115 g/sm³ (0.1–50 gr/scf) (Turner 1999; Dixit 1999).	In situations where waste gas contains both particulates and gases to be controlled, venturi scrubbers are sometimes used as pretreatment devices, removing PM to prevent clogging of a downstream device such as a packed-bed scrubber, which is designed to collect primarily gaseous pollutants.

(continues)

Table 3.2 Continued

Control Technology	Collection Efficiency	Emission Stream Characteristics			Other Considerations
		Airflow	Temperature	Pollutant Loading	
Orifice scrubbers	Orifice scrubber collection efficiencies range from 80% to 99%, depending upon the application and scrubber design. This type of scrubber relies on inertial and diffusional interception for PM collection. Some orifice scrubbers are designed with adjustable orifices to control the velocity of the gas stream.	Typical gas flow rates for an orifice scrubber unit are 0.47–24 sm³/sec (1,000–50,000 scfm).	In general, orifice scrubbers can treat waste gases up to approximately 150°C (300°F).	Orifice scrubbers can accept waste flows with PM loadings up to 23 g/sm³ (10 gr/scf) or even higher, depending upon the nature of the PM being controlled.	Orifice scrubbers generally do not require precleaning, unless the waste gas contains large pieces of debris. Precooling may be necessary for high-temperature waste gas flows, which increase the evaporation of the scrubbing liquid.
Condensation scrubbers	Collection efficiencies of greater than 99% have been reported for particulate emissions, based on study results. See the list of references cited at the end of this chapter for detailed studies.	Typical airflows are on the order of 10 sm³/sec (21,000 scfm).	The waste gas entering a condensation scrubber is generally cooled to saturation conditions, approximately 20°–26°C (68°–78°F).	Pollutant loading is dependent upon the control effectiveness for fine PM of the primary PM control system. Fine PM may, in some cases, comprise up to 90% of the total mass of PM emissions from a combustion source, and many primary control technologies have relatively low collection efficiencies for fine PM.	The fraction of fine PM emissions from a combustion source often contains cadmium and other metals. Use of a condensation scrubber to capture fine PM may provide an effective method of reducing the emission of metals.
Packed tower wet scrubbers	*Inorganic gases:* Control device vendors estimate that removal efficiencies range from 95 to 99%. *Volatile organic compounds (VOCs):* Removal	Typical gas flow rates for packed-bed wet scrubbers are 0.25–35 sm³/sec (500–75,000 scfm).	Inlet temperatures are usually in the range of 4°–370°C (40°–700°F) for waste gases in which the PM is to be con-	Typical gaseous pollutant concentrations range from 250 ppmv to 10,000 ppmv. Packed-bed wet scrubbers are	For organic vapor HAP control applications, low outlet concentrations will typically be required, leading to im-

	efficiencies for gas absorbers vary for each pollutant–solvent system and with the type of absorber used. Most absorbers have removal efficiencies in excess of 90%, and packed-tower absorbers may achieve efficiencies greater than 99% for some pollutant–solvent systems. The typical collection efficiency range is from 70% to greater than 99%. *PM*: Packed-bed wet scrubbers are limited to applications in which dust loading is low, and collection efficiencies range from 50% to 95%, depending upon the application. Condensation scrubbers potentially offer a means of extending the removal efficiency of PM.	trolled, and for gas absorption applications, 4°–38°C (40°–100°F). In general, the higher the gas temperature, the lower the absorption rate and vice versa. Excessively high gas temperatures also can lead to significant solvent or scrubbing liquid loss through evaporation.	generally limited to applications in which PM concentrations are less than 0.45 g/sm³ (0.20 gr/scf) to avoid clogging.	practically tall absorption towers, long contact times, and high liquid–gas ratios that may not be cost-effective. Wet scrubbers will generally be effective for HAP control when they are used in combination with other control devices such as incinerators or carbon adsorbers.
Impingement plate/tray tower scrubbers	*PM*: Impingement plate tower collection efficiencies range from 50% to 99%, depending upon the application. This type of scrubber relies almost exclusively on inertial impaction for PM collection. Therefore, collection efficiency decreases as particle size decreases. Short residence times will also lower scrubber efficiency for small particles. Collection efficiencies for small particles (less than 1 μm in aerodynamic diameter) are low for these scrubbers; hence, they are not recom-	Typical gas flow rates for a single impingement plate scrubber unit are 0.47–35 sm³/sec (1,000–75,000 scfm).	Inlet gas temperature is limited to 4°–370°C (40°–700°F) for PM control. For gaseous pollutant control, the gas temperature typically ranges between 4°C and 38°C (40°F and 100°F). In general, the higher the gas temperature, the lower the absorption rate and vice versa. Higher temperatures can lead to loss of scrubbing liquid or solvent through evaporation.	For organic vapor HAP control, low outlet concentrations will typically be required, leading to impractically tall absorption towers, long contact times, and high liquid–gas ratios that may not be cost-effective. Wet scrubbers will generally be effective for HAP control when they are used in combination with other control devices such as incinerators or carbon adsorbers.

(continues)

Table 3.2 Continued

Control Technology	Collection Efficiency	Emission Stream Characteristics			Other Considerations
		Airflow	Temperature	Pollutant Loading	
	mended for fine PM control. *Inorganic gases:* Control device vendors estimate that removal efficiencies range from 95% to 99%. For SO_2 control, removal efficiencies vary from 80% to greater than 99%, depending upon the type of reagent used and the plate tower design. Most current applications have an SO_2 removal efficiency greater than 90%.				
Fiber-bed scrubbers	Fiber-bed scrubber collection efficiencies for PM and VOC mists generally range from 70% to greater than 99%, depending upon the size of the aerosols to be collected and the design of the scrubber and the fiber beds.	Fiber-bed scrubbers can treat flows from 0.5 sm³/ sec to 47 sm³/sec (1,000–100,000 scfm).	The temperature of the inlet waste gas flow is generally restricted by the choice of materials. Plastic fiber beds are generally restricted to operate below 60°C (140°F).	Inlet flow loadings can range from 0.2 g/sm³ to 11 g/sm³ (0.1–5 gr/scf).	Waste gas streams are often cooled before entering fiber-bed scrubbers to condense as much of the liquid in the flow as possible and to increase the size of the existing aerosol particles through condensation. A prefilter is generally used to remove larger particles from the gas stream prior to its entering the scrubber.

^a sm³/sec = standard cubic meters per second
^b scfm = standard cubic feet per minute
^c g/sm³ = grams per standard cubic meter
^d gr/scf = grains per standard cubic foot

Table 3.3 Comparative cost data

Technology	Capital Cost		O&M		Annualized		Cost-Effectiveness	
	$ per sm³/sec[a]	$ per scfm[b]	$ per sm³/sec	$ per scfm	$ per sm³/sec	$ per scfm	$ per metric ton	$ per short ton
Gravity settling chambers	330–10,900	0.16–5.10	13–470	0.01–0.22	40–1,350	0.02–0.64	0.01–3.90	0.01–3.50
Cyclone separators	4,200–5,100	2.00–2.40	2,400–27,800	1.10–13.10	2,800–28,300	1.30–13.40	0.45–460	0.41–420
Fabric filter, pulse-jet cleaned type	13,100–54,900	6–26	11,200–51,700	5–24	13,100–83,400	6–39	46–293	42–266
Extended media cartridge filters	15,000–27,000	7–13	20,000–52,000	9–25	26,000–80,000	13–38	94–280	85–256
Dry electrostatic precipitator: Wire-pipe type	65,000–400,000	30–190	10,000–20,000	5–10	20,000–75,000	10–35	55–950	50–850
Wet electrostatic precipitator: Wire-pipe type and others	125,000–640,000	60–300	15,000–25,000	7–12	32,000–125,000	15–60	90–950	80–850
Venturi scrubbers	6,700–59,000	3.20–28	8,700–250,000	4.10–119	9,700–260,000	4.60–123	84–2,300	76–2,100
Orifice scrubbers	10,000–36,000	5–17	8,000–149,000	3.80–70	9,500–154,000	4.50–73	88–1,400	80–1,300
Condensation scrubbers	13,000	6	5,300	2.50	7,000	3.40	65	59
Packed tower wet scrubbers	22,500–120,000	11–56	33,500–153,000	16–72	36,000–166,000	17–78	0.24–1.09	0.21–0.99
Impingement plate/tray tower scrubbers	4,500–25,000	2.10–11	5,200–148,000	2.50–70	5,900–151,000	2.80–71	51–1,300	46–1,200
Fiber-bed scrubbers	2,100–6,400	1.00–3.00	3,500–76,000	1.60–36	4,300–77,000	2–37	40–710	36–644

[a] sm³/sec = standard cubic meters per second
[b] scfm = standard cubic feet per minute

References and Recommended Resources

AAF International, Inc. 1999. Core products information. Available at http://www .aafintl.com/equipmentl (last updated December 1999).

Air and Waste Management Association. 1992. *Air pollution engineering manual.* New York: Van Nostrand Reinhold.

Avallone, E., and T. Baumeister, eds. 1996. *Marks' standard handbook for mechanical engineers.* 10th ed. New York: McGraw-Hill.

Bakke, E. 1974a. The application of wet electrostatic precipitators for control of fine particulate matter. Paper presented at the Symposium on Control of Fine Particulate Emissions from Industrial Sources for the Joint U.S.-U.S.S.R. Working Group on Stationary Source Air Pollution Control Technology, San Francisco, Calif., 15–18 January.

———. 1974b. On the application of wet electrostatic precipitators for control of emissions from Soderberg aluminum reduction cells. Presentation at A.I.M.E. meeting, Dallas, Texas, February.

Barry, H. N. 1960. Fixed-bed adsorption. *Chemical Engineering* (Feb. 8): 105.

Campbell, J. M., F. E. Ashford, R. B. Needham, and L. S. Reid. 1963. More insight into adsorption design. *Hydrocarbon Processing and Petroleum Refiner* 42 (12): 89–96.

Cheremisinoff, N. P. 1998. *Pressure safety design practices for refinery and chemical operations.* Westwood, N.J.: Noyes.

Cheremisinoff, P. N., and R. A. Young. 1977. *Air pollution control and design handbook.* Vols. 1 and 2. New York: Marcel Dekker.

Cooper, D., and F. Alley. 1994. *Air pollution control: A design approach.* 2nd ed. Prospect Heights, Ill.: Waveland Press.

Corbitt, R. A., ed. 1990. *Standard handbook of environmental engineering.* New York: McGraw-Hill.

Croll Reynolds Company. 1999. Available at http://www.croll.com (accessed 19 May 1999).

Danielson, J. A. 1967. *Air pollution engineering manual.* Cincinnati, Ohio: U.S. Department of Health, Education, and Welfare.

EC/R. 1996. Evaluation of fine particulate matter control technology. Final draft, September. Prepared for USEPA Integrated Policy and Strategies Group. Durham, N.C.: EC/R, Inc.

Enviro-Chem. 1999. Monsanto Enviro-Chem Systems. Available at http:// www.enviro-chem.com (accessed 24 May 1999).

Fair, J. R. 1969. Sorption processes for gas separation. *Chemical Engineering* (July 14): 90–110.

Heumann, W. L. 1997. *Industrial air pollution control systems.* Washington, D.C.: McGraw-Hill.

Hougen, A., and K. M. Marshall. 1967. Adsorption from a fluid stream flowing through a stationary granular bed. *Chemical Engineering Progress* 43:197.

Humphries, C. L. 1966. Now predict recovery from adsorbents. *Hydrocarbon Processing* 45 (12): 88.

Industrial Filter Fabric (IFF), Inc. 1999. Cartridge filters. Product Bulletin 003, December. Available at http://www.filters.com.

Institute of Clean Air Companies (ICAC). 1999a. Control technology information: Electrostatic precipitator. Available at http://www.icac.com (page last updated 11 January 1999).

———. 1999b. Control technology information: Fabric filters. Available at http://www.icac.com (page last updated 11 January 1999).

Levenspiel, O. 1962. *Chemical reaction engineering.* New York: John Wiley & Sons.

Lewis, W. K., E. R. Gilliland, B. Cherton, and W. P. Cadogan. 1950a. Adsorption equilibria: Hydrocarbon gas mixtures. *Industrial and Engineering Chemistry* 42:1319.

———. 1950b. Adsorption equilibria: Pure gas isotherms. *Industrial and Engineering Chemistry* 42:1326.

Mattia, M. M. 1970. Process for solvent pollution control. *Chemical Engineering Progress* 66 (12): 74–79.

Merrick, D., and J. Vernon. 1989. Review of flue gas desulphurization systems. *Chemistry and Industry* (Feb. 6).

Michaels, A. S. 1952. Simplified method of interpreting kinetic data in fixed-bed ion exchange. *Industrial and Engineering Chemistry* 44 (8): 1922.

Moore, A. D. 1973. *Electrostatics and its applications.* New York: John Wiley & Sons.

Mycock, J., J. McKenna, and L. Theodore. 1995. *Handbook of air pollution control engineering and technology.* Boca Raton, Fla.: CRC Press.

Oglesby, S., and G. B. Nichols. 1970. *A manual of electrostatic precipitator technology.* Part 1, *Fundamentals.* Contract CRA 2269–73 for NAPCA. Cincinnati, Ohio.

Othmer, D. F., and S. Josefowitz. 1948. Correlating adsorption data. *Industrial and Engineering Chemistry* 40:723.

Perry, R., and D. Green, eds. 1984. *Perry's chemical engineer's handbook.* 6th ed. New York: McGraw-Hill.

Rawson, H. M. 1963. Fluid bed adsorption of carbon disulfide. *British Chemical Engineering* 8 (3): 180.

Ray, G. C., and E. O. Box, Jr. 1950. Adsorption of gases on activated charcoal. *Industrial and Engineering Chemistry* 42:1314.

Smith, J. M. 1970. *Chemical engineering kinetics.* 2d ed. New York: McGraw-Hill.

Sondreal, E. A. 1993. Clean utilization of low-rank coals for low-cost power generation. In *Clean and efficient use of coal: The new era for low-rank coal.* Paris: Organization for Economic Co-Operation and Development/International Energy Agency.

Soud, H. N., I. Takeshita, and I. M. Smith. 1993. FGC systems and installations for coal-fired plants. In *Desulfurization*, vol. 3. Warwickshire, England: Institution of Chemical Engineers.

STAPPA/ALAPCO (State and Territorial Air Pollution Program Administrators and the Association of Local Air Pollution Control Officials). 1996. *Controlling particulate matter under the clean air act: A menu of options.* July. Washington, D.C.: STAPPA/ALAPCO.

Sun, J., B. Liu, P. McMurry, and S. Greenwood. 1994. A method to increase control efficiencies for wet scrubbers for submicron particles and particulate metals. *Journal of the Air and Waste Management Association* 44:2.

Torit Products. 1999. Industrial dust collection systems. Available at http://www.torit.com/Products (last updated December 1999).

USEPA (U.S. Environmental Protection Agency). 1981. *Control technologies for sulfur oxide emissions from stationary sources.* 2d ed. April. Research Triangle Park, N.C.: Office of Air Quality Planning and Standards.

———. 1982. *Control techniques for particulate emissions from stationary sources.* Vol. 1. September. EPA/450/3–81–005a. Research Triangle Park, N.C.: Office of Air Quality Planning and Standards.

———. 1991. *Control technologies for hazardous air pollutants.* June. EPA/625/6–91–014. Washington, D.C.: Office of Research and Development.

———. 1992. *Control technologies for volatile organic compound emissions from stationary sources.* December. EPA/453/R-92–018. Research Triangle Park, N.C.: Office of Air Quality Planning and Standards.

———. 1993. *Chromium emissions from chromium electroplating and chromic acid anodizing operations: Background information for proposed standards.* July. EPA/453/R-93–030a. Research Triangle Park, N.C.: Office of Air Quality Planning and Standards.

———. 1995. *Survey of control technologies for low concentration organic vapor gas streams.* May. EPA/456/R-95–003. Research Triangle Park, N.C.: Office of Air Quality Planning and Standards.

———. 1996a. *OAQPS control cost manual.* 5th ed. February. EPA/453/B-96–001. Research Triangle Park, N.C.: Office of Air Quality Planning and Standards.

———. 1996b. *Chemical recovery combustion sources at Kraft and soda pulp mills.* October. EPA/453/R-96–012. Research Triangle Park, N.C.: Office of Air Quality Planning and Standards.

———. 1997. *Compilation of air pollutant emission factors.* Vol. 1, 5th ed. October. Research Triangle Park, N.C.: Office of Air Quality Planning and Standards.

———. 1998a. *Stationary source control techniques document for fine particulate matter.* October. EPA/452/R-97–001. Research Triangle Park, N.C.: Office of Air Quality Planning and Standards.

———. 1998b. *OAQPS control cost manual.* 5th ed. Chapter 5. December. EPA/453/B-96–001. Research Triangle Park, N.C.: Office of Air Quality Planning and Standards.

Vatavuk, W. M. 1990. *Estimating costs of air pollution control.* Chelsea, Mich.: Lewis.

Wark, K., and C. Warner. 1981. *Air pollution: Its origin and control.* New York: Harper-Collins.

Western Precipitation Company. 1952. Precipitation pamphlet. C-103R-1. Los Angeles, Calif.: Western Precipitation.

White, H. J. 1963. *Industrial electrostatic precipitation.* Reading, Mass.: Addison-Wesley.

Chapter 4
Thermal Methods of Waste Treatment

Philip Rutberg

Waste treatment is one of the most ancient—and at the same time, most up-to-date—areas of human knowledge. In various phases of history, many generations of experts in the field of public health services have concentrated on the elimination of unsanitary conditions. It became especially important at the end of the nineteenth century, when the theory and practice of public health services reached an understanding of the necessity to improve people's living conditions and household activities.

Problems connected with a radical modification of waste character have arisen. A growing amount of paper, plastic, and glass packing is appearing in waste composition. Some waste products are formed in huge amounts by industrial processes. This pollution causes immediate disastrous effects and justified anxiety regarding its far-reaching consequences. Even with the strict standards adopted in the developed countries, the waste disposal at landfills, in the ground, and under water does not guarantee neutralizing of its harmful action.

Industrial development has been accompanied by magnification of the negative effects of diverse toxic chemical substances, formed at various stages of production as wastes. It has become obvious that toxic agents survive for a long time, accumulate, migrate, and are transformed. Especially troubling are synthesized organic substances having no analogs in nature whose numbers steadily grow with developments in chemistry, chemical technologies, and related branches.

The problem of waste treatment has become very important in the last decades. Numerous studies on waste decontamination and utilization have been carried out during that period.

The whole field of waste treatment has to be transformed. Globally, waste generation is growing rapidly (see table 4.1) (Joos 2002). In the United States, for example, 230 million tons of municipal waste are produced each year, that is, roughly two kilograms of waste per person (USEPA 2004). Russia generates 185 million tons of garbage and industrial and agricultural waste and about 140 million tons of toxic waste (Al'kov and Koroteev 2000; Russian Federation 2002, 216).

It is possible to classify the following categories of waste:

- Household
- Municipal
- Agricultural
- Industrial
- Mountain (metallurgical)
- Special (radioactive, military)

At the moment in world industrial practice, various designs and technological schemes of waste processing are used. Waste parameters and national requirements to environmental protection determine their design philosophy. The whole variety of thermal processes, used at waste treatment, are presented in figure 4.1.

1   Combustion

The most commonly used devices for waste treatment are furnaces and combustion installations (see table 4.2) (Hester and Harrison 1994, 31; Lee and Huffman 1989). Combustible wastes of all types of aggregative state burn down in an airstream at temperatures of 850°–1,200°C, depending on the chemical nature of the compounds. Various kinds of fuel are supplied into the furnaces for maintaining the necessary temperature, if the internal energy of the waste itself is insufficient. As a rule, combustion takes place in bulky, capital-intensive buildings that are rather economic in main-

Table 4.1   Total amount of waste formed annually in the countries of Europe

Country	Annual amount of waste (mln. t)	Country	Annual amount of waste (mln. t)
Austria	47	Ireland	41
Belgium and Luxembourg	30	Spain	223
Denmark	12	Sweden	66
Finland	61	Great Britain	422
France	601	Norway	35
Germany	277	Switzerland	9
Greece	33	Czech Republic	89
Italy	107	Hungary	69
Netherlands	53	Poland	137
Portugal	5	Turkey	57

Total: 2,374

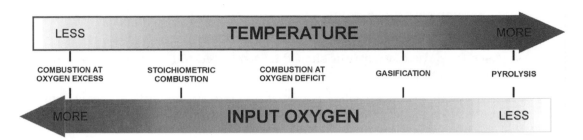

Figure 4.1 The comparative diagram of thermal processes
Combustion: *the process is exothermal, the oxidation is stechiometric with air excess, waste has a great amount of organic.*
Gasification: *the process is self-sustaining, partial combustion at oxygen deficit.*
Pyrolysis: *the process is endothermic with an exterior power source at lack of oxygen.*

Table 4.2 Waste combustion in the developed countries

Country	Number of factories	Burning waste (%)	Wastes, burning with power generation (%)
Japan	1,900	75	almost completely
France	170	42	67
United States	168	16	—
Italy	94	18	21
Germany	47	35	—
Denmark	38	65	almost completely
Great Britain	30	7	33
Sweden	23	55	86
Spain	22	6	61
Canada	17	9	7
Netherlands	12	40	72

tenance, especially when the heat productivity of the waste is sufficient for auto-thermal mode and flue gases do not contain secondary toxic substances.

Until the mid-1980s, the combustion of solid waste was considered a basic method of environmental protection, leading to the creation of the incineration industry. The average productivity of one factory in the United States, for example, is about 500 tons per day.

Waste combustion, depending on local conditions, is carried out both with heat use—that is, using a waste-heat recovery boiler—and without heat use, which has an effect on technical and economic indices (Villevold 1982).

The organization of the combustion process at such an installation should provide:

•   good interaction of waste with air in a layer
•   required temperature mode
•   quality afterburning of flue gases and necessary dwell time, preventing carryover of big unburned particles by flue gases

- combustion of waste with different (both high and low) calorific value
- wide range of operation modes

Exhaust gases from the reactor unit should be subjected to multistage cleaning from hazardous admixtures.

The schematic diagram of a waste combustion plant, taking into consideration the newest requirements on process organization and environmental emission, is shown in figure 4.2 (Piel and Platiau 1998, 2–15).

In layer furnaces, the velocity of layer movement in all zones of combustion is achieved by use of cooling mechanical grates. In rotary furnaces (see figure 4.3), the zone of combustion is motionless, and the waste moves at the expense of rotation of a cylindrical drum, located under an angle of $1 \div 3°$ to the horizon. Torches, working on natural gas or diesel fuel, are installed in furnaces for heating the furnace space up to a temperature of 800°–1,200°C. Temperature is maintained at the expense of waste combustion.

Recently, an industry connected with waste treatment and power generation has developed (D'yakov 2002). These technologies are developed as an alternative to land-fill disposal. Steam turbines are used for power generation through waste combustion. In view of the low heat coefficient of steam, such installations are characterized by low efficiency. The most frequent customers of generated thermal energy are local thermal networks.

Furnaces with a pseudo-fluidized bed are used in the industry, except in the afore-mentioned combustion methods. Thus, thorough combustion is achieved with mini-mum formation of harmful substances. An example of such method is a factory for combustion of wastewater sludge constructed by the French Company OTV (OTV

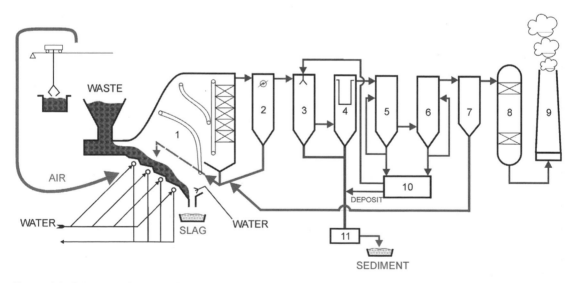

Figure 4.2 Schematic diagram of the plant for solid waste combustion
1—furnace with cooling grate; 2—electric filter; 3—vaporizer; 4—dryer; 5—preliminary cleaning device (acid gases); 6—basic cleaning device; 7—adsorber; 8—catalytic filter Nox; 9—stack; 10—cleaning of scrubber solution; 11—accumulating tank.

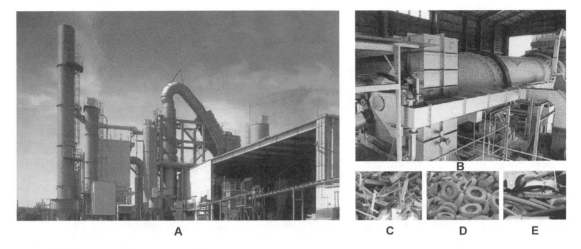

Figure 4.3 A modern factory for solid waste treatment
Productivity is 200 t/d. It was built and owned by Actree Corporation (Japan).
a—general view of the factory; b—drum-type reactor; c—wood waster; d—automobile tires; e—plastic waste.

2004). A plant using such technology is equipped with furnaces with a pseudo-fluid-ized bed, gas cleaning system, and a system of heat exchanging and provides combustion of dehydrated sludge at the expense of its own combustion heat. Natural gas is used for the furnace ignition. The ash is used as a cement additive in the manufacture of reinforced concrete constructions. The productivity on sludge is 50–60 tons per day, with a temperature of 780°–800°C. A partial process flowsheet of a sludge combustion plant is represented in figure 4.4.

The Ebara Corporation of Japan is a leader among the companies manufacturing furnaces and equipment for waste combustion (Ebara Corporation 2004). In the last 20 years, the company has developed a number of technologies using furnaces with fluidized bed and layer furnaces.

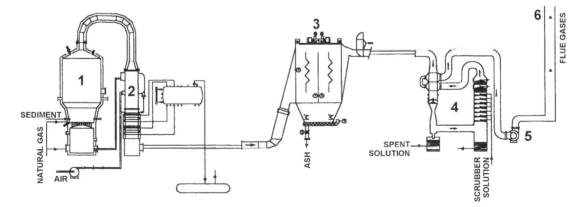

Figure 4.4   The flowsheet of a factory on sludge combustion
1—furnace with pseudo-fluidized bed; 2—heat exchanger; 3—filter; 4—gas cleaning system; 5—smoke exhauster; 6—stack.

2 Gasification

Gasification is a high-temperature thermochemical process of interaction of an organic mass with gasifying agents, resulting in formation of combustible gases. Air, oxygen, steam, and carbon dioxide and mixtures thereof are used as gasifying agents.

The process of filtration combustion in conditions of superadiabatic heating is one example of the gasification method use for garbage treatment. Such modes appear in the presence of a heat source and the heat exchange between streams of solid and gaseous substances, moving toward each other relative to the front of a chemical reaction.

Technologies created on this basis have a two-stage scheme (see figure 4.5) (Manelis, Polianchik, and Fursov 2000). At the first stage, the treated material is subjected to the steam–air gasification. The gas produced is burned at the second stage in a heat-recovery boiler, generating thermal and electrical power.

Gasification takes place in a shaft-type reactor-gasificator. The specific peculiarity of the process is that the heat released at combustion is not removed from the reactor, but rather concentrated in the gasification zone and used for generation of hydrogen from water and partially from carbon oxide.

The processing raw material is fed into the reactor from the top through the airlock chamber. Air and water vapor are supplied from the bottom. The produced gas is removed at the top part of the reactor, and ash slag is discharged from the bottom. The working mass moves in the reactor by gravity. Gasification is characterized by high power efficiency (up to 95 percent) and allows the processing of materials with a small content of combustible components or high humidity.

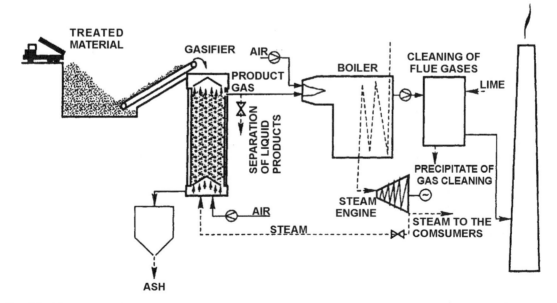

Fig. 4.5   Flowsheet of thermal treatment of combustible waste with generation of thermal and electrical energy

3   Pyrolysis

Pyrolysis of waste is one of the most promising ways of solving the ecological and power problems involved in waste disposal (Von Roll 2004; BS Engineering 2004; Thermoselect Südwest 2004). The technology of pyrolysis consists of irreversible chemical modification of waste under the action of temperature without access to oxygen. The temperature level of the process in the reactor determines the yield and quality of pyrolysis. It is characterized by production of thermodynamically stable substances: solid residue, resin, and gas. Pyrolysis can be divided into low-temperature (up to 900°C) and high-temperature (more than 900°C) types.

In low-temperature pyrolysis, shredded waste material is subjected to thermal destruction at temperatures of 500°–900°C. An increase of temperature results in an increase of pyrolytic gas yield and a decrease in the yield of liquid and solid products.

High-temperature pyrolysis is a method to produce a secondary synthesis gas, with the purpose of using it for production of steam, hot water, and electric power. It results in gasification of organic components and melting of mineral ones.

These processes, being endothermic, require input of a significant amount of heat into the reactor, and therefore an increased power input. Recuperation of this energy is possible only after combustion of generated syngas.

Figure 4.6 shows the scheme of garbage utilization (Rosdon 2004) by high-temperature pyrolysis, whose basic principles eliminate the disadvantages of classical methods of pyrolysis. The problem of maintaining such high temperatures (1,650°–1,700°C) in the zone of slag melting is solved by application of stabilizing electrical heat. Contamination of pyrolytic gas at the reactor outlet by toxic hydrocarbon, heavy metal compounds, or dioxins is eliminated by high-temperature (1,700°–1,800°C) treatment of gas in a reduction atmosphere and the discharge of the final product from the reactor without contact with the initial components.

The pyrolysis reactor has an adsorber-gas generator and a micro-blast furnace, supplied with electrodes for stabilization of the thermal conditions with variation of the waste composition.

4   Waste Treatment in Melts

The use of metallurgical smelting furnaces is a basis for the technology of waste treatment in a liquid slag melt. A number of tuyeres are located below the level of a flux bath and serve to feed an air or air–oxygen mixture into the furnace under pressure. This allows *barbotage* (intermixing), improving conditions of waste combustion and increasing the furnace productivity. The essence of the technological process consists of high-temperature decomposition of waste components in a layer of bubbling melted slag at temperatures of 1,350°–1,400°C and dwelling times of 2–3 seconds. Waste processing is carried out without preliminary sorting and drying, and also with significant variation of chemical and morphological compositions.

Oxygen is fed above the level of melt, and afterburning of exhaust gases takes place above the bath. This allows reducing the flue gas volume, ensures the gases' cleaning,

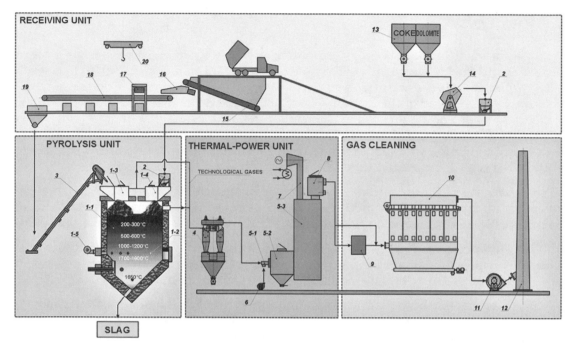

Figure 4.6 Flowsheet of a complex for solid garbage utilization
1.1—reactor; 1.2—filter—gas generator; 1.3—solid waste feeder; 1.4—coke feeder; 1.5—blow fan;
2—backit; 3—skip lifter; 4—cyclone battery; 5.1—gas-torch block; 5.2—furnace; 5.3—convection
recuperator; 6—blow fan; 7—steam-turbo-generator plant; 8—mixer; 9—electric heater; 10—bag filter;
11—smoke exhauster; 12—stack; 13—bin trestle; 14—blend mixer; 15—apron conveyer; 16—feeder;
17—iron separator; 18—sorting conveyer; 19—bunker for solid garbage; 20—electrical travelling crane.

and reduces the emission of toxic substances, including nitrogen oxides. The slag is periodically discharged and is suitable for production of building materials. The metal is collected at the furnace bottom, periodically merges, and is used as a secondary raw material. A process flowsheet of a plant (Institut Gintsvetmet 2004) for solid waste treatment is shown in figure 4.7.

Due to the high temperatures, the flue gases do not contain organic compounds, so there are no dioxins or furans. Alkaline and alkaline earth metals contained in the waste are transformed into gas at flux bath blowing. This promotes the binding of chlorine, fluorine, and oxides of sulfur into safe compounds, caught by the gas cleaning as solid dust particles. The installation's profitability can be increased by using a heat exchanger, which allows receiving hot water or steaming on an industrial scale.

There are techniques of waste processing using standard electric arc furnaces. The flux bath is formed at the expense of components and mineral part of waste. Melting takes place under the action of the electrical arc, originating between electrodes after voltage supplying. Air or an air–oxygen mixture is used to intensify the process.

5   Plasma Methods of Waste Treatment

Widely used methods of thermal waste processing have a number of basic disadvantages. First of all, there is the formation and emission into the atmosphere of a great

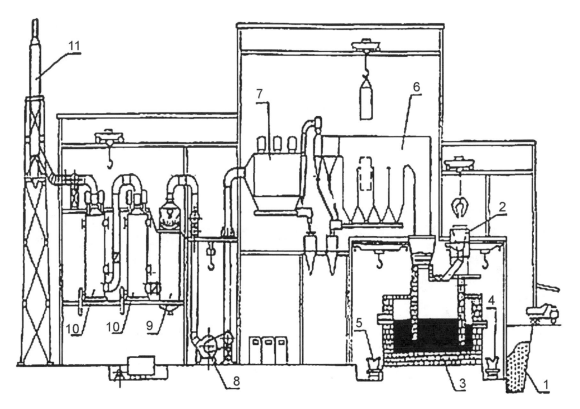

Figure 4.7 Apparatus-process flowsheet of solid waste treatment
1—raw material storage; 2—feeding unit; 3—furnaces with a melt; 4—alloys for metallurgical factories;
5—slag for building material production; 6—recovery boiler; 7—dry electrostatic cleaner; 8—smoke exhauster;
9—scrubber; 10—wet wet-stage precipitator; 11—stack.

amount of toxic substances: drag-out ash containing heavy metals, soot, carbon monoxide, oxides of sulfur and nitrogen, chlorine compounds, and also such supertoxic substances as dioxins and furans. Slag also contains unburned carbon and polyaromatic substances.

Existing techniques produce a great amount of toxic substances. That is why it is necessary to meet the strict emission requirements using extremely expensive purification plants, which considerably increase operating costs. The poor ecological and economic performance of such techniques has resulted in the search for and creation of new methods of waste treatment. Processes of high-temperature mineralization of waste under the action of isothermal plasma, obtained by gas passing through an electrical arc, is one such method. The range of temperature variation in this process is rather wide, stretching from 1,700°C to almost 10,000°C. The time needed for complete waste transformation in conditions of plasmochemical process is 0.01–0.5 seconds, depending on the nature of the waste and the process temperature.

The effectiveness of thermal energy input for simultaneous stimulation of chemical and physical modifications in a substance is depicted in figure 4.8. The diagram shows the difference of plasma heat in comparison with heat from natural fuel. The plasma processes ensure high and effective temperatures of processing, which cannot be reached by other methods of heat.

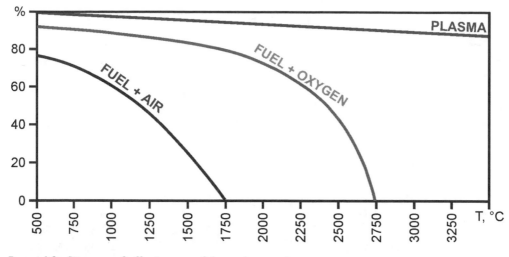

Figure 4.8   Diagram of effectiveness of thermal energy input

Various stages of substance transformation under heat action are presented in figure 4.9. Dissociation of combustion products gains practical significance only at temperatures above 1,800°C, which is connected with energy consumption, necessary for molecular decay into atoms. The most widespread components of gases of industrial technology are CO_2, CO, H_2O, O_2, N_2, and H_2. Of these, carbon dioxide dissociates most easily, and atomic hydrogen is the hardest one to form. At temperatures above 1,500°C, processes of dissociation and their heat consumption assume greater importance. In particular, the dissociation determines a limiting temperature of about 2,700°C, which is accessible at combustion of cold fuel in oxygen. Thus, the dissociation process limits the technological possibilities of the chemical energy of fuel, as it does not allow especially high temperatures.

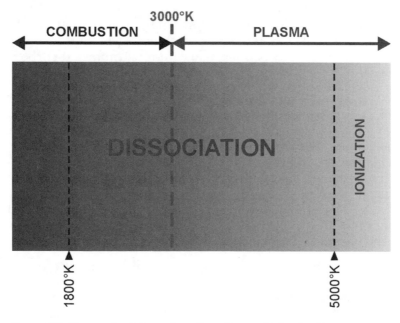

Figure 4.9   Stages of substance transformation under heat action

The line at 2,700°C is an upper limit of accessible temperature in industrial waste processing using traditional kinds of fuel. Traditional combustion cannot decompose products into elementary molecules; thus, resin and unburned carbon are formed in such combustion systems. The realization of technological processes with higher temperatures is possible only using the electrical power.

The plasma dissociation allows the complete decomposition of practically all known organic and many inorganic compounds into elementary atoms and molecules of the elements. In these conditions, thermodynamically stable two- or three-atom compounds—oxides, hydrides, halogenides—can be formed, depending on the chemical nature of the waste and plasma-forming gas.

Different amounts of energy consumption are necessary for the ionization of atoms of various substances. The degree of ionization at a given temperature depends mainly on the content of the elements with low ionization potential in a gaseous medium. Sensible thermal ionization can take place at temperatures of 5,000°–6,000°C.

The high-temperature gas stream, at the expense of dissociation and ionization processes, has a high energy content. The heating ability of such a stream on a substance's surface reaches values one to two orders of magnitude higher than that from fuel. This allows an acceleration of many technological processes connected with handling of materials.

Compared with traditional combustion, plasmochemical installations have a number of essential advantages (Rutberg 2003; Rutberg et al. 2002):

- The possibility of temperature regulating in the basic reactor in the range of 750°–10,000°C
- Deep destruction of waste with a simultaneous decrease of volume of exhaust gases
- Smaller weight and dimension characteristics of the reactor and the installation as a whole
- The possibility of complete automatic control of the process
- Minimum expenses of time and means on remedial maintenance of high-temperature plasma generators
- More complete conversion of carbon into oxides of carbon, CO and CO_2
- Use of already available, inexpensive electrical power
- By-products of plasma processing are easily predicted, harmless, and acceptable from the point of view of environment protection

It is expedient to use plasmochemical installations for combustion of relatively small amounts of concentrated supertoxic substances of hazard classes I and II, hazardous medical waste, and halogen-containing gaseous, liquid, and solid waste. It is especially expedient to use such installations at a site of formation or storage of supertoxic waste. In these conditions, it is possible to use both stationary and mobile plasmochemical installations.

At the same time, as evaluations show, plasma technologies appear to be economically more attractive, in comparison with other methods, for treatment of a wide range of waste (see table 4.3) (Joos 2002).

Table 4.3 Economic indices of various methods of waste treatment

Method of treatment	Cost of treatment (Euro/ton)	
	Min	Max
Disposal at disposal areas	105	160
Traditional combustion	100	140
Pyrolysis, thermolysis	90	150
Plasma method without syngas production	100	120
Plasma method with syngas production	70	80

Use of plasma technologies allows the realization of a more purposeful and pure process for reaching the following results:

• In syngas generation, there is much higher conversion of carbon into CO and CO_2 (owing to the higher temperature of the process) and consequently higher effectiveness of syngas production from primary raw material

• A much higher velocity of chemical processes in the reactor at the expense of high thermal and chemical activity of low-temperature plasma, due to a rather high density of energy

• The possibility of an essential decrease of toxic substances at the expense of singularities of proceeding physicochemical processes and the possibility of controlling these processes (Rutberg 2003).

As the physical-chemical processes require the input of a significant amount of energy at a high specific density into the reactor, the realization of these processes has become possible due to successes in creation of reliable and cheap generators of dense plasma as a necessary unit of the technological process.

5.1   Generators of Low-Temperature Plasma

Strong current arc discharges and plasma generators, developed on their base, intended for treatment of various types of waste are used in practically all plasma systems (Rutberg 2003). The existing designs of plasma generators can be divided into two basic groups: DC plasma generators and AC plasma generators.

The internal energy of the electric arc transforms into internal gas energy in a plasma generator, caused by the gas heat exchange with the arc discharge column. The flow of the arc column by the stream of plasma-forming gas is used to intensify the heat exchange process. At the same time, the gas flow thermally insulates the chamber walls from heat flows, yet does not prevent heat transfer by radiant flux from the arc column toward the plasma generator walls. Complete thermal insulation of electrodes is impossible, owing to the electrical contact between the arc and the electrode surface. In contact area, especially at the presence of arc spots and significant current values (greater than 100 Amps), the density of heat flow reaches 10^9–10^{10} W/m^2. The long-duration operation of electrodes under these conditions is possible only if they are made of refractory materials (tungsten, molybdenum, hafnium, etc.) and cooled at the site of the electrode joint with the chamber case of the plasma generator. Electrodes are made of copper or steel if oxidizing media (air, etc.) are used as a plasma-forming gas and they need strong cooling. In this case, the increase of the electrode lifetime is necessary to provide conditions under which arc sports will continuously move along the electrode surface. Since it reduces the need for the electrode cooling, the average temperature of their surface is lower than the melting point. Electric arc erosion is insignificant (10^{-6}–10^{-4} g/C) at the locations of arc sports due to the short lifetime (less than 10^{-4} seconds). There are two forms of arc sport movement in DC plasma generators: gas-dynamic and at the effect of the magnetic field action on the current.

Physical processes in the positive arc column (electric conductivity, thermal conductivity, radiation, etc.) determine heat transfer into the plasma-forming gas blowing the arc. The erosion and the lifetime of electrodes are mainly determined by near-electrode phenomena. The basic section of the column is blown by the transverse flow of the plasma-forming gas. Small sections of the arc near the electrode surface are not blown. Complex, nonequilibrium processes in thin electrode layers cannot be controlled by any modern diagnostics. That is why the theoretical methods would be of great use. The volt-ampere characteristic is an integral property of the arc in the plasma generator as in the free-burning arcs.

The fundamental difference of volt-ampere characteristic of the plasma generator electric arcs from the free-burning arcs or arcs stabilized with the walls is that their character and the value of the average intensity of the electric field E depend on total gas flow rate. In most cases, they are low dropping. This is true for the range of the current strength I of 200–600 A. At higher currents the dropping character of the curves becomes better defined (see figure 4.10).

5.1.1 DC Plasma Generators

The arc column is optically transparent and radiates as a volumetric emitter at a plasma temperature of around 10,000°C and a particle density of about 10^{-19} cm^{-3} (see figure 4.11). Under the stable condition of cylindrical column shape and intensive turbulent heat exchange, the share of radiant convective heat exchange to the channel walls is insignificant, and the plasma generator has an efficiency of 0.6–0.8. This mode of arc burning takes place at relatively low currents and arc lengths for DC plasma generators with a power of 100–200 kW. That is why the peculiarities of DC plasma generators are:

*   The distinguishing characteristic of all DC powerful plasma generators is a relatively long discharge chamber with a comparatively small internal diameter. In this configuration, the working gas is fed along the whole length of the working chamber, causing arc contraction and a surface temperature as a rule higher than 10,000°–15,000°C. That is why the basic heat exchange between the arc and the working gas is at the expense of radiation, and energy losses are insignificant.

*   DC use, as a rule, requires ballast resistance connected in series with the arc. It is used for stable arc burning with a dropping volt-ampere characteristic and arc current control. It causes additional active power losses.

*   The power supply system for a DC plasma generator is comparatively expensive (expensive thyristor equipment).

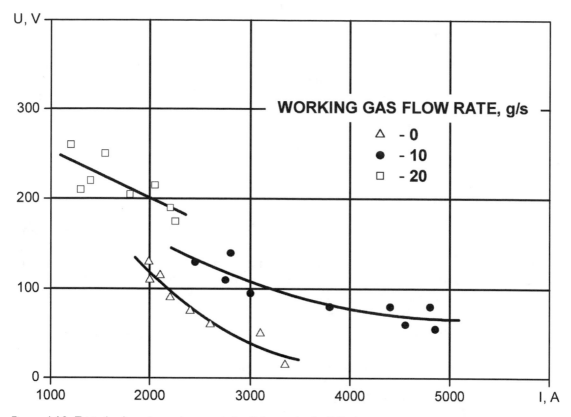

Figure 4.10 Typical volt-ampere characteristic of the arc in the DC plasma generator

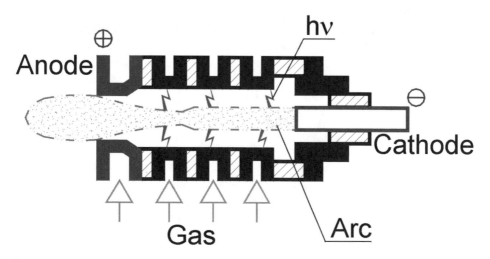

Figure 4.11 Schematic of DC plasma generator

5.1.2   AC Plasma Generators

Single-phase and multi-chamber AC plasma generators (see figure 4.12) have the same specific features as DC plasma generators, but in addition, they require a secondary ignition source for the arc after current passing through zero. In this situation, multiphase plasma generators are preferable. Single-chamber multiphase AC plasma

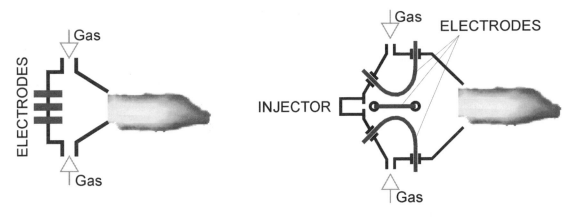

Figure 4.12   Schematic of AC plasma generator

generators provide simultaneous burning of several arcs in one chamber. The chamber can have a spherical form, which is optimal from the point of view of minimizing energy loss on the chamber walls.

Several arcs can burn simultaneously in one chamber, and at the expense of electron diffusion (also due to the additional injection into the chamber volume) a sufficiently vast volume with a high concentration of current carriers can be formed. This results in the diffusion character of arc burning; in this case, the temperature is rather lower than at the contracted mode of burning. The diffusion mode of arc burning provides a smooth current transition through zero, and the current curve form is close to sinusoidal.

There are two basic mechanisms for gas heating in the diffusion mode: forced convection and blowing of some gas through the arc. Energy-heated gas, at the expense of excitation energy relaxation and atom recombination, is redistributed among the rest of the gas, which causes it to heat up to average mass temperature. In this case, the losses are rather low (there are practically no radiation losses) and the transformation coefficient of the discharge into gas energy is high.

When the current value varies, which is connected with arc instability, the AC power supply system automatically reacts by supplying additional electromagnetic force and thereby suppresses the instability.

A power supply system for a multiphase plasma generator using a commercial frequency (50 or 60 Hz) is significantly cheaper and simpler than power supply systems of DC arcs. Furthermore, it can be made of standard industrial blocks.

The maximal parameters reached for AC plasma generators operating in a continuous mode are up to 3 MW in the United States (Westinghouse) and up to 2 MW in Russia—in discontinuous modes, up to 80 MW in Russia—at a thermal efficiency of 60–90 percent. A description of typical power plasma generators is presented below.

Plasma Generators with Rod Electrodes

Plasma generators with rod electrodes are intended for heating of inert gases: nitrogen and hydrogen at pressures of 0.1–6 MPa. The flow rate for nitrogen is 0.01–10 kg/sec. Heating is carried out to temperatures of 1,700°–6,000°C with a thermal efficiency of 60–85 percent. The power range of these plasma generators is from 100 kW up to 80

MW. The design of such a plasma generator is shown in figure 4.13 (Glebov and Rutberg 1985).

Plasma Generators with Rail Electrodes

Three-phase plasma generators with rail electrodes are intended for operation with air and other oxidizing media. Their power ranges from 0.1 to 1 MW. One type of plasma generator with rail electrodes (in this case, delivering 500 kW of power) is shown in figure 4.14.

The use of electrodynamic movement of the electrical arc in a field of its own current (the rail-gun effect) is the basis of this type of plasma generator operation. Arcs after the interelectrode breakdown move along the tubular electrodes up to their end. A plasma injector is used for repeated ignition of the arc. It provides concentration of electrons n_e in the interelectrode gap of 10^{-14}–10^{-16} cm^{-3}. That is sufficient for ignition of basic arcs at the comparatively low voltage of the power supply, approximately 300–500 V.

Rising arcs move along the electrodes with velocity of 10–30 m/sec, depending on the current magnitude and inclination angle of the electrodes. A three-phase system of electrodes and their configuration avoid arc extinction and current interruption. Thus, the rail-gun effect provides distribution of the thermal load along the electrode length. Intensive water-cooling of tubular electrodes, made of copper alloys, allows strong currents to pass, increasing the lifetime of the electrodes and minimizing any

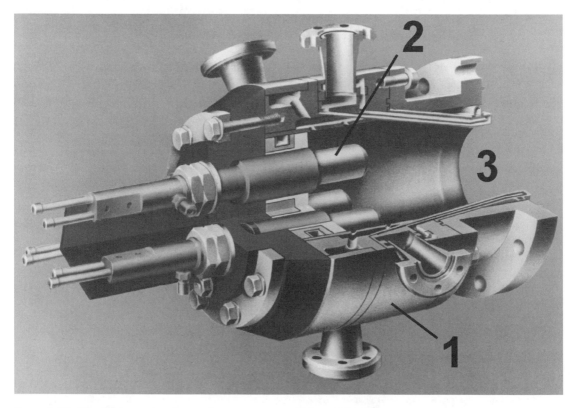

Figure 4.13   Three-phase electric-arc plasma generator with rod electrodes
1—case, 2—electrodes, 3—nozzle.

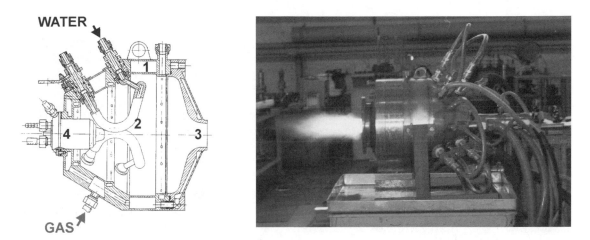

Figure 4.14 Plasma generator with rail electrodes
1—case, 2—electrode, 3—nozzle, 4—injector.

gas medium contamination. The arcs, with a diffuse combustion regime, occupy a large share of the discharge chamber. Such a plasma generator system can produce thermal efficiency of up to 90 percent, depending on operation conditions.

The volt-ampere characteristics of the system drop at low currents and increase as the current increases (above 10 kA). The change in volt-ampere characteristics at strong currents is caused by plasma temperatures at which the electric conductivity does not significantly depend on temperature (Coulomb scattering) (see figure 4.15). In all cases, the increase in flow rate causes a voltage increase. The increase of gas flow rate and power boost efficiency.

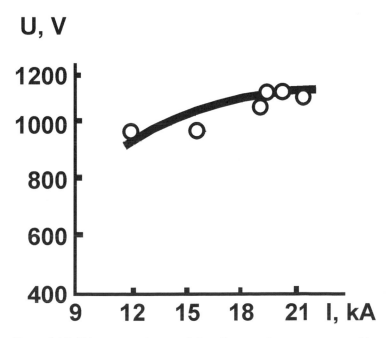

Figure 4.15 Volt-ampere characteristics of power plasma generator with rod electrodes

5.2   Plasma Combustion

The most advanced technologies make it possible not only to destroy wastes or reduce their volume but also to receive products for further commercial use. These technologies include combustion, gasification, and pyrolysis (Rutberg 2003; Choi and Park 1997).

Plasma combustion is a process that takes place at high temperatures (above 600°C) and results in complete decomposition of organic compounds. But not all combustion technologies can be recommended for some types of waste. Required temperatures cannot be achieved in single-chamber furnaces, even with low-temperature pyrolysis. Some wastes (military, medical, chemical, etc.) require temperatures higher than 1,200°C. Their calorific value cannot provide such high temperatures solely through waste combustion. That is why it is necessary to use additional fuel at the initiation of combustion. Plasma technology solves these problems in such processes (Rutberg 2002, 2003; Carter and Tsangaris 1995a).

5.2.1   High-Temperature Plasma Oxidization of Hazardous Medical Waste

An installation for plasma combustion of hazardous medical waste has been developed and constructed at the Institute of Problems of Electrophysics of RAS (IPE RAS) in Russia (see figure 4.16) (Rutberg 2002; Rutberg et al. 1999; Rutberg, Bratsev, et al. 2001). Solid medical waste is supplied to the installation in boxes and placed in a temporary storage site or in a special room equipped with a refrigerator.

Waste combustion is performed in a rotary furnace, with waste supplied to the furnace periodically through a gas-tight gate. The combustion in the furnace happens in oxygen-deficit conditions. The air excess factor is 1.0–1.2.

The temperature of flue gases at the furnace outlet is supported at the level of 1,100–1,250°C by the heat of waste combustion and hot air from plasma torches installed at the end face of the furnace feed head. Plasma torches are used also for initial heating of the furnace at start-up and maintainance of the necessary temperature in the furnace during interruptions of the combustion process.

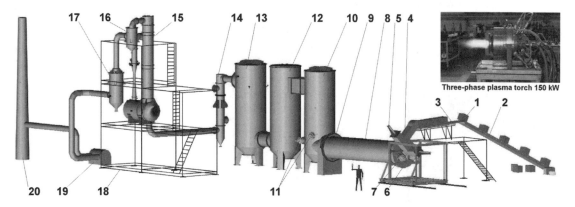

Figure 4.16 General view of the installation for plasma treatment of hazardous medical waste
1—box with waste; 2—conveyor belt; 3—roller; 4—feeding shaft; 5—pneumatic pusher; 6—loading chamber; 7,11—plasma torches; 8—rotary furnace; 9—unloading chamber; 10—afterburner; 12—quenching chamber; 13—quencher; 14—recuperator; 15—wet scrubber with Venturi; 16—maerosol filter; 17—adsorber; 18—support structure; 19—gas blower; 20—stack.

Slag is removed from the furnace into a device for slag quenching filled with water. After quenching and cooling, the slag is sent for burial.

Flue gases from the rotary furnace go into the afterburner, which provides combustion of carbon oxides contained in the flue gases, residuals of hydrocarbons, mechanical carbon in a form of coke particles, and soot removed from the furnace. The high temperature in the afterburner (1,200°–1,300°C) and extended dwelling time of the flue gases (on the order of two seconds) promote decomposition of stable organic compounds and prevent the exit of supertoxic compounds such as dioxins.

The temperature in the afterburner is achieved through the heat content of the flue gases themselves as they come in from the furnace, as well as the feed of hot (1,700°C) air from the plasma torch and the heat of combustion of the fuel residuals contained in the flue gases. A supply of additional air and water into the chamber is provided to increase the quality of the afterburning and ensure the possibility of temperature regulation.

Then the flue gases are sent to the cooling and cleaning systems. Abrupt cooling of flue gases to temperatures of 900–950°C takes place in the quenching chamber. In parallel, there is the process of reduction of nitrogen oxides, which are formed in great quantities by operation of the plasma torches. A solution of urea and water is fed into the chamber as part of this process. The urea–water solution decomposes at temperatures higher than 132°C, releasing ammonia, which acts as a reducing agent for the nitrogen oxides. In addition, water from the solution adds significant cooling of the flue gases.

Temperature of flue gases is reduced to 500–550°C in the quencher through this process of water injection. The recuperator then cools the flue gases to temperatures of 300–320°C by transmission of a part of heat to cold air, which is hereinafter used at installation for the technological purposes.

Cleaning volatile ash, acid gases, and some heavy metal vapors from the flue gases is carried out in a wet scrubber with a venturi—initially in the venturi tube, then in a packed-bed scrubber. Both devices are mounted at a storage tank for spraying solution, which combines the functions of a gas-liquid separator and is sprayed through the injectors by a circulating sodium solution.

The aerosol filter at the outlet of the wet scrubber and venturi cleans the flue gases of fine dust, aerosols, and traces of acid gases. It also catches droplets of the neutralizing solution.

At this point in the process, the flue gases are diluted by air, heated in the recuperator, and supplied into the adsorber at a temperature of 90°–95°C. The adsorber is filled with activated carbon and is for extraction of residual compounds of heavy metals and dioxins. After passing through the adsorber, the flue gases are once again diluted by heated air and, with a temperature of 120°–150°C, discharged into the atmosphere through the stack.

5.2.2 Plasma Combustion of Liquid Supertoxic Agents

An installation for plasma combustion of liquid toxic waste has also been developed and constructed at the IPE RAS (see figure 4.17) (Rutberg 2002; Rutberg et al. 2000; Rutberg et al. 2002). Plasma treatment of CA simulators is carried out in two chambers

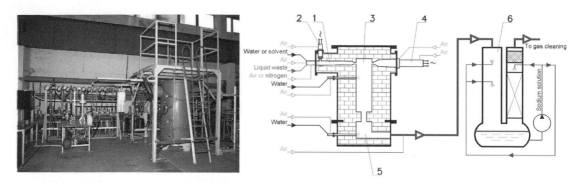

Figure 4.17 The flowsheet of installation for plasma destruction of liquid toxic substances
1—reactor-primary furnace; 2—single-phase plasma torch (up to 10 kW); 3—rector-afterburner; 4—three-phase plasma torch (up to 3 kW); 5—quenching chamber; 6—gas cleaning system.

mounted in series (a two-reactor design)—a tunnel primary furnace and cyclone reactor-afterburner.

Ignition of liquid simulators of chemical warfare agents in the primary furnace is carried out with the use of a twisted plasma jet, into which the simulators, dispersed beforehand by compressed air, are injected. The input of combustible solvents into the plasma jet is designed to improve the process of igniting difficult combustible and incombustible simulators. The process of combustion or gasification depends on the proportions of the supplied fuel; air and water can be used in the primary furnace. The processes proceed at temperatures of 1,200°–1,300°C.

Gases from the primary furnace move with a high velocity into the cyclone reactor-afterburner. Here, at temperatures of 1,400°–1,500°C, the final oxidation of the initial substances and the products of their incomplete oxidation occurs. For this purpose, a tangential stream of hot air from the plasma torch is fed into the reactor. A significant (two-second) dwelling time of the flue gases in the afterburner ensures full mineralization of any organic substances.

After passing through the reactor-afterburner, the flue gases are quickly cooled to temperatures of 380°–400°C by water injection and the input of cold air. At the gas cleaning installation, the flue gases are subjected to further cooling and a three-stage cleaning to rid them of acid gases and aerosol particles. A wet alkaline method is adopted as the gas cleaning method, using a filamentary filter for removing aerosol particles. The cleaned flue gases are fed into a sanitarium column and then discharged into the atmosphere.

5.3   An Experimental Plant for High-Temperature Plasma Pyrolysis

The processes at each stage of plasma treatment are extremely complicated and diverse. Most of them are not well investigated, which is why in a number of cases the technological and construction parameters of the installation cannot be chosen on the basis of strict mathematical calculation (Carter and Tsangaris 1995b). A pilot laboratory-scale plant has been built at the IPE RAS to study the mechanism of plasma processes in the plasma treatment of waste (see figure 4.18) (Rutberg 2003; Rutberg, Safronov, et al. 2001).

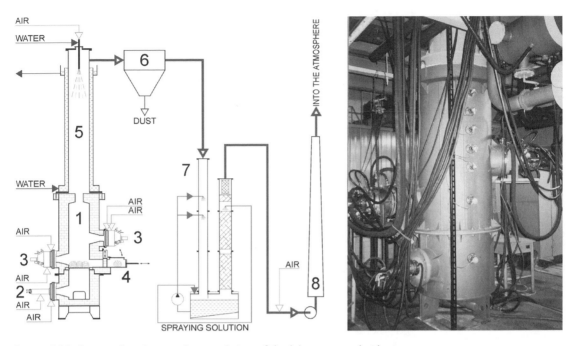

Figure 4.18 Process flowsheet and general view of the laboratory-scale plant
1—reactor unit; 2—plasma torch 30 kW; 3—plasma torch 150 kW; 4—lock loading chamber; 5—quencher;
6—filter; 7—wet scrubber with Venturi; 8—stack.

In this experimental facility, waste, packed in bags, is periodically fed through the lock chamber into the shaft reactor, which consists of two chambers. The lining of the reactor unit allows the performance of experiments at temperatures as high as 1,300°–1,500°C and in the presence of HCl and HF acid gases. Waste is heated in a layer at the motionless grate dividing the reactor unit. AC plasma torches of two types (30 kW and 150 kW) are used to heat the reactor and maintain the required temperature. Flue gases from the reactor go into a quencher, where they are cooled by water injection and then are directed to the gas cleaning system, which consists of a filter and wet scrubber with venturi.

In one experiment, wood blocks were used as a fuel to model wood waste. The initial moisture was 13 percent. The average amount of wood fed into the reactor was about 25 kilograms.

Samples of flue gases were taken from the gas duct and the content of nitrogen oxide and carbon oxides was analyzed (see figure 4.19). The most interesting fact obtained during the experiment is that nitrogen oxide (NO) was absent practically from the beginning of the process, during the stage of volatile-matter yield and the reduction stage of carbon combustion. The NO content, before feeding fuel, was about 4,000 ppm, which corresponds, with regard to dilution, to a temperature of plasma-forming gas at the plasma torch outlet of approximately 1,900°C. In figure 4.19, the NO content in the flue gases is the lower curve; the higher one is carbon monoxide (CO) content. At that content of CO, not more than 3,000–7,000 ppm NO was found in the flue gases. With further decrease of the CO content, the NO concentration increases, and at the minimum content of CO, it achieves the initial level. This fact indicates that, at the reduction zones in the reactor, the amount of nitrogen oxides introduced with the plasma-forming air can be decreased to a minimum.

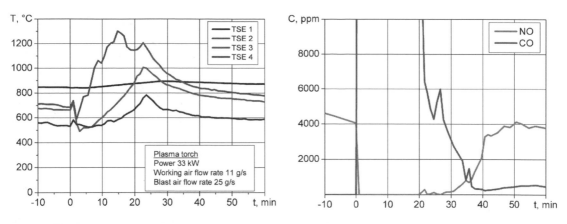

Figure 4.19 Temperatures in the reactor at layer combustion of model wood waste

5.4   Plasma Pyrolysis with Gradual Melt Formation

Plasma pyrolysis with gradual melt formation cannot be considered to be absolutely satisfactory because it requires significant energy consumption for melt formation. Nevertheless, its use is expedient in some cases.

The Startech Environmental Corporation in the United States has developed a new technology for treatment of liquid and solid waste (Startech Environmental 2004). Figure 4.20 represents a general view of this installation, which is called a "plasma converter of wastes." The installation processes municipal and industrial wastes, as well as paints, solvents, and oils. It can process up to 200 kilograms per hour.

The technology of the Startech installation is based on the process of plasma pyrolysis, with the use of a 200-kW DC plasma torch. The plasma-forming gas used is air. Waste is fed from the stockhouse into the lock feeder and then by small portions into the reactor. If it is necessary to supply liquid, metering pumps and injectors are used. A plasma torch with an arc of indirect influence is used at the installation. The plasma torch can be fixed at the cover or can move under the action of an external mechanism inside the reactor. Waste is fed into the chamber type reactor (see figure 4.21) and subjected to the action of high-temperature plasma. The pyrolysis of the organic part

Figure 4.20 General view of Startech installation

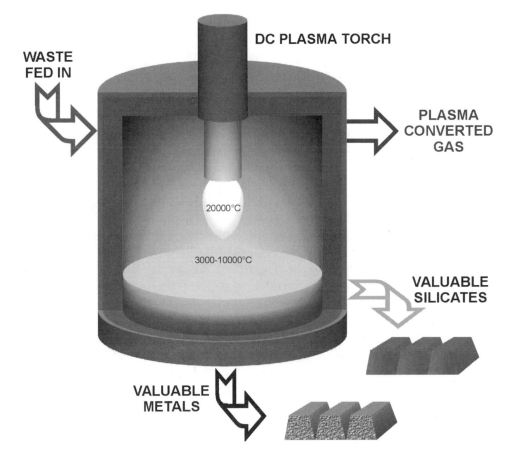

Figure 4.21 Diagram of Startech reactor

of the waste takes place at a temperature of 1,600°C, with generation of high-calorific syngas in off-gases. The melt forms at the bottom: accumulated silicates as slag, with metals in the waste accumulating at the surface of the melt.

Syngas is a product of the treatment (see table 4.4). It passes from the reactor at a temperature of 1,200°C and goes into a gas cleaning system. After that, it is possible to use it as a fuel for electric power generation. Pure syngas is used as a chemical raw material. Metals and ceramic silicates are periodically discharged from the reactor.

Table 4.4 Typical composition of syngas

Gas	Percentage (%)
CO	22–32
CO_2	2–5
H_2	20–45
O_2	3–5
N_2	15–35
Total CH	5–10

When processing rigid scraps, the volume of the solid waste decreases 300-fold after treatment.

5.5   Plasma Pyrolysis in the Reactor with a Molten Bed Formed by Joule Heat and a Freely Burning Arc

Integrated Environmental Technologies in the United States has developed a technology known as a plasma enhanced melter which can process solid industrial, medical, and cold wastes (Integrated Environmental Technologies 2004). The installation for 4 tons per day, put in operation in 1999, is shown in figure 4.22. The facility serves a large hospital and other medical establishments situated nearby.

The method uses a chamber-type reactor (see figure 4.23), inside which two groups of graphite electrodes are located. Two power sources are used: an AC power source and a group of electrodes create a zone of Joule heat, which forms a bath of melt at the bottom of the reactor. Energy is introduced immediately into the melt through dropped electrodes. The melt is a glassy conducting mass, formed by the inorganic part of the waste. A DC power source and electrodes of different polarity form a zone of arc plasma. A stable plasma arc is formed between the surface of the melt bath and the rod electrodes.

The waste is fed directly into a zone of plasma-arc heat, where the organic part dissociates in an oxygen deficit and forms CO, H_2, N_2, HCl, and H_2S. The inorganic oxides are dissolved in a melt bath, and molten metals precipitate at the bottom of the reactor.

The electrodes are made of graphite sections 0.6 m long and 0.15 m in diameter. Every two to three weeks it is necessary to extend the electrodes supporting the temperature of the melt bath. DC electrodes powering the zone of plasma-arc heat require extension every two or three days.

The process of treatment proceeds at a temperature of 1,100°–1,400°C and is close to pyrolysis. Syngas, vitrified slag, and reduced metals are formed at the outlet of the reactor. The process is shown in figure 4.24.

A wet method of gas cleaning is used. At the first stage, acid gases and particles of dust are removed, and at the second stage, traces of acid gases and aerosols are taken

Figure 4.22 General view of 4 t/day installation for medical waste treatment, Honolulu (USA)

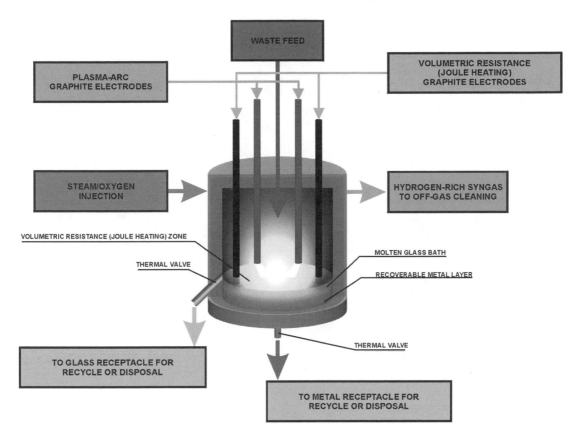

Figure 4.23 Diagram of PEM reactor

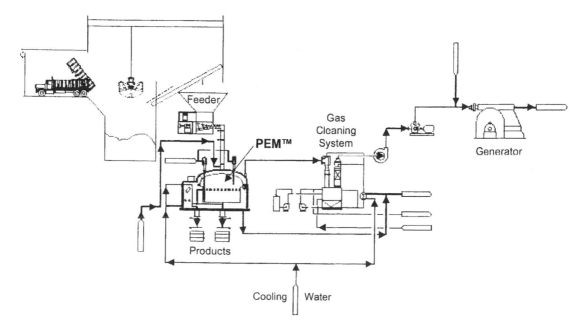

Figure 4.24 PEM process flowsheet

out. Cleaned syngas consists basically of H_2 and CO and is used for further generation of electrical power.

It might be well to point out that this method has the same disadvantages as the plasma pyrolysis with gradual melt formation described above. They are clearly defined because energy liberation during the heating of the melt is released mainly in a zone of anode voltage drop in the arc.

5.6   Plasma Pyrolysis with Issue of Vitrified Slag

Vanguard Research, Inc. (VRI) in the United States has developed the plasma energy pyrolysis system (PEPS), a technology for treatment of solid garbage and industrial waste (Vanguard Research 2004). The company constructs stationary plants for plasma pyrolysis with syngas generation capable of processing 5–15 tons per day of waste; they also produce a mobile module capable of processing 3–5 tons per day.

A 10-ton/day installation is represented in figure 4.25. In this version (see figure 4.26), a 500-kW DC plasma torch with an arc of indirect operation works on air or nitrogen, powered by a 1-MW diesel generator. A reversible system of waste supply eliminates blockage and bridging that makes it extremely safe and reliable. Packed or in bulk, the waste is fed into the screw feeder by the mechanical pusher. The chamber-type reactor is heated by the plasma torch to a temperature of 1,650°C, at which point dissociation of the organic part of the waste and vitrification of slag take place.

Syngas exits the reactor at a temperature of 1,040°C and is quickly cooled by water in order to prevent reactions that can create complicated organic compounds such as dioxins. The gas produced as a result of pyrolysis of, for example, medical waste, contains carbon, particles of metals, and acid gases that must be removed before use. The

Figure 4.25 General view of the installation for treatment of solid garbage for 10 t/day. Lorton, Virginia, USA

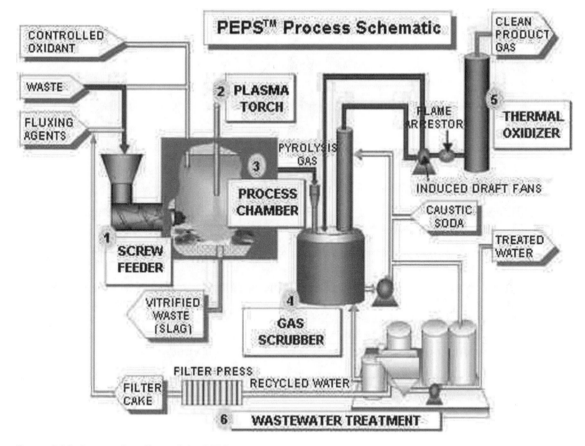

Figure 4.26 Process flowsheet of the PEPS process

volume of exhaust gases is 10 times less than that from combustion of the same amount of waste. The air-exhauster supplies the gas from the reactor into the quencher and scrubber for cleaning. After cleaning, the gas saturated with water mixes with natural gas or methane and is transformed in a thermal oxidizer to meet a nominal atmospheric composition. A comparison of the harmful emissions from a PEPS installation and U.S. standards for combustion, using the example of medical waste, is shown in figure 4.27.

The high temperature in the reactor and high-energy composition of the obtained gas enables electrical power generation and decreases the capital and operating costs. At the second stage, the gas that is produced is subjected to pyrolysis in a plasma reactor. Syngas can be used in a gas turbine for rotation of an electrical generator to supply power into the PEPS system or power network. Alternatively, syngas can be used for steam production at combustion in a recovery boiler, or in a heat exchanger mounted in a stream of exhaust gas.

5.7   Combined Gasification and Plasma Pyrolysis

The Swedish company ScanArc has developed a technology called PYROARC for the treatment of solid, liquid, or gaseous waste (ScanArc Plasma Technologies 2004). A peculiarity of the process is that it proceeds in two stages. Solid waste is fed into a

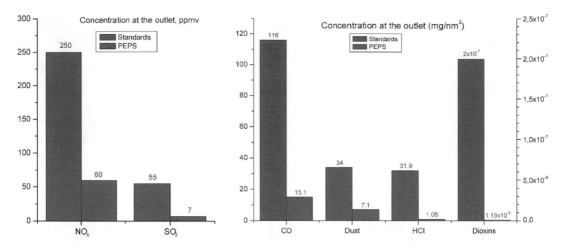

Figure 4.27 Emission Indexes of hazardous substances of PEPS installation for medical waste processing

shaft furnace, where gasification of the organic part and melting into slag of the inorganic part takes place. The gasification of solid waste is based on thermal energy partially created by combustion of carbon and carbon monoxide. It takes place in a reverse-flow gasifier with a heat exchange between the supplying product and exhaust gases. Preliminary heated air or oxygen is used as a gasifying agent.

The gas obtained in this first stage is then subjected to pyrolysis in the plasma reactor. Syngas, steady to nonleachible slag, melted metal, and some secondary dust are formed (see figure 4.28).

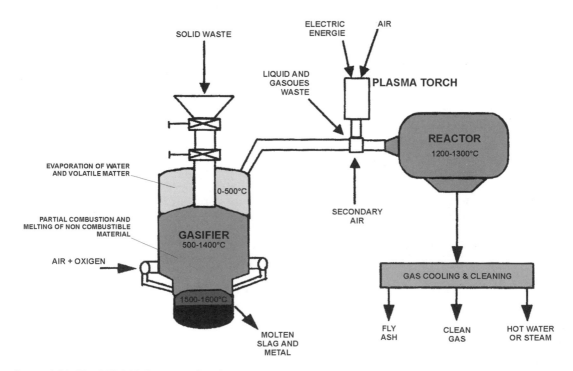

Figure 4.28 The PYROARC process flowsheet

The basic part of the process is a reactor (see figure 4.29). Fuel gas, produced in a shaft reactor and containing resins and hydrocarbons, is fed into a mixing zone before the plasma generator. Secondary gas (air, steam) is introduced to control the oxidation level of the produced syngas. After the mixing zone, there is a zone of extension, where temperatures are equalized at 1,200°–1,300°C and the decomposition reaction is initiated.

There is a variant of this system, in which liquid and gaseous wastes are fed into the mixing zone bypassing the gasifier, mixed with air plasma, and decomposed in the reactor forming the syngas.

A pilot installation for 300–700 kilograms per hour of solid waste, depending on structure, and for 50–500 kilograms per hour of liquid or gaseous waste has been built in Hofors, Sweden. It uses a DC plasma torch (with 1-MW power using air as the working gas) with an indirect operation arc. The produced syngas has a calorific value, about 4 MJ/m³, and contains about 35–40 percent carbon and molecular hydrogen. This gas is used in a gas turbine for electrical power generation or for production of vapor in a boiler.

5.8 Plasma Gasification with Adding of Solid Fuel, Formation of Inert Mass Melt, and Residual Heat Utilization

The Westinghouse Plasma Corporation of the United States, together with the Japanese metallurgical company Hitachi Metals, has constructed in Yoshii, Japan, a factory for treatment of 24 tons of garbage per day using a new technology (see figure 4.30) (Hitachi Metals 2002). The facility has a vertical shaft-type reactor and two MARC-3 DC plasma torches of 300 kW each. Air is the plasma-forming gas.

Solid waste is fed from the top simultaneously with limestone and coke, forming a "coke pudding" on which the waste lies. Plasma torches heat up the bottom section to a temperature of 1,500°C. Partial combustion of the coke, with formation of CO_2 and CO, takes place under the action of air heated up in plasma torches. These gases react with the excess coke and substances contained in the solid waste, forming CO, H_2, and CH_4, which are the basic components of fuel gas. A zone of melt formed from the inorganic part of the waste (slag) is created by the action of heat. As slag accumulates,

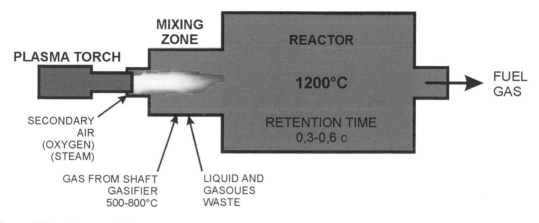

Figure 4.29 Diagram of the reactor

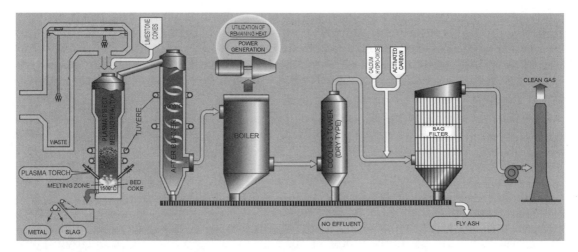

Figure 4.30   Schematic process flowsheet

it is removed from the reactor. Thus, syngas and vitrified slag are formed after treatment.

The fuel gas then goes into the afterburner, where it is burned with additional air. The dwelling time of the gases in the afterburner is 2.5 seconds, at a temperature of more than 1,000°C. The energy obtained from this combustion is used for production of steam in a boiler. The steam is then fed into the turbine for electric energy generation. The effectiveness of the generated energy is more than 20 percent, and the obtained electricity is used at the same plant, creating a closed cycle of processing.

Flue gases from combustion go through a cleaning process and after filtration are discharged in the atmosphere. The composition of the discharged gases is:

Dioxins	<0.01 ng/nm³
Dust	<0.01 g/nm³
Sulfur oxide	<20 ppm
HCl	<30 ppm
NO	<50 ppm

In 2002 in Sapporo, Japan, a factory for processing of municipal solid waste and milled residuals of automobiles was constructed using this technology. The factory has two plasma reactors with productivity of 4 tons per hour each. Four DC plasma torches made by Westinghouse with power of 300 kW each are installed on each reactor.

6   References and Recommended Resources

Al'kov, N. G., and A. S. Koroteev. 2000. Complex technology of multistage utilization of solid garbage with electrical power generation. *Izvestia Akademii Nauk. Energetika* (2000, no. 4): 21–33.

BS Engineering. 2004. http://www.bseri.com.

Carter, G. W., and A. V. Tsangaris. 1995a. Plasma gasification of biomedical waste. In *Proceedings of the International Symposium on Environmental Technology*

Plasma Systems and Applications, 8–11 October 1995, Atlanta, Georgia, USA, 1:239–50.

———. 1995b. Plasma gasification of municipal solid waste. In *Proceedings of the International Symposium on Environmental Technology Plasma Systems and Applications, 8–11 October 1995, Atlanta, Georgia, USA,* 2:321–32.

Choi K.-S. and Park D.-W. 1997. Pyrolysis of waste tires by thermal plasma. In *Thirteenth International Symposium on Plasma Chemistry (ISPC 13), 1997,* 4:2447–51. Peking: Peking University Press.

D'yakov, A. F. 2002. State and prospects of development of non-tradition electric power engineering in Russia. *Izvestia Akademii Nauk. Energetika* (2002, no. 4): 26–29.

Ebara Corporation. 2004. http://www.ebara.co.jp.

Glebov, I. A., and Ph. G. Rutberg. 1985. *Powerful plasma generators.* Moscow: Energoatomizdat.

Hester, R. E., and R. M. Harrison. 1994. Waste incineration and the environment/eds. Manchester, England.

Hitachi Metals. 2002. Reports. Plasma direct melting plant. Hitachi Metals Report no. E-321–03–2002. Available at http://www.hitachi-metals.co.jp.

Institut Gintsvetmet. 2004. http://www.gintsvetmet.ru.

Integrated Environmental Technologies. 2004. URL http://www.inentec.com.

Joos, M. G. 2002. Plasma-based total treatment of waste and renewable energy sources: Ecologic and economic aspects. Page 91 of the proceedings of Colloquium, Ghent University, Belgium, 21 March.

Lee, C. C., and G. L. Huffman. 1989. Incineration of solid waste. *Environ. Progr.* 8 (3).

Manelis, G. B., E. B. Polianchik, and V. P. Fursov. 2000. Energotechnologies of combustion on the basis of phenomena of superadiabatic heatings. *Chemistry in the Interests of Stable Development* (2000, no. 8): 537–45.

OTV. 2004. http://www.otv.fr.

Piel, A., and D. Platiau. 1998. Etude des technologies de thermolyse des dechets menagers. Final report, June. Liege, Belgium: Institut Scientifique de Service Public.

Rosdon. 2004. http://www.rosdon.com.

Russian Federation. 2002. About state and environment protection of Russian Federation in 2001. Available at http://www.eco-net.ru.

Rutberg, Ph. G. 2002. Some plasma environmental technologies developed in Russia. *Plasma Sources Science and Technology* 11:159–65.

———. 2003. Plasma pyrolysis of toxic waste. *Plasma Physics and Controlled Fusion* 45:957–69.

Rutberg, Ph. G., A. N. Bratsev, A. A. Safronov, V. E. Popov, B. M. Laskin, and M. Caplan. 2001. Installation on plasmachemical disinfection of hazardous medical waste. In *Progress in plasma processing of materials,* 751–60. Begell House.

Rutberg, Ph. G., A. A. Safronov, A. N. Bratsev, and B. M. Laskin. 2000. Treatment of Cl-F organic toxic compounds. *Proceedings of the First IAEA Technical Committee Meeting on Applications of Fusion Energy Research to Science and Technology, October 30–November 3, 2000, Chendu, P.R. China*, 22–27.

———. 2002. Scientific-engineering foundation of plasma-chemical technological treatment of toxic agents (TA) and industrial super-toxic agents. In *Environmental Aspects of Converting CW Facilities to Peaceful Purposes*, ed. R. R. McGuire and J. C. Compton, 211–22. The Netherlands.

Rutberg, Ph. G., A. A. Safronov, A. N. Bratsev, V. E. Popov, S. D. Popov, A. V. Surov, and M. Caplan. 2001. Plasma furnace for treatment of solid toxic wastes. In *Progress in plasma processing of materials*, 745–50. Begell House.

Rutberg, Ph. G., A. A. Safronov, A. N. Bratsev, and A. A. Ufimtsev. 1999. Plasma installations of the destruction of high toxic medical waste. Paper presented at the First International Symposium on Nonthermal Medical/ Biological Treatments Using Electromagnetic Fields and Ionized Gases, 12–14 April 1999, Norfolk, Virginia.

ScanArc Plasma Technologies. 2004. URL http://www.scanarc.se.

Startech Environmental. 2004. URL http://www.startech.net.

Thermoselect Südwest. 2004. http://www.thermoselect.de.

USEPA (U.S. Environmental Protection Agency). 2004. http://www.epa.gov/epaoswer/non-hw/muncpl/facts.htm.

Vanguard Research. 2004. http://www.vriffx.com.

Villevold, R. S. 1982. Design and maintenance of incineration plants. *Stroyizdat* (1982): 5–20.

Von Roll. 2004. http://www.vonrollinc.com.

Chapter 5
Advanced Wastewater Treatment Technologies

Lidia Szpyrkowicz, Neti N. Rao,
and Santosh N. Kaul

Research on the use of electrochemical methods for the treatment of industrial waste-water has demonstrated that many pollutants, organic and inorganic, can be effectively removed by anodic oxidation or cathodic reduction or in a bulk solution by electro-generated media. Electrochemical techniques are more reliable than many conventional processes and are gaining popularity for the treatment of industrial wastewater. They have proven efficient in destroying a variety of pollutants, including ammonia (Marinec and Lectz 1978), nitrites (Lin and Wu 1998), benzoquinone (Feng et al. 1995), benzene (Fleszar and Ploszynska 1985), dyes (Szpyrkowicz, Juzzolino, et al. 2000; Szpyrkowicz, Juzzolino, and Kaul 2001), thiourea dioxide (Szpyrkowicz, Juzzolino, et al. 2001), phenols (Comninellis 1992; Comninellis and Nerini 1995), chlorophenols (Polcaro and Palmas 1997), formaldehyde (Do and Yeh 1995), cyanides (Szpyrkowicz et al. 1998; Szpyrkowicz, Kaul, et al. 2000), toluene (Bejan et al. 1999), alcohols (Kowal, Port, and Nichols 1997; Burstein et al. 1997; Do and Yeh 1996), and hydrocarbons (Otsuka and Yamanaka 1998). A possibility of water reclamation by electrochemical means also is currently being explored for controlled ecological environments as lunar and Martian habitats (Tennakoon 1996).

Electrochemical techniques, when applied for wastewater treatment, are generally being operated at high cell potentials; thus, the anodic process occurs in the potential region of water discharge. In the anodic discharge of water, adsorbed hydroxyl radicals (•OH) can be generated which are strong oxidizing agents of the majority of the organic pollutants (Simond, Schaller, and Comninellis 1997). If chloride is present, an indirect oxidation via formation of active chlorine is usually the predominant mechanism (Do and Yeh 1995; Czarnetzki and Janssen 1992; Chiang, Chang, and Wen 1995).

Electrochemical reactors have been successfully used since the early 1960s for the recovery of pollutants from metal plating factories, for silver recovery from photographic baths, and in other processes of metal recovery (Rajeshwar, Ibanez, and Swain 1994). Among various applications of electrochemical processes, those in which oxidation of pollutants can be achieved is particularly important, as it can offer an alternative to traditional chemical oxidation. Thus electrochemical processes have been considered recently for other environmental applications, ideally requiring total combustion of organic pollutants to CO_2 and H_2O or modification of their structure, resulting in improved biodegradability (Comninellis 1992; Simond, Schaller, and Comninellis 1997; Foti et al. 1999). Application of electrochemical reactors has proven feasible for the purification of wastewater from the tanning industry (Naumczyk et al. 1996), the metal plating industry (Ho, Wang, and Wan 1989), the textile industry (Szpyrkowicz, Juzzolino, and Kaul 2001), dye production (Abdo, Rasheed, and Al-Ameeri 1987), and other sources (Murphy et al. 1992).

It turns out that the composition of the wastewater, and in particular the type of pretreatment it has undergone before electro-oxidation, can influence strongly the performance of electrochemical oxidation. For example, studies conducted at the University of Venice (Szpyrkowicz, Naumczyk, and Zilio Grandi 1994, 1995) with graphite, Ti/Pt, Ti/Pt-Ir, Ti/RuO$_x$, Ti/PbO$_2$, and other anode materials proved that purifying tannery wastewater by an electrochemical process can be efficient, particularly if the process is applied for the pretreated effluents. The rate of removal of a single polluting species also proved to be a function of the electrocatalytic properties of the anode material.

Equally interesting environmental applications can be offered by photocatalysis. In photocatalysis, acceleration of a photoreaction, caused by the presence of a catalyst, occurs. The catalyst may accelerate the photoreaction by interaction with the substrate in its ground state or excited state or with a primary photoproduct, depending upon the mechanism of photocatalysis. In photocatalysis, no energy is stored; there is merely an acceleration of a slow event by a photon-assisted process. There are many types of well-characterized solid-state materials that belong to the metal oxide, chalcogenide, and perovskites families that act as photocatalysts.

The study of photocatalysis began in the 1970s. Titanium dioxide (TiO_2), a simple oxide material, has emerged as an excellent photocatalyst material for environmental purification. Many scientific studies have proven titanium dioxide to be a chemically stable, UV-responding material of great potential to decontaminate water, air, and solids. The photocatalytic process is employed to break down and destroy many types of organic pollutants. It has been used to purify drinking water and air streams, destroy bacteria and viruses, remove metals from waste streams, and break down organics into simple components of water and carbon dioxide.

1   Electrochemical Processes

1.1   The Main Types of Electrochemical Processes Applicable for Wastewater Treatment

1.1.1   Electro-oxidation

The type of the oxidative process involved in the removal of a pollutant by electrochemical processes is a function of the material of the anode, and application of different anodic materials with different electrocatalytic properties can affect reactor treatment efficiencies. Two different processes can occur at the electrode as a function of the anodic material (Simond, Schaller, and Comninellis 1997; Foti et al. 1999). On anodes having a high electrocatalytic activity, such as platinum, direct oxidation can take place. Other materials can be involved in the production of redox mediators that, in the bulk solution, can undergo a reaction with a pollutant, setting forth an indirect electrochemical oxidation. It has been shown that on dimensionally stable anodes (DSA), oxidation of organics can occur only under conditions of simultaneous oxygen evolution, which can be accomplished either by electrochemical oxidation of physically adsorbed hydroxyl radicals or by electrochemical oxidation of the surface itself, followed by the release of oxygen in a chemical decomposition step.

Anodes on which electrocatalytic mineralization is favored are those at which the formation of hydroxyl radicals can be fast, thus having many active sites and low oxygen vacancies in the oxides. Anodes composed of metallic oxides at the highest possible oxidation level and/or which contain an excess of oxygen (a condition that can be obtained, for example, by doping certain materials with metallic oxides) can be considered suitable to conduct mineralization, and thus may be suitable for application during industrial wastewater treatment for the complete destruction of organic substances. However, if only a modification of the structure of the organics is needed, for example, to improve their biodegradability or change the properties, noble metal anodes can be used.

For electrochemical processes occurring via a direct reaction of the pollutant at the electrode surface, the term *direct electrolysis* is generally used. *Indirect electrolysis* refers to those processes in which a redox couple, whose oxidized form is being continuously renewed by an anodic heterogeneous reaction, is involved in a homogeneous phase reaction with the pollutant. Indirect electro-oxidation of pollutants can be conducted when chloride, ferric, or silver ions are present, the first being the most important from the point of view of practical applications, as chlorides are a common constituent of several industrial wastewaters.

It should be noted that the performance of reactors that operate by applying direct or indirect electrochemical oxidation (or reduction) is influenced by mass transfer in different ways.

Direct Electro-oxidation

Figure 5.1 explains the concept of a direct electrochemical reaction applied for the pollutant abatement, using the anodic process as an example. As can be seen, oxidation occurs directly on an electrode by the application of appropriate potential, without the involvement of other substances. The interplay of electron transfer and mass transfer is fundamental: mass transport and the electron transfer are linked in series. At low overpotentials, transport rates are normally fast, compared with the electrochemical reaction kinetics, and electron transfer is the rate-determining step (cf. Pletcher and Walsh 1993, fig. 2). Current density (reaction rate) increases exponentially with overpotential, which leads to the reactant being progressively depleted at the interface and a concentration gradient in the boundary layer is developed. As an overpotential increases, the rates of electron transfer and mass transport are comparable and both processes control the rate of the overall reaction (the region of the mixed control). As the overpotential increases further, the surface concentration of electroactive species tends to zero and the current density becomes independent of the poten-

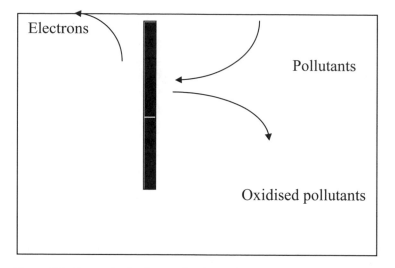

Figure 5.1 Removal of pollutants by direct electro-oxidation

tial; the reaction rate is wholly determined by the mass transport rate. An increase of the mass transfer rate by stirring or turbulence promoters results in the increase of the removal of the contaminants.

Electrochemical reactors designed to process single substances are commonly operated near the limiting current density j_L in order to maximize the reaction rate, while minimizing energy losses to parasitic reactions such as solvent decomposition.[1] The exact real wastewater composition is seldom known, so the value of limiting current density with respect to the direct oxidation of pollutant cannot be defined and the process is operated under some arbitrarily chosen current density.

Under the conditions of the complete mass transport control, assuming that migration of the active species in the electric field is negligible (an assumption valid when

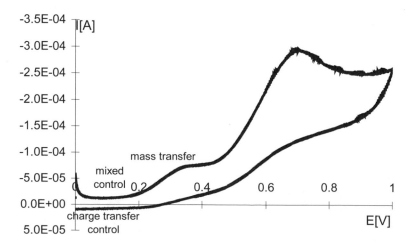

Figure 5.2 An example of the current–potential curve for the positive region, evidencing the region of the charge transfer control, mixed control, and mass transfer control, respectively, for the potential ranges (roughly) 0–0.25, 0.25–0.35, and 0.35–0.45 V

the concentration of the base electrolyte is high), the limiting current I_L expresses the duty of the reactor:

$$I_L = k_m AnFC^\infty$$

where:

> I_L = limiting current (A)
>
> k_m = mass transport coefficient (m/sec)
>
> A = electrode surface area (m²)
>
> n = number of participating electrons
>
> F = Faraday constant: 96,485 C/eq
>
> C^∞ = bulk concentration of the reacting species (mol/m³)

Interesting considerations can be made based on cyclic voltammetric experiments. As an example, figure 5.3 shows the results of investigations performed at the Ti/Pt anode for thiourea dioxide (TUD) with sodium chloride as a supporting electrolyte (Szpyrkowicz, Juzzolino, et al. 2001). The curve with the full line corresponds to the voltammogram obtained in the solution with the base electrolyte alone. During the forward scan, a wave of around 1.5 V due to chlorine evolution can be seen. On addition of TUD, another wave is observed in the forward scan around 0.7 V. The height of this peak is proportional to the TUD content. This suggests that the direct oxidation of TUD occurs at the Ti/Pt anode. Confirmation of this fact was obtained by performing cyclic voltammetric measurements in sodium sulfate electrolyte (not here presented), during which the anodic wave at around 0.7 V was again observed.

If the potentials applied during exhaustive electrolysis are higher than those for direct electro-oxidation and allow generation of redox mediators, several parallel reactions contribute to the removal of the contaminant. This was the case, for example, of electro-oxidation of TUD, described in Szpyrkowicz, Juzzolino, et al. 2001, when the anode potential oscillated between 2 and 3 V and, as the solution contained chlorides, an indirect electrochemical oxidation of TUD also occurred. This occurred due to the oxidation of chloride and water, reactions in which active chlorine and oxygen, respectively, are produced, giving surface mediators (e.g., $Cl_{ads}\bullet$ or $OH_{ads}\bullet$); the mediators were continuously regenerated as soon as they were consumed by the TUD.

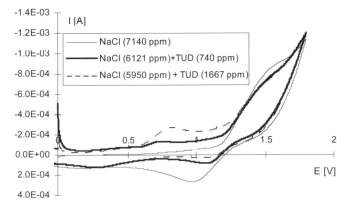

Figure 5.3 Cyclic voltammograms of the NaCl and the
NaCl + TUD solution; Ti/Pt anode

Cyclic voltammograms performed on an aqueous sulfite solution, recorded with a Ti/Pt anode, are shown in figure 5.4 (Szpyrkowicz, Juzzolino, et al. 2001). In this case, an anodic wave at 0.5–0.6 V appears when Na_2SO_3 is added to a base electrolyte. This wave increases as the concentration of sulfites increases, which indicates that the wave belongs to sulfite oxidation. Since the sulfite oxidation wave occurs at less-positive potentials than the TUD wave, it follows that, at the oxidation potential of TUD, SO_3^{2-} is concomitantly oxidized when both pollutants are present in the solution.

In the tests on electro-oxidation of sulfites in a reactor equipped with a Ti/Pt material as an anode, agitation had a noticeable influence on the rate of the process. Experiments of electro-oxidation of sulfites performed in a reactor equipped with this anode at the hydrodynamic conditions characterized by Re = 250 resulted in a lower rate of electrochemical reaction, by comparison to the tests at Re = 5,000. The mass transfer coefficient was in this case equal to 25.2×10^{-6} m/sec. An aspect of the influence of the mass transfer on direct electrode reactions is further discussed in section 1.4.1.

Indirect Electro-oxidation

In the process of indirect electro-oxidation of pollutants, an electrochemically generated redox reagent is used as a chemical reagent (or catalyst) to convert pollutants to less-harmful products or even to CO_2 and H_2O. The redox reagent acts as an intermediator to transport electrons between the pollutant and the electrode. The general scheme of this reaction is:

$$C \rightarrow C^+ + e^-$$
$$C^+ + R \rightarrow O + C$$

where C is the mediator, R is the pollutant in the reduced form, and O is the pollutant in the oxidized form. Figure 5.5 depicts schematically the electrode and bulk reactions involved.

The following conditions should be assured for the indirect electro-oxidation to proceed with a high efficiency:

- The rate of production of the mediator should be high.
- The rate of the reaction of the mediator with a given pollutant should be higher than the rate with other compounds eventually present.

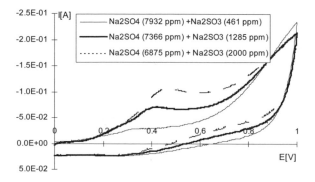

Figure 5.4 Cyclic voltammetric study of a sulfite solution in a sulfate medium

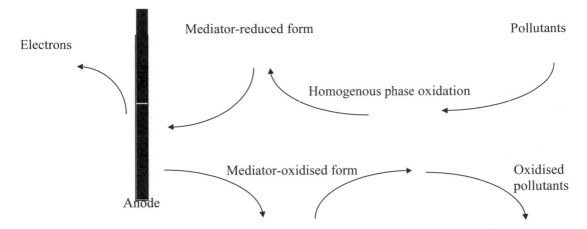

Figure 5.5 Removal of pollutants by indirect electro-oxidation with an anodic regeneration of the mediator

- The potential at which electrogeneration of the mediator occurs should be distant (lower) from the potential of solvent electrolysis, to prevent loss of energy for an unproductive process.
- The electrode should not be prone to adsorption of the pollutant, to avoid occupation of the active sites otherwise free for the reaction of the generation of the active form of the mediator.

In fact, indirect electrochemical processes offer a promising alternative solution for destruction of pollutants in industrial wastewaters. They can be particularly attractive if conducted in undivided electrochemical reactors, which can operate at lower cell voltages and are simpler and cheaper to engineer than their counterparts incorporating ion-permeable membranes, separating anolyte and catholyte. Especially for treatment of industrial wastewaters with numerous pollutants present at low concentrations, indirect electrochemical oxidation, mediated in bulk solution by a redox couple with electrochemical regeneration of its oxidized form, is generally preferred to direct electro-oxidation. The most often applied mediators are listed in table 5.1.

Among the redox couples indicated in table 5.1, the chlorine species are of particular importance, due to the presence of chloride in several types of industrial wastewa-

Table 5.1 Redox mediators that can be used for indirect electro-oxidation of pollutants and their standard potentials

Mediator couple	Standard Reduction Potential (vs SHE)
Cl (0/-I)	1.36
F (0/I)	2.87
$S_2O_8^{2-}/SO_4^{2-}$	1.96
Co (III/II)	1.92
Ce (IV/III)	1.72
Ag (II/I)	1.98
Fe (III/II)	0.77

Source: Rajeshwar, Ibanez, and Swain 1994

ter and the lack of need for its recovery from the purified effluents. This latter advantage makes this mediator particularly competitive with silver and cobalt and offers a possibility of conducting the process at low running costs.

Indirect electro-oxidation of pollutants in wastewater by electrogenerated chlorine has been used widely (Szpyrkowicz, Naumczyk, and Zilio Grandi 1994, 1995; Do and Yeh 1995; Comninellis and Nerini 1995; Bonfanti et al. 2000; Rao et al. 2001), because of the low cost of chloride, the relatively high solubility of chlorine (approximately 0.1 M; see Boxall and Kelsall 1992), and its strongly oxidizing properties.

The influence of hydrodynamic conditions in a mediated electro-oxidation of pollutants is much more complex than during direct electro-oxidation. The dependence of the reaction rates on stirring is not straightforward and cannot be obtained from the first principles. Studies on the influence of stirring during chlorine mediated electro-oxidation of dyes have shown that the magnitude of the effect of agitation in an isothermal reactor is comparable with the temperature effects. The pseudo-first-order rate constant observed for electro-oxidation of the Procion Red Rective Dye HEXGL, normalized by the specific surface area of the anode (A / V_R), is shown in figure 5.6 as a function of agitation (internal recycle).

At this point, it is interesting to compare the influence of stirring on the abatement, by chlorine-mediated electrolysis, of pollutants present in a mixture in real wastewater. The time dependence of several pollutants such as sulfides, along with their chemical oxygen demand (COD), TOC, TKN, and ammonium concentration, are well described by pseudo-first-order kinetics, their depletion rate being strongly dependent on the current density and significantly influenced by the hydrodynamic conditions in the reactor (Szpyrkowicz, Kelsall, et al. 2001).

The apparent pseudo-first-order kinetic constant k_{obs} for the removal of sulfides (see figure 5.7) indicates that the rate of sulfide depletion depends both on current density and on agitation rate, for the synthetic and real wastewaters, which behaved similarly. The positive influence of stirring can be explained by a contribution of direct anodic oxidation of sulfides, proved to be possible at a Ti/Pt-Ir anode material by Kelsall and Thompson (1993) and Rajalo and Petrovskaya (1996).

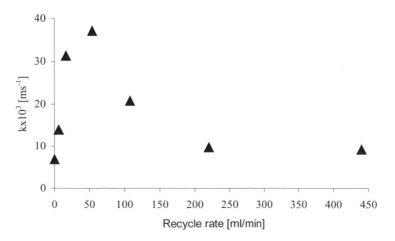

Figure 5.6 Variation of the kinetic rate constant (normalized by the A/V$_R$ ratio) in function of the internal recycle rate in the reactor

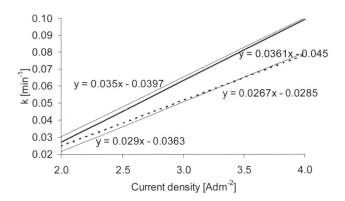

Figure 5.7 *Influence of stirring on the apparent rate constant for sulfide depletion in real and synthetic wastewater. Solid lines: conditions with mixing; dashed lines: without mixing. Bold lines indicate real wastewater.*

In contrast to sulfide, the depletion rate of ammonium ions was decreased by an increase in agitation rate, as shown in figure 5.8. This may have been due to slower homogeneous kinetics for their reaction with chloric (I). In such a case, stirring enhances the electrodes' loss reactions, and thus its effect on the overall observed rate is negative. This effect was particularly evident when the reactor was treating real wastewater at higher current densities.

1.1.2 Electro-reduction

Electro-reduction is applied for the electrochemical removal of metal ions from industrial streams. In this process, metal ions are electrodeposited onto a suitable cathode. Large electrode area–cell volume ratios and high effective mass transfer conditions are needed. Therefore, the use of wire gauzes, foils, packed beds, fluidized beds as electrodes, and rotating electrodes under vigorous stirring is common. High cell resistances mean wasted power and undesirable heat generation in the cell. Finally, the use of separators that do not allow intermixing of anolyte and catholyte during the

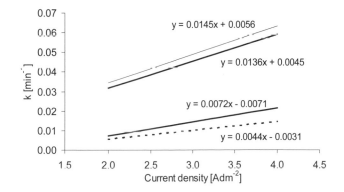

Figure 5.8 *Influence of stirring on the apparent rate constant for ammonium ion depletion in real and synthetic wastewater. Solid lines: conditions without mixing; dashed lines: with mixing. Bold lines indicate real wastewater.*

electrolysis procedure, and yet do not contribute appreciably to the cell resistance, is of paramount importance.

Consider the deposition of a solid product (e.g., copper) from its solution precursor, O (e.g., Cu^{2+}) (Rajeshwar, Ibanez, and Swain 1994). This can be represented as:

$$O + ne^- \leftrightarrow R \text{ (solid)}$$

When several monolayers of R are deposited on an inert electrode or an electrode made of R (Cu), the activity of R, a_R, is constant and equal to 1 at the completion of electrolysis. Hence, the Nernst equation takes the form

$$E = E^o + RT / nF\{\ln(\gamma_o C_o(1 - x))\},$$

where C_o is the initial concentration of O, and x is the fraction of O reduced to R.

For less than one-monolayer coverage, $a_R \sim 1$, and it can be assumed that a_R is proportional to the fractional coverage of the deposit on the electrode surface, θ:

$$a_{R = \gamma R\theta = \gamma R} A_R / A = (\gamma_R n_R A_a) / A,$$

where A_R is the area occupied by R, A_a is the cross-sectional area of a molecule of R (cm²), and n_R is the number of molecules of the deposit. At equilibrium,

$$n_R = V_s C_o N_A,$$

where V_s is the solution volume and N_A is the Avogadro number. Therefore,

$$E = E^o + (RT/nF)\ln\{\gamma_o A / (\gamma_R V_s N_A A) + (RT/nF)\} ((1 - x)/x)$$

For environmental applications, a variety of electrode material choices exist, including reticulated vitreous carbon (RVC), carbon foams, and various forms of graphite.

1.1.3   Electro-coagulation

Numerous pollutants, disperse dyes being one example, are present in wastewater in a colloidal form, which precludes their removal by simple sedimentation. This derives from the fact that colloidal particles are electrically charged and a repulsion force (zeta potential) contrasts natural attraction forces, which exist between any two masses. Dosing of electrolytes into such a system causes a compression of the diffusive layer, allowing the particles to be brought together, considering that at a short distance Van der Waals forces prevail over repulsive electrostatic forces.

Removal of colloids can be obtained by coagulation/flocculation, which is a two-step process. The first step (coagulation) comprises an initial destabilization of the surface charge, accomplished by adding coagulants. This leads to the adsorption of ions of different charge on the surface of the particles, reducing the net surface charge and consequently the repulsive force. The second step (flocculation) consists of formation of bigger, easily sedimentable or floatable agglomerates. The coagulating compounds generally used for wastewater treatment are inorganic salts, particularly those of aluminum and iron. The kinetics of coagulation is a function of the rate of hydrolysis of a coagulant, the rate of destabilization of colloids, and the rate of transport of ions and particles in the system. The rate of flocculation depends on the nature of the coagulating species and an eventual presence of flocculating agents (generally chain-structured molecules, acting as bridges between coagulated colloids).

The first step of coagulation, the hydrolysis of coagulants, is a multiphase process, which leads to a production of several monomers and polymers. In the case of aluminum, five monomers can be formed: Al^{3+}, $Al(OH)_2{}^+$, $Al(OH)^{2+}$, $Al(OH)_3$, and $Al(OH)_4{}^-$; there are three possible polymers: $Al_2(OH)_2{}^{4+}$, $Al_3(OH)_4{}^{5+}$, and $Al_{13}(OH)_{24}{}^{7+}$. If aluminum is added in the form of a salt, its principal hydrolyzed forms are monomers. During coagulation with metals in the form of a salt, their hydrolysis causes consumption of alkalinity, leading to lowering of the pH and also to an eventual production of free CO_2, which interferes with sedimentation. To obviate these problems, prehydrolyzed compounds such as poly-aluminum chloride (PAC) can be used (van Benschoten and Edzwald 1990; Jain et al. 2001). Other advantages of PAC are its flocculating effect on the particles and a wider range of optimal pH levels suitable for good sedimentation of the sludge (Lu and Tang 1999).

An alternative method of the removal of colloids is electro-coagulation. Electro-coagulation is a process during which destabilization of colloids occurs both as a consequence of the electrical field generated between the electrodes (polarization coagulation) and due to the action of coagulating compounds, produced in situ by oxidation of the anode. It has been observed that while using aluminum as a sacrificial electrode, the products of the hydrolysis of the aluminum contain both monomers and polymeric compounds. Their characteristics are somehow comprised between the products of hydrolysis of inorganic metal salts and prehydrolyzed compounds (Lu and Tang 1999).

Coagulating compounds destabilize the colloidal system, forming flocks of particles, which can be separated easily by flotation due to the formation of gas bubbles during the electrolysis. In fact, during the electrolysis a constant flux of gas bubbles having dimensions as small as 20–100 μm is produced (Fukui and Yuu 1984; Gawronski 1996). The diameter of aggregates of the particles increases with an increase in the rate of agitation, reaching a plateau at a certain value of the rotation speed of the impeller that can be considered optimal (Fukui and Yuu 1984).

The main reactions occurring at the electrodes during electro-coagulation using aluminum as a sacrificial anode are:

anode: $Al \rightarrow Al^{3+} + 3e^-$

cathode: $2H_2O + 2e^- \rightarrow H_2 + 2OH^-$

In case chlorides are present and with anode potentials exceeding those for oxygen and chlorine evolution, these reactions also occur to some extent. Thus, for example, for the alkaline bulk pH and with chloride ions present in solution, other anodic reactions can be:

$$4OH^- \rightarrow O_2 + 2H_2O + 4e^-$$
$$2Cl^- \rightarrow Cl_2 + 2e^-$$

The quantity of aluminum produced during the first phase of coagulation can be calculated according to:

$$\Delta(Al) = \eta I \Delta t M_{Al}/nFV \qquad (1)$$

where:

η = anodic current efficiency

I = applied current (A)

M_{Al} = molecular weight of aluminum (g/mol)

Δt = time of electrolysis (sec)

F = Faraday constant: 96,485C/eq

V = electrolyzed volume (dm³)

n = number of exchanged electrons

Figure 5.9 reports the total quantity of aluminum determined in the flocculated sludge and in the solution, sampled at different time intervals during electro-coagulation of dyes containing wastewater (Szpyrkowicz 2002). For comparison, the quantity calculated using equation (1), assuming 100 percent current efficiency, is also given in figure 5.9. As can be seen in the figure, the experimental value exceeded the theoretical one, indicating an efficiency of the anodic process higher than 100 percent. Even as high an efficiency as 190 percent was obtained after 6.5 minutes of electrolysis (an apparent efficiency higher than 100 percent was obtained also by Nameri and colleagues [1998] in the tests of electro-coagulation of fluoride containing water). This difference in mass can be explained by the contribution of corrosion pitting to aluminum dissolution.

1.1.4   Fenton and Electro-Fenton Methods

Fenton's reagent–based oxidation constitutes oxidation of organic compounds using ferrous iron (Fe^{2+}) and hydrogen peroxide (H_2O_2). The high removal efficiencies of this method (see, e.g., Kuo 1992) can be explained by the fact that oxidation reactions are coupled to coagulation occurring due to the presence of ferrous/ferric cations; thus, these metallic ions play a double role as catalyst and coagulant in the process.

A process of coagulation that occurs simultaneously with oxidation involves the formation of hydroxy-complexes of iron:

$$(Fe(H_2O)_6)^{3+} + H_2O \leftrightarrow (Fe(H_2O)_5OH)^{2+} + H_3O^+$$

$$(Fe(H_2O)_5OH)^{2+} + H_2O \leftrightarrow (Fe(H_2O)_4(OH)_2)^+ + H_3O^+$$

The products of the above reactions polymerize when the pH of the solution is kept between 3.5 and 7, following the reactions:

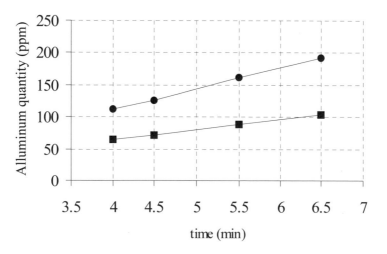

Figure 5.9 Theoretically produced (squares) and experimentally determined (dots) quantity of aluminum released during electro-coagulation

$$2(Fe(H_2O)_5OH)^{2+} \leftrightarrow (Fe_2(H_2O)_8(OH)_2)^{4+} + 2H_2O$$

$$(Fe_2(H_2O)_8(OH)_2)^{4+} + H_2O \leftrightarrow (Fe_2(H_2O)_7(OH)_3)^{3+} + H_3O^+$$

$$(Fe_2(H_2O)_7(OH)_3)^{3+} + (Fe(H_2O)_5OH)^{2+} \leftrightarrow (Fe_3(H_2O)_5(OH)_4)^{5+} + 2H_2O$$

The performance of the Fenton process depends heavily on factors such as pH of the solution, the quantity of hydrogen peroxide added, and the dose of $FeSO_4$. In the chemical Fenton process, the hydrogen peroxide reacting with ferrous ions forms a strong oxidizing agent (hydroxyl radical), whose oxidation potential is higher than that of ozone (2.8 V against 2.07 V). The main reactions that occur in the solution during the Fenton process are:

$$H_2O_2 + Fe^{2+} \rightarrow Fe^{3+} + OH^- + HO\bullet \qquad (2)$$

$$HO\bullet + RH \rightarrow H_2O + R\bullet$$

$$R\bullet + Fe^{3+} \rightarrow R^+ + Fe^{2+}$$

$$R^+ + H_2O \rightarrow ROH + H^+$$

where R indicates a generic oxidizable substance. Other reactions can also occur:

$$HO\bullet + H_2O_2 \rightarrow H_2O + HO_2\bullet$$

$$HO_2\bullet + Fe^{3+} \rightarrow Fe^{2+} + O_2 + H^+$$

$$HO\bullet + Fe^{2+} \rightarrow Fe^{3+} + OH^-$$

$$HO\bullet + HO\bullet \rightarrow H_2O_2$$

Anodic Fenton's Reagent

It is possible to produce hydroxide peroxide at the anode, in order to realize an electro-chemical "anodic" Fenton. At the anode, water can be directly oxidized to H_2O_2 (electro-chemical potential E = 1.77 V) as follows (Merli at al. 2003):

$$H_2O \rightarrow {}^1/_2 H_2O_2 + H^+ + e^-$$

The presence of Fe^{2+} ions, in combination with H_2O_2, may lead to bulk generation of HO• radicals, according to the Fenton reaction (equation [2] above), which is quite effective at the low pH values (about 2) reached at the anodic compartment of the cell. This reaction can be achieved together with direct substrate oxidation at the anode.

Nevertheless, this process presents some disadvantages typical of other anodic processes, such as expensive materials and rapid deactivation. Above all, with anodic generation of Fenton's reagent, electro-regeneration of ferrous ions is not possible.

Cathodic Fenton

Chemical and anodic Fenton systems present two main limitations:

1. A continuous supply of H_2O_2 is required (in the first case).
2. Regeneration of Fe^{2+} via reaction ($R + Fe^{3+} \rightarrow Fe^{2+}$ + product) is less efficient than with electrochemical reduction of Fe^{3+} (in the second case).

A Cathodic Fenton process (Alvarez-Gallegos and Pletcher 1998; Oturan et al. 2000; Oturan et al. 2001; Panizza and Cerisola 2001; Qiang, Chang, and Huang 2002), on the other hand, can electrochemically generate H_2O_2 by reduction of dissolved oxygen in acidic solution at selected cathodes, following the reaction:

$$O_2 + 2H_3O^+ + 2e^- \rightarrow H_2O_2 + 2H_2O$$

Oxygen is constantly sparged in the solution and it progressively adsorbs on the electrode. The reduction yield depends on the medium and on oxygen solubility: being apolar, O_2 is more soluble in organic solvents than in water; at 25°C, 1 atm, its saturation concentration is 1mM. The limiting step of this reaction is the transfer of the first electron, which generates $O_2^{\bullet-}$, a strongly alkalinic radical (following Broensted definition), by the process:

$$O_2 + e^- \rightarrow O_2^{\bullet-} \qquad E_o = -0.33 \text{ V vs NHE}$$

If electrons are supplied, it is possible to generate (electrochemically) all the other reduced forms of oxygen (HO_2^-, H_2O_2, $HO^\bullet$, $HO_2^\bullet$, H_2O, HO^-) through different reactions. In aqueous solution $O_2^{\bullet-}$ is rapidly converted to O_2 and HO_2^-:

$$2\,O_2^{\bullet-} + H_2O \rightarrow O_2 + HO_2^- + HO^- \qquad E_o = 0.2 \text{ V vs NHE}$$

On an inert electrode and with a proton excess, a more rapid chemical reaction can occur:

$$O_2^{\bullet-} + H_3O^+ \sim HO_2^\bullet + H_2O$$

The hydroperoxide radical is a weak acid ($pK_a = 4.89$), which can give homolythic disproportion:

$$HO_2^\bullet + HO_2^\bullet \rightarrow H_2O_2 + O_2$$
$$k = 8.6 \times 10^5 \text{ M}^{-1}\text{S}^{-1} \qquad K_{(25°C)} = 10^{25} \text{ aim} \qquad E_o = 1.44 \text{ V vs NHE}$$

Hydroperoxide radicals may also react with $O_2^{\bullet-}$ (at a pH lower than 5):

$$HOO^\bullet + O_2^{\bullet-} \rightarrow HOOH + O_2 + H_2O \qquad E_o = 1.44 \text{ V vs NHE (a pH 0)}$$

In the Cathodic Fenton system, the reduction of Fe^{3+} to Fe^{2+} also occurs at the cathode:

$$Fe^{3+} + e^- \rightarrow Fe^{2+} \qquad E_o = +0.771 \text{ V vs SHE}$$

The H_2O_2 generated at the cathode can be coupled with Fe^{2+} in order to produce the Fenton reagent. The cathodic regeneration of Fe^{2+} in acidic solution allows this reagent to be provided continuously to the system, thus preventing the rapid decrease in iron ions that is observed in the chemical Fenton method.

Though a few researchers have studied H_2O_2 electro-generation in acidic solutions, there are contrasting results. Sudoh and colleagues (1986) investigated the decomposition of aqueous phenol by electro-generated Fenton's reagent, obtaining the highest current efficiency (85 percent) with a cathodic potential of -0.6 V vs. a saturated Ag/AgCl electrode (SSE) and pH 3. Tzedakis, Savall, and Clifton (1989) reported that the electrolysis of an oxygen-saturated H_2SO_4 solution (0.6 M) using a stirred mercury pool electrode yielded a current efficiency of 55 percent at a cathodic potential of -0.30 V vs. SCE. Chou and colleagues (1999) used the electro-Fenton process to remove aqueous chlorophenols at a cathodic potential of -0.6 V vs. SCE and reported that the current efficiency increased with decreasing pH. Hsiao and Nobe (1993) studied the oxidative hydroxylation of phenol and chlorobenzene using electro-generated Fenton's reagent, thus attaining an optimal cathodic potential of -0.55 V vs. SCE and a H_2O_2 generation favored at low pH values.

1.2   The Mechanism of Electro-oxidation Mediated by Chlorine and Other Species

When chlorides are present in the wastewater, an indirect anodic oxidation occurs, with strong oxidants such as free chlorine generated on the anode surface during electrolysis. The pollutants are then destroyed by chlorine in the bulk solution. Since anodic oxidation of chloride is readily achieved, wastewater that can be effectively purified by electrolysis is that which contains a high concentration of this ion and pollutants that are readily oxidized by chlorine.

The real wastewater subjected to electrolysis is generally a multicomponent cocktail of various compounds, of which only few can be determined quantitatively. For this reason, TOC and COD, an aspecific parameter that reflects the presence of organic and inorganic substances that can be oxidized by $K_2Cr_2O_7$ under well-defined conditions, are often used as figures of merit. Most of the organic substances present in the wastewaters contribute to the detected COD value, together with sulfides or other inorganic compounds present in the reduced form.

During treatment of polluted effluents, a reactor equipped with anodes that have electrolytic properties toward Cl_2 evolution is generally operated under galvanostatic conditions, at which anode potentials exceed the reversible potentials for reactions (3) and (4).

$$2H_2O \leftrightarrow O_2 + 4H^+ + 4e^- \qquad E_o = 1.23 \text{ V} \qquad (3)$$

$$2Cl^- \leftrightarrow Cl_2(el) + 2e^- \qquad E_o = 1.36 \text{ V} \qquad (4)$$

As a consequence, both reactions can occur, along with the direct anodic oxidation of some pollutants (R) to oxidized species (O):

$$R \rightarrow O + ne^-$$

These anodic reactions occur simultaneously with the following primary cathode reaction:

$$2H_2O + 2e^- \rightarrow 2OH^- + H_2 \qquad (5)$$

The reaction mechanism of chlorine oxidation depends on the type of anode material. For example, on nonpassive platinum and iridium, it can be represented by (Trasatti 1987):

$$Cl^- \leftrightarrow Cl_{(ad)} + e^-$$

$$2Cl_{(ad)} \rightarrow Cl_2$$

The process becomes much more complicated on the oxide film anodes, at which it is believed that Cl^- ions are discharged on oxidized surface sites following the process (Trasatti 1987):

$$(S^z) \leftrightarrow (S^{z+1}) + e^-$$

$$(S^{z+1}) + Cl^- \rightarrow (S^{z+1}) Cl_{(ad)} + e^- \qquad (6)$$

$$(S^{z+1}) Cl_{(ad)} + Cl^- \leftrightarrow (S^z) + Cl_2$$

where (S^z) is an active site of the anode.

The rate of reaction (4) may be kinetically or transport controlled, depending on the applied current density, chloride ion concentration, and prevailing hydrodynamic conditions, which may also depend on the rate of anodic gas evolution (Gijbers and Janssen 1989); hence, these parameters also control the current efficiency of reaction (4).

This anode reaction is followed by the diffusion of molecular chlorine away from the electrode (Trasatti 1987):

$$Cl_{2(el)} \rightarrow Cl_{2(sol)}$$

If its solubility is exceeded locally at the electrode, then bubbles may form. Whereas aqueous chlorine predominates at pH levels below about 3.3, at high bulk pHs its diffusion away from the electrode may be coupled to its hydrolytic disproportionation by reaction (7) to form hypochlorous acid at pHs less than about 7.5 and hypochlorite ions at higher pH levels.

$$Cl_{2(sol)} + H_2O \leftrightarrow HClO + H^+ + Cl^- \tag{7}$$
$$HClO \leftrightarrow H^+ + OCl^-$$

Chlorine, hypochlorous acid, and hypochlorite ions are collectively often referred to as "active chlorine." All three species can be reduced at the cathode of an undivided reactor by:

$$ClO^- + H_2O + 2e^- \rightarrow Cl^- + 2OH^- \tag{8}$$

Hydrogen evolution by reaction (5) generates a high local pH at the cathode, so that hypochlorite ions predominate. Loss reaction (8) limits the maximum active chlorine concentration that can be achieved in such reactors to about 0.1 M (Boxall and Kelsall 1992; Czarnecki and Janssen 1992; Hu, Lee, and Wen 1996). In addition, further loss reactions such as chlorate formation may occur in the bulk solution, but they may be assumed to have insignificant rates at ambient temperatures because the steady state hypochlorite concentration may be kept low by homogeneous reactions with the various pollutants (R):

$$R + HClO \rightarrow products + Cl^- \tag{9}$$

Chlorate may also be formed by oxidation of hypochlorous acid (Krasnobrodko 1988; Boxall and Kelsall 1992), mass transport usually controlling the rate, depending on the anode material:

$$6HClO + 3H_2O \rightarrow 2ClO_3^- + 4Cl^- + 12H^+ + 3/2O_2 + 6e^- \tag{10}$$

As reactions (3) and (10) acidify the anode diffusion–reaction layer, while reactions (5) and (8) generate alkaline conditions at the cathode, a large pH gradient exists across the anode–cathode gap, with possible kinetic consequences for oxidation of pollutants, as the reduction kinetics of active chlorine species decreases with increasing pH.

A scheme of some processes and reactions involved in the removal of pollutants in the electrochemical reactor is depicted in figure 5.10. Active chlorine is a mediator, constantly renewed by the charge transfer reaction at the anode, involving the mass transfer of the chloride ion from the bulk to the electrode, and the transport of the hypochlorite to the bulk. Various pollutants compete for chlorine, and it is reasonable to postulate that relatively fast reactions (e.g., those with sulfides and ammonium) occur preferentially near the anode surface, while slower reactions (e.g., COD oxidation) occur predominantly in the bulk.

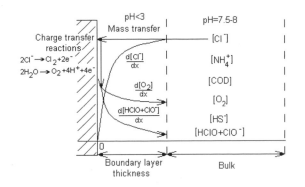

Figure 5.10 A scheme of the reactions and processes
involved (Szpyrkowicz, Kelsall, et al. 2001)

To explain the results of the removal of pollutants during electrolysis of chloride containing wastewater, it is generally hypothesized that the indirect electro-oxidation, involving various forms of chlorine, can be a predominating process, as suggested in the literature (Bonfanti et al. 2000). In principle, however, anodic production of ozone can also occur (Foller and Tobias 1982) when the anode potential exceeds the normal potential $E^\ominus$ of ozone evolution, equal to 1.51 V, as shown:

$$3H_2O \leftrightarrow O_3 + 6H^+ + 6e^- \tag{11}$$

or by oxidation of oxygen dissolved in the electrolyte:

$$H_2O + O_2 - 2e^- \rightarrow O_3 + 2H^+ \tag{12}$$

In solutions containing active chlorine compounds in the presence of a catalyst, several radical species can also be formed, including ClO•, Cl•, OH•, and H• (Krasnobrodko 1988). The source of these radical species can be HClO and ClO$^-$:

$$HClO + ClO^- \rightarrow ClO\bullet + Cl\bullet + OH\bullet \tag{13}$$

This reaction is explained by the theory of heterogeneous catalysis (Krasnobrodko 1988):

$$HClO + ClO^- + K \rightarrow K\begin{bmatrix} ClO^- \\ \\ HClO \end{bmatrix}^\bullet \tag{14}$$

where K is a catalyst. Decomposition of the above complex produces hypochloric radicals and the free catalyst:

$$\left[K\begin{matrix} ClO^- \\ \\ HClO \end{matrix}\right]^\bullet \rightarrow ClO\bullet + ClO\bullet + H^+ + K \tag{15}$$

Active ClO• radicals can take part also in the reaction of generation of atomic oxygen and OH• radicals:

$$ClO\bullet + ClO^- + OH^- \rightarrow 2Cl^- + 2O + OH\bullet \tag{16}$$

The next step in the chain of reactions is:

$$OH\bullet + OCl^- + OH^- \rightarrow ClO\bullet + OH^-\bullet \tag{17}$$

At the same time, the atomic oxygen generated by this reaction recombines to form molecular oxygen:

$$O + O \rightarrow O_2$$

The loss of $ClO\bullet$ radicals can proceed via a bimolecular reaction:

$$2ClO\bullet \rightarrow 2Cl^- + O_2$$

This last reaction leads to a slow die-out of the whole chain of the reactions begun by reaction (12).

Another oxidizing species that can be formed in the reactor is the hydrogen peroxide produced by the cathodic reduction of dissolved oxygen:

$$O_2 + H_2O + 2e^- \rightarrow HO_2^- + OH^- \tag{18}$$

The above reactions show that the scenario describing the presence of different oxidizing species can be much more complex than depicted in figure 5.10, involving chlorine and eventually ozone, chemically active atomic oxygen, relatively poorly soluble active molecular oxygen, and various radicals, including hydroxyl radicals. The $OH\bullet$ radicals are the most powerful oxidants among the species cited above ($E_o = 2.80$ V) and lead to a complete combustion of pollutants to CO_2 and H_2O. They are nonspecific oxidants that react at significant rates with a variety of organic contaminants (Westerhoff et al. 1999).

The scenario of possible reactions and formation of different oxidizing species is principally governed by the anode potential, which thus has a fundamental role in the performance of the reactor. As the anode potential during exhaustive electrolysis of polluted effluents often exceeds the reversible potentials for discharge of water molecules and chloride ions or is higher than this value, oxygen evolution can be expected to occur in parallel with generation of chlorine and other oxidizing species and radicals mentioned above. Moreover, direct anodic oxidation of certain pollutants to oxidized species cannot be excluded. When, for example, free cyanides are present in the wastewater cocktail, they can be readily oxidized (assuming the anode material to have electrocatalytic properties toward this reaction) at low anode potentials, around 0.6 V (Szpyrkowicz, Ricci, and Daniele 2003):

$$CN^- + 2OH^- \rightarrow CNO^- + H_2O + 2e^-$$

In all types of electrolyzed wastewater that contains chlorides in a high concentration, their discharge on the anode contributes significantly to maintaining the current. As demonstrated by voltammetric investigations (not presented here) performed with different anode materials in effluents rich in Cl^-, chlorine evolution was a well-evident anodic process for several tested electrode materials: Ti/Pt, $Ti/RuO_2\text{-}TiO_2$, $Ti/SnO_2\text{-}Sb_2O_5$, $Ti/Pt\text{-}Ir$, $Ti/MnO_2\text{-}RuO_2$, $Ti/RhO_x\text{-}TiO_2$, and $Ti/PdO\text{-}Co_3O_4$.

The mass transport–limited current density for oxidation of Cl^- can be estimated from:

$$j_L = nFk_m(Cl^-)$$

where:

j_L = mass transport–limited current density (A/m²)

n = number of electrons involved (n = 1 for Cl⁻)

F = Faraday constant: 96,500 C/eq⁻¹

A = surface of the anode (m²)

k_m = mass transport rate coefficient (m/sec)

(Cl⁻) = concentration of chloride ions in the solution (mol/m³)

This equation can be used to evaluate whether the reactor operates under mass or charge transfer conditions for chlorine evolution. Assuming k_m for the quasi-quiescent conditions is approximately equal to 10^{-5} m/sec and the concentration of chlorides is around 160 mol/m³, the mass transport anodic current density can be roughly estimated equal to 160 A/m². Considering that mixing due to the evolution of gas bubbles causes a rise in the mass transfer coefficient value, the total current attributed to chloride discharge may be even slightly higher.

Examining figure 5.10, under the conditions where active chlorine is the main mediator, two different reaction regions can be distinguished: the hydrodynamic boundary layer where, due to reactions (3) and (10), the pH is low; and the other region (the bulk solution), where the pH is neutral or slightly alkaline. It is likely that when the predominant mediator is the "active chlorine," removal of pollutants having instantaneous or fast-reaction kinetics with chlorine could be mediated mostly by hypochlorous acid, while other pollutants are eliminated by reaction with hypochloric ions. As already reported, the redox potentials of these mediators are different, and thus the effect due to the variation of pH in the two regions would be reflected by still lower rates of the reactions occurring in the bulk. The half-reactions with the mediators can be represented as:

$$HClO + H^+ + 2e^- \rightarrow Cl^- + H_2O$$
$$ClO^- + H_2O + 2e^- \rightarrow Cl^- + 2OH^-$$

Both these reactions lead to the "recovery" of chloride ions, which are in turn continuously oxidized at the anode to form chlorine/hypochlorite again.

If a certain pollutant can also react with some other mediators produced by reactions (9)–(18), the resulting rate of the process will depend on the relative contribution of single reactions with oxidizing agents, whose concentrations in turn will be strongly dependent on the operational conditions of the reactor.

It was observed that during the treatment of many kinds of industrial wastewaters, various pollutants are simultaneously removed from the very first moment of the process. This means that there is a competition between the single pollutant and the mediators produced by the anodic discharge of chloride and water molecules and the cathodic reduction of oxygen. Considering a wide variation of the type of compounds generally present in the wastewater, it cannot be excluded that some substances may also be oxidized directly at the anode surface, if the standard potential for oxidation of a given pollutant is situated in the potential window below chloride oxidation. Subsequently, on exceeding the chloride oxidation potential, the value of which differs slightly for different anode materials due to the variation of the overpotential for chlorine evolution, chlorine generation by an anodic oxidation of chlorides would predominate, as already stated, over other anodic reactions, owing to a high concentration of chloride ions as compared to other species present in the electrolyzed solution, until the limiting current density for Cl_2 evolution is reached.

1.3   The Kinetics of Indirect Electro-oxidation for Depletion of Pollutants

Electrode reactions (3) and (5) (in section 1.2) are charge transfer controlled, as is reaction (4) at low current densities and high chloride concentrations, but the hypochlorite loss reactions at the anode (reaction [10]) and the cathode (reaction [8]) are mass transfer controlled. Thus, increasing turbulence in the reactor promotes the mass transport of Cl^- ions to—and $Cl_2(aq)$ away from—the anode, but it also enhances the rate of hypochlorite reduction at the cathode. However, with other oxidizable species in the bulk solution, their homogeneous oxidation by electrogenerated chlorine will prevent it from accumulating in the bulk solution, so minimizing its cathodic reduction rate.

Fast homogeneous reactions will affect the concentration profiles in the anode diffusion–reaction layer, enhancing the chloride flux to the anode, whereas slow reactions will occur predominantly in the bulk solution. Hence, a mass balance on active chlorine species formed in a solution of volume V with current efficiency ϕ_e at an anode of area A_a and mass transport rate coefficient $k_{m,a}$—but consumed by further oxidation of hypochlorous acid at the anode, by reduction of hypochlorite ions at the cathode of area A_c and mass transport rate constant $k_{m,c}$, and by chemical reaction with pollutant R and second-order rate coefficient k—can be represented by:

$$v\frac{d['Cl_2']}{dt} = \frac{\phi_e j_a A_a}{nF} - A_a k_{m,a}[HOCl] - Vk['Cl_2']x[R]$$

The anodic current efficiency Φ_e depends on the anode current density j_a, $k_{m,a}$, and the chloride concentration.

Considering that a pollutant can be removed by several concomitant reactions, its depletion in a batch reactor may be described by:

$$-\frac{dC}{dt} = \sum_i k_i[C]^{m_i}[C]^{m_i}['Cl_2'] + \frac{I\phi_e(\text{pollutant})}{nFV} \tag{19}$$

where the terms in the sum describe the decrease of the concentration C of the pollutant due to the indirect electro-oxidation by a number i of chemical reactions involving active chlorine (Cl_2), and the second term indicates direct oxidation of the contaminant at the anode by a charge-transfer reaction that involves n electrons.

During indirect electro-oxidation mediated by electrogenerated chlorine, the rate of removal of the pollutants is proportional to the rate of chlorine generation. Thus, it is generally found to rise with the increase of applied current, for currents lower than the limiting value. For this reason, the electrodes at which chlorine evolution occurs and adsorption of single pollutants or intermediates of their oxidation is not observed give better performance of the reactor. This observation is consistent with the achievement of the best results obtained using the Ti/Pt-Ir anode by comparison with several other tested materials, at which some contaminants were found to react directly (Szpyrkowicz, Ricci, and Daniele 2003). Moreover, the chlorine evolution at a Ti/Pt-Ir electrode can be considered a fast reaction (Foti et al. 1999). The presence of iridium in the coating prevents oxidation of the platinum (Foti et al. 1999), the latter being a better catalyst for chloride oxidation than the oxidized surface.

It is clear that it is impossible to determine experimentally the rate constants k_i (equation [19]) and the reaction order m for i parallel reactions occurring simultaneously. While the contribution of direct anodic oxidation can be estimated for synthetic wastewater, it is indeterminable for real effluents, due to the presence of chloride. Thus, to define quantitatively the rate of electro-oxidation of a pollutant present in a wastewater cocktail, some assumptions leading to a necessary simplification of the scenario have to be made and a simplified approach is generally used.

Commonly, it is assumed that the oxidation reaction occurs mostly via an indirect electro-oxidation and that the reduced mediators "recovered" after the reaction with a pollutant are, in turn, continuously reoxidized at the anode. Assuming stationary conditions, under which there is no accumulation of mediators in the solution and the rate of their production (which is proportional to the applied current I) and rate of their consumption are equal, the concentration of mediators during electrolysis can be assumed to be constant. Equation (19) then simplifies to:

$$- \frac{d[C]}{dt} = k_{obs}[C]^m \qquad (20)$$

where k_{obs} is the observed rate constant, which reflects the contributions of several reactions, including direct and mediated electro-oxidation, and m is the pseudo-order of reaction—both of which can be measured experimentally.

Experimental results indicate that the removal of the pollutants can, in the majority of cases, be described by the pseudo-first-order reaction. The values of the kinetic rate constants as a rule increase with the current density and with the degree of pretreatment of the wastewater (as an example, see table 5.2 for kinetic rate constants for the removal of ammonium from tannery wastewater). One reason for this probably lies in the competition for the anode active sites between the compounds that can be directly oxidized on the electrode surface and those whose reactions lead to a production of bulk oxidation mediators, the former having often a more sluggish kinetics, leading to a temporary block of active sites. In fact, while the reaction rate constant, for example, of the first step of chlorine evolution (reaction [6]) is in the range of 1 sec^{-1} (Tomcsanyi

Table 5.2 The kinetic rate constants for the removal of ammonium from the wastewater of different strengths

Type of wastewater	Applied current density (A/dm²)	Pseudo-first-order rate constant ($\times 10^4$ min^{-1})
Raw wastewater (type A—heavily polluted with organic compounds)	2	1.9
	4	13.1
Biologically treated wastewater (type A)	3	4.1
	4	7.8
	5	11.6
	6	16.7
Biologically treated wastewater (type B— less polluted than type A)	3	12.0
	4	22.6
	5	40.6

Source: Naumczyk et al. 1996a

et al. 1999), the rate constant for a direct oxidation, for example, of sulfides, is 10^3 times lower (Rajalo and Petrovskaya 1996).

According to the simplified model described by equation (20) for $m = 1$, the time needed to eliminate pollutants by electro-oxidation, from their initial concentration (C_1) to the concentration defined by the standards (C_2) would be:

$$t = \frac{\ln \dfrac{[C_1]}{[C_2]}}{k_{obs}}$$

An analogy can be made between chlorine absorption in a water solution followed by a homogeneous phase reaction and the indirect electrochemical oxidation with active chlorine as a mediator (Szpyrkowicz, Cherbanski, et al. 2001). Assuming the gas-liquid interface to be approximated by the electrode-liquid interface, a difference between conventional fluid–fluid reaction and an indirect electro-oxidation is a lack of the gas film resistance. By analogy with the fluid–fluid reactions, information regarding localization of the zones of the reactor where the reactions occur may be of fundamental importance. It can be obtained by evaluating the Hatta number Ha that, for a first order reaction, for example, is given by (Levenspiel 1999):

$$Ha^2 = \frac{\max \ possible \ conversion \ in \ the \ film}{\max \ difusional \ transport \ through \ the \ film} = \frac{kCD_{Cl_2}}{k_l^2}$$

where:

k = a first-order reaction rate constant (sec^{-1})

D_{Cl_2} = diffusivity of chlorine (m²/sec)

k_l = mass transport rate coefficient (m/sec)

C = concentration of a pollutant (mol/m³)

For a Hatta number higher than 1.4, the reaction between chlorine and the pollutant occurs within the boundary film near the anode and a high anode surface will be beneficial for indirect electro-oxidation under these conditions. For Hatta numbers between 0.14 and 1.4, the reaction occurs in the film and in the bulk solution and its rate is intermediate with respect to the mass transport rate.

1.4   Design of Electrochemical Reactors

1.4.1   Direct Electro-oxidation (Reduction)

Assuming the reactor is a batch or plug-flow-type reactor, the observed change of the concentration of the electroactive species, obtained from the mass balance, is given by:

$$\frac{dC_t}{dt} = I_t \, / \, (nFV)$$

where I_t is the instantaneous current (used for a given reaction) at time t and V is the working volume of the reactor.

Under the complete mass transport control, I_t is the same as I_L, which equals $k_m An$-FC_t. Substituting, the following expression is obtained (Pletcher and Walsh 1993):

$$\frac{dC}{dt} = -k_m A C_t / V \tag{21}$$

which, after integration, gives an expression for the concentration of a pollutant in the outlet of these reactors (Pletcher and Walsh 1993):

$$C_{out} = C_{in} exp(-k_m A t / V) \tag{22}$$

On the basis of the mass and charge balances, in analogy to the batch and plug-flow reactors, a design equation for the single-pass CSTR reactor can be obtained (Pletcher and Walsh 1993):

$$C_{out} = C_{in}/(1 + k_m A t/V)$$

From equations (21) and (22), the importance of the mass transfer and of the highly developed electrode surface area in electrochemical reactors operating with direct electrochemical oxidation of pollutants is evident. One way to fulfill this requirement is to operate with three-dimensional electrodes. This solution, however, has also several drawbacks, such as:

- the difficulty of obtaining a uniform potential and current distribution
- an easy entrapment of gas bubbles (H_2 or O_2), which are likely to evolve simultaneously with the main reactions, while dilute solutions are treated (as is often the case during treatment of wastewater), leading to an increase of the cell resistance and electrical energy consumption
- the high pressure drop and tendency to clog

Consequently parallel plate-type cells often are used to assemble the electrochemical reactor. The preference to use plate electrodes derives from:

- versatility of the reactor for single-phase or multiphase processing and adaptability to monopolar or bipolar operation
- relative uniformity of current distribution
- ease of manufacture and assembly
- flexible modularity (additional cells can be easily added to the stack)
- facility to incorporate membranes, if required

Current–mass transfer correlations for different types of reactors can be defined from the first principles. The general form of this correlation is:

$$Sh = a Re^b Sc^c$$

where:

> Sh = Sherwood number ($Sh = k_m d/D$)
> Re = Reynolds number ($Re = vd_e/\eta$)
> Sc = Schmidt number ($Sc = \nu/D$)

Symbols in the dimensionless groups are:

> d_e = equivalent diameter (m)
> D = diffusion coefficient of electroactive species (m²/sec)
> v = mean linear velocity of electrolyte (m/sec)
> ν = kinematic viscosity of the electrolyte (m²/sec)

The a, b, c, and d are coefficients whose values differ in function of the hydrodynamic regime and the cell geometry.

It can be shown (Pletcher and Walsh 1993) that the limiting current and the Sherwood number are related via the mass transfer coefficient, so we have:

$$Sh = j_L l \, / \, nFADc^\infty$$

where l is the characteristic length parameter.

As a direct consequence of this relation, the duty of the reactor is supposed to increase exponentially with turbulence. In particular, for the plug-flow reactor, the duty would be a function of the volumetric flow rate of electrolyte, with the exponent equal to b. Figure 5.11 gives an example of the increase of the performance of the electrochemical batch reactor with parallel plate anodes operating under a complete recycle under different volumetric flow rate conditions for the removal of cyanides (Szpyrkowicz, Ricci, and Daniele 2003).

Recently, an increase in the number of commercial electrochemical reactors based on the parallel plate geometry has been seen. The filter-press arrangement in particular achieves great attention (Picket and Wilson 1982; Schwager, Robertson, and Ibl 1980; Stork and Hutin 1981; Trinidad and Walsh 1996), due to a possibility of realizing high mass transfer rates, enhanced by an implementation of the promoters of turbulence. The enhancement of the mass transport rate can be of a factor of two, depending on the geometry of the promoters (Ralph et al. 1996). However, while the expanded plastic mesh turbulence promoters do provide a simple and cost-effective method of enhancing mass transport, they also increase the pressure drop over the reactor, thereby increasing the pumping costs, and can shield much of the electrode surface, producing a nonuniform potential distribution. Furthermore, they can give rise to gas blinding and cause flow segregation. As a consequence, the mass transfer rates can even be decreased with respect to those for the unmodified channels, as reported by Wragg and Leontaritis (1997).

If the electrodes' working potentials are close to those used for water electrolysis, differences in local current-voltage distributions, existing over the electrode surface area, can result in gas evolution (O_2 and H_2), which in turn can produce a modification

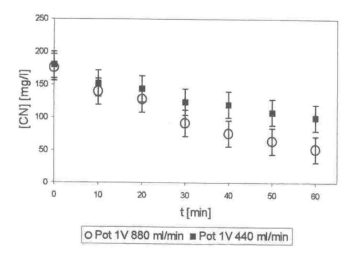

Figure 5.11 Time trend of cyanide concentration during exhaustive electrolysis of copper complex cyanide solution in a batch reactor

of the hydrodynamic boundary layer at the electrode surface, due to the rising motion of the bubbles (Wragg and Leontaritis 1997).

1.4.2 Mediated Electro-oxidation (Reduction)

A simplified procedure for developing a reactor for mediated electro-oxidation (reduction) of polluted effluents is based on an assumption that the reactor will perform isothermally and that the rates of depletion of contaminants will not change with conversion. In practice, these two hypotheses may not hold, as it may be difficult to control the rise of the reactor temperature due to the Joule effect, particularly if long treatment times are necessary. Regarding the second assumption, it should be verified experimentally if the rates of pollutant depletion do not change with conversion due to the presence of increasingly higher concentration of intermediates as the reaction advances (Szpyrkowicz, Zilio Grandi, et al. 2000), which can compete for the mediator with the pollutant itself.

Once the validity of the rate expressions is determined, the expressions for homogeneous phase reactors can be used in developing the continuous flow electrochemical reactor. Thus, the necessary volume of the reactor in which mediated electro-oxidation will proceed at a rate equal to r_A can be calculated (Scott and Fogler 1992). For a CSTR reactor, this volume would be:

$$V = F_o X/(-r_{A exit})$$

where F_o is the entering molar flow of a pollutant (mol/sec) and X is the conversion of the pollutant, equal to $(C_{in} - C_{out})/C_{in}$. For the plug-flow reactor, the volume can be calculated by:

$$V = F_o \int \frac{dX}{-r_A}$$

The necessary area of the electrodes can be estimated by considering that the ratio between the electroactive area and the reactor volume should be maintained, in the continuous flow reactor, the same as the value that characterized the batch experiments performed to define r_A.

1.5   Current Efficiency and Energy Consumption

As several reactions can occur simultaneously at the electrode surface, including the parasite reactions of solvent decomposition, the current delivered to the reactor will be used only in part for a reaction in which a pollutant of interest is involved. The ratio between the charge used for a given transformation and the total charge applied determines the current efficiency Φ_e, which can be determined, according to Faraday's laws of electrolysis, from:

$$\Phi_e = \frac{mnF}{Q}$$

where m is the number of moles of electroactive species converted in the process. A value of Φ_e below 100 percent indicates the occurrence in the reactor of a back-reaction or, more commonly, a parasite reaction.

In case the process of electrochemical treatment of polluted effluents is followed by monitoring of an aspecific parameter of chemical oxygen demand (COD), the current efficiency Φ_e can be evaluated, assuming a four-electron reaction during COD destruction, by (Comninellis 1992):

$$\Phi_e = \frac{FV}{8I}\frac{[(COG)_{in} - (COG)_{out}]}{\Delta t}$$

where Δt is the time of electrolysis (sec) and the other variables have the meanings indicated previously.

The applicability of the electrochemical process for treatment of polluted effluents is generally conditioned by its running costs, which in turn are a function of the energy consumption. The energy requirements of the electrochemical process depend on the duration of electrolysis and the applied cell voltage, according to:

$$EC = IUt$$

where:

> EC = energy consumption (kWh)
>
> t = time needed for the removal of a given pollutant (hr)
>
> I = total current applied (A)
>
> U = overall cell voltage (V)

Because of the necessary time of electrolysis and the cell voltage related to the electrocatalytic properties of the electrode material, the energy requirements EC for the treatment of wastewater differ as a function of operational conditions.

It has been observed that the energy consumption for effective wastewater electrolysis decreases with the degree of the pretreatment, as a result of the progressive lowering of the quantity of pollutants present in the reactor feed and, consequently, a diminishing of the necessary time of treatment. For example, an effective removal of ammonium from raw tannery wastewater requires 46.9 kWh/m³, while only 4.0 kWh/m³ is needed when the wastewater undergoes a deep pretreatment (Szpyrkowicz, Naumczyk, and Zilio Grandi 1995).

Often, particularly in case of the removal of metal ions by electro-reduction, it appears more meaningful to calculate energy consumption in reference to the quantity of the removed pollutant, rather than the wastewater volume. In this case, the specific unit energy consumption (SEC), expressed in kWh necessary to remove 1 kg of pollutant, can be calculated according to:

$$SEC = 10^{-3}\,UIt/m$$

where:

> U = cell potential (V)
>
> I = applied current (A)
>
> t = reaction time (hr)
>
> m = mass of the removed pollutant (mol)

Substituting for m with the calculated quantity of eliminated metal (Me_o), provided the assumption of a pseudo-first-order kinetics is valid, the equation for the specific energy consumption becomes:

$$E = 10^{-3}\,UIt(Me_o)^{-1}\,(1 - \exp(-k_{obs}t))$$

where k_{obs} is the rate constant for the overall process of pollutant depletion and (Me_o) is its initial concentration (mol/dm³).

From this equation, it is apparent that two parameters can influence energy consumption, namely, cell potential and the rate constant of the reaction. Lowering the cell potential can be achieved by the addition of strong electrolytes to the solution. However, under some conditions, negative changes in the kinetics of the reaction caused by the addition of phosphate and chloride may lead to a net increase of the unit energy consumption by lowering the reaction rate (Szpyrkowicz, Zilio Grandi, et al. 2000). Thus, the opportunity of adding other electrolytes should be always carefully examined.

The energy consumption is strongly influenced by the Ohmic drop in the cell (the last parameter is, in turn, related to the inter-electrode gap) and should be considered to be specific to the experimental setup used.

1.6   Examples of Application to Wastewater Treatment

1.6.1   Direct Electro-oxidation

Cyanides

The electro-oxidation of cyanides using Pt, Ni, Ti, PbO_2, and graphite has been extensively studied for the removal of simple cyanides (Arikado et al. 1976; Kelsall, Savage, and Brandt 1991; Hine et al. 1986; Szpykowicz 1998; Szpyrkowicz, Kaul, et al. 2000; Szpyrkowicz, Zilio Grandi, et al. 2000; Szpyrkowicz, Juzzolino, and Kaul 2001). For the removal of cyanides complexed with copper, SS/Pt, Pt, and SS fiber anodes have been tested (Ho, Wang, and Wan 1989; Wels and Johnson 1990; Lin, Wang, and Wan 1992; Hwang, Wang, and Wan 1987). These studies were mainly focused on the destruction of cyanides, paying less attention to the electrodeposition of copper on the cathode for its subsequent reuse in the plant. It was shown that the direct oxidation is very slow on platinum (Tamura et al. 1974), Ti/PbO_2 (Hine et al. 1986) and graphite (Arikado et al. 1976), indicating a kinetically limited reaction, inhibited by adsorbed species, proceeding with low current efficiency. Better results have been obtained using nickel anodes (Kelsall, Savage, and Brandt 1991), a PbO_2-coated anode (Hine et al. 1986), or a reticulated three-dimensional electrode (Tissot and Fragniere 1994). Trials to increase the specific surface area were also performed, using packed-bed (Lin, Wang, and Wan 1992) or trickling towers (El-Ghaoui, Jansson, and Moreland 1982).

The pathway of the direct electrochemical oxidation of copper-complexed cyanides depends on the pH of the medium. At a pH above 12, oxidation leads to formation of CO_2 and N_2 and deposition of CuO on the anode (Hwang, Wang, and Wan 1987). At pHs of 7.0–11.7, cuprocyanide ions first dissociate to free cyanide ions and then are electro-oxidized. At pHs of 10.5–11.7, cyanide ion and brown azulmin from the polymerization of CN^- are produced. At a pH between 7 and 8.6, carbonate, ammonium, and azulmin are formed. In the weak acidic solutions (pH 5.2–6.8), oxalate, oxamide, and ammonium are produced. Current efficiency of oxidation proved to be dependent on pH. There are no literature data, however, on the energy consumption at different pHs.

Studies have also been carried out on the indirect oxidation of cyanides in a chloride-rich medium (El-Ghaoui and Jansson 1982), a method indicated especially suitable for diluted solutions. There are no quantitative data, however, on the electro-deposition of metallic copper under these conditions.

Particular problems arise when dealing with the cyanide complexed by such metals as iron, copper, or zinc, due to the high chemical stability of these complexes. The electro-oxidation of complexed cyanide was found feasible, but it manifests a slower kinetics as compared to the electro-oxidation of free cyanide.

Several researchers have observed that the rate of cyanide decomposition can be favorably influenced by the presence of copper in the solution (El-Ghaoui and Jansson 1982; Ho et al. 1990; Szpykowicz et al. 1998), postulating that the film of copper oxides deposited on the anode exhibits electrocatalytic properties. The mechanism of this process was studied by Wels and Johnson (1990) in cyanide decomposition experiments carried out on the copper oxide film obtained by electrodeposition from an Na_3-$Cu(CN)_4$ solution. They postulated that anodic conversion of Cu^{III} to Cu^{II} provides a favorable mechanism for the oxygen transfer step, which occurs concomitantly with electron transfer in the oxidation of CN^- to CNO^-. The oxygen transfer step occurs via generation of hydroxyl radical $(\bullet OH)_{ads}$ at the Cu^{III} site during anodic discharge of H_2O.

In alkaline conditions, cyanide complexed with copper can be electrochemically oxidized to cyanate, which is finally converted to carbonate and nitrogen gas. It was confirmed that the rate of cyanide decomposition is favorably influenced by the presence of copper in the solution, with the film of copper oxides, deposited in situ on the anode, exhibiting electrocatalytic properties (Szpyrkowicz, Kaul, et al. 2000). Because of in-situ formation of the electrocatalytic film, the anode substrate material becomes unimportant, as it acts solely as a mechanical support.

For the electro-oxidation of cyanide complexed with copper, occurring in strongly alkaline solutions (pH > 12), Hwang, Wang, and Wan (1987) proposed the following reaction sequence:

$$CN^- + 2OH^- \rightarrow CNO^- + H_2O + 2e^- \tag{23}$$
$$2CNO^- + 8OH^- \rightarrow 2\,CO_3^{2-} + N_2 + 4H_2O + 6e^-$$
$$Cu(CN)_n^{(n-1)-p} + 2nOH^- \rightarrow Cu^+ + nCNO^- + nH_2O + 2ne^-$$
$$2Cu^+ + 2OH^- \rightarrow Cu_2O + H_2O$$
$$Cu_2O + 2OH^- \rightarrow 2CuO + H_2O + 2e^-$$

The following (improbable) mechanism was proposed recently to explain the electrochemical oxidation of copper cyanide complexes (Casella and Gatta 2000):

$$Cu(CN)_n^{(1-n)} + 2H^2O \rightarrow CuOOH + 3H^+ + nCN^- + 2e^-$$
$$CuOOH + H^+ + e^- \rightarrow Cu(OH)_2$$
$$CuOOH + H^+ + e^- \rightarrow CuO + H_2O$$

In addition to reaction (23), cyanate can also decompose, depending on the potential, producing ammonium ions (Kelsall, Savage, and Brandt 1991):

$$CNO^- + 2H_2O \rightarrow NH_4^+ + CO_3^{2-}$$

During the destruction of cyanide complexed with copper, simultaneous removal of copper by electrodeposition and elimination of COD were obtained in an electrochemical reactor equipped with a Ti/Pt anode or simply an SS anode. The COD removal was well described by the pseudo-first-order kinetics, the composition of the solution strongly influencing the reaction rates. Under very alkaline conditions, which appeared to be the most indicated, the rate constant for COD removal was 0.01 hr^{-1}. The process performed better when no additional electrolytes were present. Under these conditions, current efficiencies reaching 100 percent or higher were obtained, indicating that chemical processes (probably the autocatalytic reaction of cyanate decomposition in the presence of carbonate ions) were occurring simultaneously to electro-oxidation. Low current efficiency for copper removal, which characterized the reactor performance particularly at the end of electrolysis, probably derived from the fact that the process became strongly mass-transfer controlled.

A comparison of the results of the study conducted under strong agitation (Szpyrkowicz et al. 1998) with the literature data regarding experiments done without mixing showed that the distribution of the eliminated copper between the anodic film and the metallic cathodic deposit may have been related to fluodynamic conditions in the cell.

A pH of 13 or above is also thought to decrease energy consumption, as the concentration of hydroxyl ions showed a strong influence on the performance of the reactor. The energy consumption for the elimination of copper also proved to be influenced by its initial concentration, showing an inversely proportional relationship. For example, for an initial copper concentration equal to 1,100 mg/dm^3, the power consumption in the reactor under study was as low as 5.46 kWh for 1 kg of removed copper.

The performance of the reactor was better, as regards both the kinetics and energy consumption, if conducted as a direct electro-oxidation with no addition of chloride and under very alkaline conditions (Szpyrkowicz et al. 1998). Electrochemical treatment of copper-complexed wastewater can thus be an alternative to chemical oxidation, the latter having a disadvantage of creating hazardous sludge.

Thiourea Dioxide

Thiourea dioxide (TUD) is a reducing agent used in the dyeing process of synthetic materials with disperse dyes (Cegarra 1976; Szpyrkowicz, Juzzolino, et al. 2001). The procedure of dyeing comprises two steps. First, the fabric undergoes dyeing in the batch-wise mode at a high temperature (135°C), or at lower temperatures with the addition of specific carriers. In the second step, the excess dye not perfectly adhering to the material is removed by a reducing agent (sodium hydrosulfite is commonly used), which reacts with the chromophore group of the dye. The unreacted TUD contained in the reducing bath contributes significantly to the organic and nitrogen load of wastewater if no pretreatment is carried out and the bath is mixed directly with fabric rinsing water.

The reaction of TUD with a dye can be sketched schematically (Heaton 1994):

This reaction generates sulfites and urea. The problem related to the presence of these compounds in the spent reducing baths and the need of their elimination was encountered during the study of the electrochemical oxidation of TUD. Both sulfites and urea are known to be inhibitors of biological processes, which are the means currently used for textile wastewater treatment.

An investigation was launched to ascertain whether the electrochemical process could remove all types of pollutants and whether it can provide a global solution to the problem of the pretreatment of a spent reducing bath. For this reason, experiments on the electro-oxidation of TUD were supplemented by separate tests to identify the kinetics of the elimination of sulfites and urea. The experiments designed to follow the removal of TUD, sulfites, and urea included both direct and mediated electrolysis. They were conducted using eight different anodes in order to evaluate the influence of the anode material on the process kinetics and the possibility of electrocatalytic reactions. The role of hydrodynamic conditions in the performance of a reactor was also considered.

The destruction of TUD was followed in time by the analysis of COD and sulfates. As noted above, the decomposition of TUD generates urea and sulfites; consequently, during the experiments with the wastewater containing TUD, along with electro-oxidation of TUD, a concomitant oxidation of in situ–generated sulfites into sulfates was also expected to occur in the electrochemical reactor. In fact, a continuous increase of the concentration of sulfates was observed in all the runs, as can be seen from figure 5.12a. The COD value attributed to TUD was obtained by subtracting the contribution due to sulfites from the total COD of the samples. Assuming that urea could be an intermediate of TUD electro-oxidation, in analogy to chemical oxidation, it would not be revealed by the analysis of COD because under the strongly acidic conditions of COD determination, it decomposes into CO_2 and NH_4^+.

The study showed that electro-oxidation is a successful method for the destruction of TUD, sulfites, and urea, which are the by-products of its decomposition. The process proved feasible to be conducted in a simple undivided cell reactor equipped with parallel plate electrodes. The performance of the reactor was a function of the electrode material, due to the materials' different electrocatalytic properties toward the removal of considered pollutants. The best results for TUD and SO_3^{2-} electro-oxidation were obtained with the Ti/Pt electrode, which showed electrocatalytic effect for both the compounds, indicating a possibility of their direct electro-oxidation on the anode. Destruction of TUD and SO_3^{2-} proceeded also via indirect electro-oxidation, mediated by chlorine evolved on the anode. No electrocatalytic effects were observed for urea with any of the tested anode materials. For elimination of urea, the Ti/Pt-Ir electrode proved to be the best anode, most probably due to its high efficiency in electro-oxidation of chlorides into chlorine.

As can be seen from figure 5.12, which, for the Ti/Pt anode, also compares the performance of the reactor under different hydrodynamic conditions, the degree of the turbulence significantly influenced the performance of the reactor. By increasing the degree of agitation, the rate of the removal of COD and the rate of formation of sulfates increased. This indicates that under these conditions the process kinetics was mass transport controlled and was described by equation (21).

The values of the mass transfer coefficient calculated for the experiments related to the removal of TUD conducted under the highest conditions of turbulence ($Re = 5,000$) are reported in table 5.3. For a Ti/Pt anode and lower Re values ($Re = 60$ and

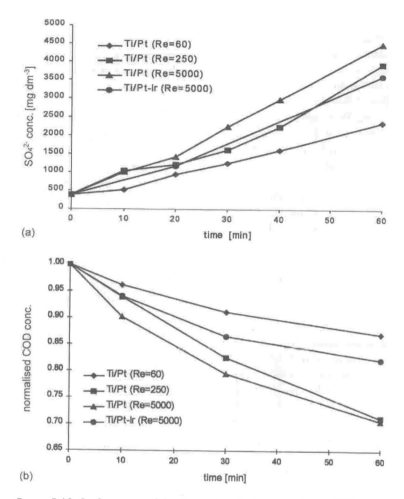

Figure 5.12 Performance of the electrochemical reactor during TUD electro-oxidation. Graphs show experiments with sodium chloride as supporting electrolyte and Ti/Pt-Ir and Ti/Pt anodes. (a) Trend of SO_4^{2-} concentration vs. time. (b) Trend of normalized COD values, attributed to TUD, vs. time.

250), the mass transfer coefficients were 2.91×10^{-6} m/sec and 6.76×10^{-6} m/sec, respectively.

1.6.2 Indirect Oxidation

Sulfides

While sulfide solutions are known to oxidize directly at anodes such as Ti/Pt-Ir (Kelsall and Thompson 1993; Szpyrkowicz, Kelsall, et al. 2001), when the electrolyzed solution contains chloride and the pH exceeds 7, the following overall reaction can occur, driven by "chlorine" reduction:

$$HS^- + 4H_2O \rightarrow SO_4^{2-} + 9H^+ + 8e^-$$

Data indicate thiosulfate as an intermediate product (Rajalo and Petrovskaya 1996), although aqueous sulfur chemistry is complex, with many other sulfoxy intermediates

Table 5.3 Results of electro-oxidation of TUD, urea, and sulfites

| Anode material | Electro-oxidation of TUD | | Electro-oxidation of Urea | | Electro-oxidation of SO_3^{2-} | |
| | COD | SO_4^{2-} | TKN | | SO_3^{2-} | |
	$k_m \times 10^{-6}$ (m/sec)	$k_m \times 10^{-6}$ (m/sec)	Φ_e (%)	$k_m \times 10^{-6}$ (m/sec)	Φ_e (%)	$k_m \times 10^{-6}$ (m/sec)
Ti/Pt	7.35	51	19	0.81	132	25.2
Ti/Pt-Ir	4.20	46	97	3.38	93	17.7
Ti/PdO-Co$_3$O$_4$	—	—	69	2.33	—	—
Ti/RhO$_x$-TiO$_2$	—	—	40	1.63	68	11.9
Ti/SnO$_2$-Sb$_2$O$_5$	—	—	0	—	—	—
Ti/RuO$_2$-TiO$_2$	—	—	49	1.75	93	16.5
Ti/PbO$_2$	—	—	104	4.20	—	—
Ti/MnO$_2$-RuO$_2$	—	—	74	3.03	96	20.1

to sulfate formation possible, in addition to elemental sulfur, which could partially passivate the anode.

Figure 5.13 shows an example of the reactor performance in terms of the time dependence of normalized sulfide concentrations for three current densities with and without additional agitation. The rate of sulfide depletion, well described by psuedo-first-order kinetics, proved to be strongly dependent on the current density and significantly influenced by the hydrodynamic conditions in the reactor. Previously shown values (cf. figure 5.7) of the apparent pseudo-first-order kinetic constant k_{obs} for the removal of sulfides indicate that the rate of sulfide depletion from the synthetic and real (tannery) wastewaters behave similarly. These results also suggest that direct oxidation of this pollutant may contribute to its elimination.

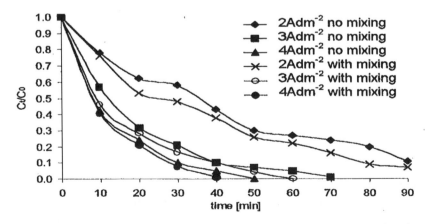

Figure 5.13 Time dependence of normalized sulfide concentrations in real and synthetic (tannery) wastewaters for various current densities and hydrodynamic conditions of the reactor (Szpyrkowicz, Kelsall, et al. 2001)

Ammonia

Ammonia is one of the most important among several different pollutants present in industrial water, for example, from coke production, tanneries, fish processing, and slaughterhouses (Szpyrkowicz, Naumczyk, and Zilio Grandi 1994, 1995; Szpyrkowicz, Kelsall, et al. 2001). In anaerobically treated tannery wastewater, ammonium ions are often considered the primary cause of the depletion of dissolved oxygen, if the effluents are discharged into water bodies. Preliminary studies by cyclic voltammetry of a Ti/Pt-Ir anode in a synthetic ammonia-bearing wastewater did not show any activity of this material toward ammonium ion oxidation even under very alkaline conditions, when NH_3 is expected to occur in free form.

Despite direct electro-oxidation can be, in principle feasible, in the presence of chloride in the solution a principal route of the removal of this contaminant is via an indirect electrochemical oxidation, in which anodic formation of chlorine is coupled to its homogenous reaction with ammonium ions near the anode and in the bulk. Though the reaction can involve chloramine intermediates, the overall reaction, analogous to breakpoint chlorination, can be sketched as follows:

$$2NH_4^+ + 3HClO \rightarrow N_2 + 3H_2O + 5H^+ + 3Cl^-$$

The depletion of ammonia is well described by the pseudo-first-order kinetics, and the value of the rate constant varies, for the same conditions of applied current, with agitation intensity. The decrease of the rate constant with stirring (cf. figure 5.8) may be due to the enhancement, with agitation, of the anodic and cathodic loss reactions involving active chlorine, without any advantage deriving from mixing for the relatively slow homogeneous kinetics of the ammonia reaction with chloric(I).

It is important to note that other mediators that have a normal potential higher than that of chlorine can also be capable of ammonia oxidation. These mediators include ozone (its capability to destroy NH_4^+ was confirmed by Singer and Zilli [1975]), generated at the anode due to reactions (11) and (12); atomic oxygen, produced by reaction (16); and hydrogen peroxide, produced by the cathodic reduction of oxygen via reaction (18). The normal potentials of these three mediators are 2.07 V, 2.42 V, and 1.78 V, respectively. Interestingly enough, the last mediator is produced at the cathode, where strongly alkaline conditions established by the main cathode reaction (5) should favor oxidation of NH_4^+ by oxygen-based mediators (Singer and Zilli 1975).

Tanneries

Electrochemical reactors, used since the early 1960s for the recovery of pollutants from metal plating factories, have been considered recently for other environmental systems applications, ideally requiring total combustion of organic pollutants to CO_2 and H_2O or modification of their structure, resulting in improved biodegradability (Szpyrkowicz, Naumczyk, and Zilio Grandi 1994, 1995; Naumczyk et al. 1996; Szpyrkowicz, Kelsall, et al. 2001; Szpyrkowicz, Ricci, and Daniele 2003). The tanning industry, one of the leading economic sectors in many countries, generates large quantities of heavily polluted wastewater containing ammonia, sulfides, and organic substances, including tannins. Treatment of such wastewater by conventional biological methods is often not adequate to remove the pollutants completely, especially ammonia and tannins, the latter being characterized by the low biodegradability common to polyphenolic compounds.

Tannery wastewaters usually contain a high concentration of chlorides, and thus are easily amenable to indirect electro-oxidation via a redox couple Cl$^-$/Cl$_2$. Previous studies conducted at the University of Venice (Szpyrkowicz, Naumczyk, and Zilio Grandi 1994, 1995) with graphite, Ti/Pt, Ti/Pt-Ir, Ti/RuO$_x$, Ti/PbO$_2$, and other anode materials, established the feasibility of purifying tannery wastewaters by an electrochemical process. The efficiency of wastewater treatment was found to be a function of the electrocatalytic properties of the anode material, Ti/Pt-Ir giving the best performance. The composition of pollutants in the wastewater, and in particular the type of pretreatment it had undergone before electro-oxidation, also influenced the performance of electrochemical oxidation.

Unfortunately, electrochemical oxidation proved uneconomic when applied to raw wastewaters, due to the very high energy consumption required in depleting pollutants to acceptable concentrations. However, the process was found competitive in costs of operation and treatment efficiency when applied as a final polishing step or a substitute for biological nitrification-denitrification. In this case, the volume of the reactor is 15 to 20 times smaller than the nitrification-denitrification unit and the removal of nitrogen can be achieved at about 15 kWh/m^3 of wastewaters containing ammonia at concentrations of 10 mol/m^3.

Figure 5.14 shows an example of the performance of the batch reactor, equipped with a Ti/Pt-Ir anode, in terms of the time dependence of the normalized COD, TOC, sulfide, tannin, and ammonium ion concentration for a current density of 400 A/m^2 and operation without any mechanical or hydraulic mixing (Szpyrkowicz, Kelsall, et al. 2001). A 60 percent depletion of sulfides and tannins can be observed to occur in the first 10 minutes of electrolysis, with the COD and TOC being eliminated at a much slower rate than other pollutants. An initial rapid decrease in sulfide and tannin concentration is reflected by the significant lowering of COD during the first 10–15 minutes of treatment. The rate of nitrogen removal was significant, with more than 40 percent depletion in 20 minutes.

Under the conditions of one study (Szpyrkowicz, Kelsall, et al. 2001), most of the pollutant depletion processes were described by an apparent first-order reaction, ex-

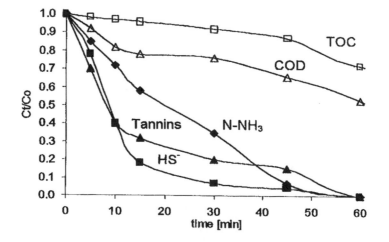

Figure 5.14 Time dependence of normalized concentrations of pollutants from tannery wastewater during chlorine-mediated electro-oxidation

cept those involving tannins, for which a second-order rate law was more appropriate. In the case of tannins, a reaction rate was independent of agitation intensity and was a function of the current density only. As implied by the data in figure 5.15, increasing current enhanced tannin depletion rates, but the gradients of plots were a nonlinear function of current density.

As reported previously, the rate of oxidation of ammonia, sulfides, and organic substances varied with agitation intensity, as their different homogeneous rate constants affected the apparent efficiency for chlorine generation. The depletion rate of sulfides and COD increased with an increase in agitation rate, by contrast to ammonia, the removal rate of which was affected negatively by stirring. These differences in performance of the reactor influenced its selectivity toward the removal of single species of contaminants.

Ascribing apparent pseudo-first-order kinetic rate constants k_1, k_2, and k_3 to sulfide, ammonium, and COD removal, respectively, and introducing a concept of a differential selectivity ratio for mass transfer with parallel first-order reactions (Westerterp 1984), the influence of hydrodynamic conditions at various current densities on the competing reactions can be expressed in terms of the ratios k_1/k_3 and k_2/k_3; some sample values are listed in table 5.4.

These data indicate that current density had little effect on reactor selectivity, while hydrodynamic conditions have been shown to strongly affect both the kinetics

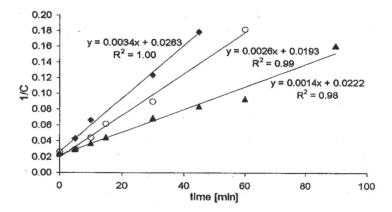

Figure 5.15 Effect of current density on the time dependence of the reciprocal of tannin concentrations in a real wastewater. Densities: ($\blacklozenge$) 400 A/m^2; ($\blacktriangle$) 300 A/m^2; (o) 200 A/m^2.

Table 5.4 Relative ratios of the apparent pseudo-first-order constants for the removal of sulfides (k_1), ammonium (k_2), and COD (k_3) under different hydrodynamic conditions and current densities

	k_1/k_3		k_2/k_3	
Current density (A/m^2)	Without mixing	With mixing	Without mixing	With mixing
200	7.14	1.93	2.00	0.36
300	7.69	2.31	2.15	0.38
400	8.30	2.36	2.20	0.27

of oxidation (see figures 5.7 and 5.8) and selectivity of pollutant depletion. On the basis of the type of pollutant whose elimination is chosen as priority, the reactor could be operated in different modes to maximize the depletion rate of the target species. If oxidation of organic substances (COD) is to be minimized and the removal of ammonium ion and sulfides maximized, then the data in table 5.4 suggest operating the reactor with no additional mixing.

Textile Wastewater from Dyeing Operations

Wastewaters that are generated at various stages of dyeing differ in strength and temperature. Their pollution load is high and is caused mainly by spent dyeing baths, composed principally of unreacted dyeing compounds, dispersing agents (e.g., polyvinyl alcohol or PVA) if disperse dyes are used, surfactants, salts, and organics, washed out from the material undergoing dyeing. The wastewaters are characterized by high color, high COD, and pHs that can vary from 2 to 12 (Gurnham 1965; Lin and Lin 1993; Szpyrkowicz and Zilio Grandi 1996; Szpyrkowicz 1996; Naumczyk, Szpykowicz, and Zilio Grandi 1996; Szpyrkowicz, Juzzolino, et al. 2000; Szpyrkowicz, Juzzolino, and Kaul 2001; Szpyrkowicz, Cherbanski, et al. 2001).

The conventional treatment of wastewaters generated in the process of dyeing consists of biochemical oxidative destruction, which often produces final effluent still exceeding standards for color and nitrogen. To improve the overall treatment efficiency, various processes are used for the pretreatment of dye-bearing wastewater, including chemical oxidation with different reagents, such as hypochlorite, ozone, ozone + UV, hydrogen peroxide, hydrogen peroxide + UV, and hydrogen peroxide + Fe^{2+} salt (Fenton's reagent) (Lin and Chen 1997; Sarasa et al. 1998; Wu et al. 1998; Saunders, Gould, and Southerland 1983; Carriere, Jones, and Broadbent 1993). Ozonation combined with chemical coagulation also proved efficient as a pretreatment step before the biological process (Lin and Lin 1993). A novel process of photocatalytic oxidation of dyes bearing wastewater by applying TiO_2 as a catalyst and UV or solar irradiation has also been recently investigated (Wang 2000; Goncalves et al. 1999; Reutergardh and Iangphasuk 1997; Kiriakidou, Kondarides, and Verykios 1999).

Relatively fewer studies have been conducted on the application of methods based on electrochemical processes. The research of Lin and Peng (1994) and Lin and Chen (1997) was conducted using steel electrodes; thus the coupled effect of electrocoagulation caused by Fe ions from the corroded anode and a possible direct oxidation on the anode and reduction at the cathode were observed. Under the best conditions of current density (92.5 A/m²), the efficiency of COD removal was around 50 percent for simple electrolysis and 60 percent when some amount of polyaluminum chloride was added. The energy consumption was equal to 0.003 kWh/dm³.

In a study by Naumczyk, Szpykowicz, and Zilio Grandi (1996) on the application of electrochemical oxidation to the treatment of full-scale textile plant wastewater containing a mixture of reactive and acid dyes, 92 percent removal of COD was obtained while using a Ti/RuO_2 anode. An investigation into the possibility of applying the electrochemical oxidation process to treat a solution bearing disperse dyes also gave encouraging results, consisting of nearly 40 percent removal of COD in 40 minutes of electrolysis (Szpyrkowicz, Juzzolino, et al. 2000). In fact, a partial solubility of disperse dyes causes solid–liquid separation processes to be not fully efficient for their removal (Do and Chen 1994) and justifies the use of oxidative methods.

The class of disperse dyes comprises organic non-ionic compounds, the solubility of which depends on their structure (Cegarra, Puente, and Valldeperas 1981). Disperse dyes deriving from aminoanthraquinone and 1,4-diaminoanthraquinone are slightly water soluble due to the presence of an ethanol group. The structure of some examples of the disperse dyes is shown below:

Dyes deriving from 1-aminoanthraquinone and 1,4-diaminoanthraquinone:

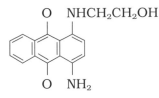

Nitro-diphenyl-amino (mostly yellow and orange) dyes:

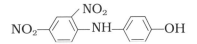

Dyes characterized by the presence of an azo-group:

$$R\text{—}\bigcirc\text{—}N=N\text{—}\bigcirc\text{—}N\begin{cases}CH_2CH_2OH \\ CH_2CH_2OH\end{cases}$$

Other compounds in this class, which have an amino group, are insoluble in water.

Szpyrkowicz, Juzzolino, and Kaul conducted a study (2001) to compare the effectiveness of chemical and electrochemical processes when applied for the treatment of textiles. The results of experiments on the destruction of disperse dyes by chemical oxidation using ozone, hypochlorite, and Fenton's reagent were compared with the data obtained by electrochemical oxidation. While hypochlorite oxidation resulted in only 35 percent reduction in color, ozonation enabled 90 percent color removal at 0.5 g O_3 dm^{-3}; however, the removal of COD was low (10 percent). Electrochemical oxidation using a Ti/Pt-Ir anode at low pH for 40 minutes of electrolysis resulted in 79 percent COD removal and 90 percent color removal. Indirect oxidation by means of chlorine-deriving compounds was the predominating process leading to the pollutants depletion. The best treatment results were obtained with the Fenton process, which with the optimal pH level of 3 and with H_2O_2 and ferrous sulfate doses of 600 and 550 mg/dm^3, respectively, resulted in a final effluent being colorless and with the residual COD equal to 100 mg/dm^3.

Figure 5.16 compares the UV–visual spectra of wastewater after electro-oxidation in a reactor equipped with a Ti/Pt-Ir anode, using sulfate and chloride as supporting electrolytes and after hypochlorite oxidation. The analysis was performed after dissolution of the samples in EtOH. The spectrum of the sample electrolyzed with the sulfate solution as electrolyte shows a nonzero baseline due to the formation of compounds that are not soluble in EtOH. A large absorption around 210 nm is the same as the one obtained for the wastewater sample before treatment. Some peroxides, which absorb in the range between 300 and 200 nm, might also have been produced by the cathodic reduction of oxygen, as the reactor was an undivided cell. Furthermore, by contrast with the spectrum of a sample obtained after electrolysis in the presence of

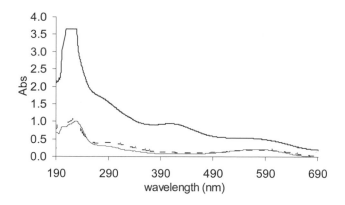

Figure 5.16 Spectra of the wastewater after chemical (—) and electrochemical oxidation with 0.05 N Na₂SO₄ as supporting electrolyte (—) and 0.1 N NaCl (– – –) Szpyrkowicz, Juzzolino, et al. 2000).

chloride, the spectrum with sulfate shows no decrease in the absorbance at the wavelength region corresponding to yellow and red. These results may suggest that the dyes were removed by indirect electrolysis when chloride was present and that the Ti/Pt-Ir anode has no selectivity for any of the three disperse dyes studied.

In order to obtain further information on the electrochemical processes occurring at the anodes, a series of cyclic voltammetric experiments was performed, using both Ti/Pt-Ir and Ti/Pt materials as the working electrodes. The potential range explored varied between − 1.5 and 2 V, in order to include, respectively, the hydrogen and oxygen evolution processes (Adams 1969). The voltammograms obtained in the presence of 186 mg/dm³ of the dyes and other organic compounds of the bath clearly indicated that no process attributable to direct oxidation of dyes or other organic substances present in the bath was evident in the different electrolyte solutions investigated. This result is in agreement with the hypothesis about the involvement of indirect electro-oxidation in the destruction of the organic matter present in the solution. In fact, it appears from figure 5.16 that the mechanism of chemical chlorination of the dyeing bath may be similar to indirect electro-oxidation, as with both the processes, peaks relative to the wavelength of yellow and red are almost completely removed.

The proven feasibility of application of electro-oxidation for the destruction of the pollutants present in dyeing baths containing partially soluble disperse dyes indicates that electrochemical oxidation, which can lead to substantial decolorization, is promising for the treatment of this kind of wastewater. The efficiency of the treatment depended on the nature of the supporting electrolyte and the bulk pH in the reactor and, to a lesser degree, on the type of the anode material. Table 5.5 shows the values of the apparent pseudo-first-order rate constant for the removal of color during electrolysis under conditions of free pH evolution. The best results were obtained in a chloride-rich medium under acidic pH conditions using the Ti/Pt-Ir anode. The control of pH at the acidic level at the value of 4.5 resulted in more than a 30-fold increase of the reaction rate. Far lower efficiencies obtained during the comparative chemical oxidation of pollutants by hypochlorite ions indicate that electrochemical oxidation is preferred to the commonly applied chemical treatment.

Table 5.5 Results of electro-oxidation of the dyeing bath using various anodes

	Anode potential (V)	Cell potential (V)	Color removal			COD removal	
Anode material			Degree of removal (%)	$k_{obs} \times 10^{-4}$ (sec^{-1})	r^2	Degree of removal (%)	Current efficiency Φ_e (%)
Ti/Pt	4.36	8.0–7.6	40	2.01	0.82	9	21.4
Ti/RuO$_2$-TiO$_2$	4.33	7.8–7.4	42	2.03	0.86	26	60.4
Ti/SnO$_2$-Sb$_2$O$_5$	4.59	8.6–7.4	45	2.24	0.80	23	61.7
Ti/Pt-Ir	4.56	8.6–7.8	50	2.54	0.89	39	104.8
Ti/MnO$_2$-RuO$_2$	4.44	8.2–7.4	46	2.36	0.90	10	23.9
Ti/RhO$_x$-TiO$_2$	4.41	8.6–7.6	47	2.43	0.89	29	77.6
Ti/PdO-Co$_3$O$_4$	4.32	8.6–7.8	48	2.43	0.90	25	57.9

Note: Electrolysis time = 40 minutes
Source: Szpyrkowicz, Juzzolino, et al. 2000

2 Photocatalysis

2.1 Principles of Photocatalysis

Crystalline solids are categorized as conductors, semiconductors, or insulators according to their ability to conduct electrical current. The valence band (VB) is the energy level occupied by the outermost electrons in a solid. The conduction band (CB) is the highest energy level in an atom. Because it is unoccupied by electrons at the ground state, electrons from the valence band can move into it and are free to flow within the solid. The valence and conduction bands are often separated by an energy barrier called the *band gap.*

In a conductor, the valence and conduction bands overlap, and electrons can easily move to the conduction band without any energy input. Within that band, they can flow freely from atom to atom in the solid. In insulators and semiconductors, the valence band (which is completely occupied by electrons) is separated from the conduction band (which is empty) by the band gap energy zone. In order for the electrons to move to the conduction band and subsequently move from atom to atom within the solid, an external energy input equal to or greater than the band-gap energy must be applied. The difference between an insulator and a semiconductor is a matter of degree. As a rule of thumb, a material is considered to be an insulator if the band-gap energy is greater than about 3.5 V and to be a semiconductor when it is less than 3.5 V.

Semiconductors are categorized as *n-type* or *p-type,* depending on whether they contain an excess of electrons or holes within their lattice. An n-type semiconductor has an excess of electrons and a p-type semiconductor has a deficiency of electrons in the crystal lattice. For example, titanium dioxide (TiO$_2$) is considered to be an n-type semiconductor because its crystal structure is oxygen deficient, causing an excess of electrons within the crystal lattice. Copper (I) oxide (Cu$_2$O) is considered p-type because it is copper deficient, causing an excess of holes within the crystal lattice. Semiconductors can be altered from n-type to p-type or vice versa by the insertion of impurities (doping) called *electron donors* or *electron acceptors* into the crystal lattice.

Photoactive catalysts are semiconductors that are primarily composed of metal oxides. Table 5.6 lists several different photoactive semiconductors; their band-gap energies vary widely, ranging from 2.2 V to 3.5 V. The *equivalent wavelength* is the wavelength of a photon that has energy equal to the band-gap energy. The table shows that photoactivation of the anatase crystalline form of TiO_2, for example, requires light of wavelength less than or equal to about 413 nm.

Semiconductors typically used in solar energy applications and oxidation processes are categorized as n-type semiconductors because the main charge carriers are electrons. TiO_2 is most commonly used because it is very photoactive and stable in comparison to other semiconductors and is very fascinating material.

A semiconductor is characterized by band structure: a filled VB (valence band) separated by the E_g (energy gap) from a vacant CB (conduction band) as shown in figure 5.17. The energy level at the bottom of the conduction band (LUMOs) reflects

Table 5.6 Band-gap and threshold wavelengths for oxide semiconductors suitable for photocatalytic environmental applications

Semiconductor	Band gap, E_g (eV)	Threshold wavelength (nm)
TiO_2	3.0	413
ZnO	3.2	385
SnO_2	3.5	346
$SrTiO_2$	3.2	385
WO_3	2.8	437
Fe_2O_3	2.2	515
In_2O_3	2.7	450
CdO	2.1	528

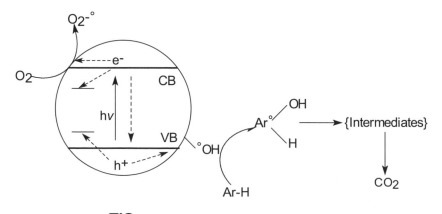

Figure 5.17 Sequence of photophysical and photochemical events taking place upon irradiation of a TiO_2 particle with hv greater than E_g together with secondary reactions to total mineralization to CO_2

the reduction potential of the photoelectrons, whereas the uppermost level of the valence band (HOMOs) is a measure of the oxidizing ability of the photo holes. The flatband potential V_{fb}, fixed by the nature of the material and by the proton exchange equilibrium, determines the energy of the two charge carriers at the interface. Hence, reductive and oxidative processes for adsorbed couples with redox potentials more positive and more negative than the V_{fb} of the conduction and valence bands, respectively, can be driven by surface-trapped e^- and h^+ carriers. Figure 5.17 illustrates a fraction of the complex sequence of events that may take place in a semiconductor photocatalyst.

The band gap defines the wavelength sensitivity of the semiconductor to irradiation. Photoexcitation with light of energy greater than or equal to E_g ($h\nu \geq E_g$) promotes an electron from VB into CB, leaving a hypothetical hole (h^+) in VB (see figure 5.17). A hole can also be identified as a chemical entity, for example, a bound O^- lattice radical or a surface-associated OH radical in a typical metal oxide semiconductor. This hole can cause oxidation of an adsorbed substrate (e.g., OH or H_2O) to OH radicals through electron transfer at the interface. On the other hand, the photogenerated electron in CB reduces a suitable acceptor (A), such as H^+ or O_2, and generates reactive H_{ads} (or H_2) or $O_2^-\bullet$ (superoxide radical anion).

It is imperative that in order to find appreciable generation of reactive chemical species (such as $\bullet OH$ or $O_2^-\bullet$), one must efficiently separate the photogenerated electrons and holes by kinetically favoring the formation of their chemical equivalents. The Schottky barrier at the semiconductor–liquid junction performs this task partially. In a typical TiO_2/UV process, the substrate to be oxidized by $O_2^-\bullet$ or $\bullet OH$ is usually an organic pollutant chemical generalized as RX. By virtue of the 3.0–3.2 eV band gap (413–380 nm), illumination of TiO_2 results in the generation of h^+ with oxidizing power of $+2.8$ V and e^- with reduction power of -0.2 V (based on the positions of VB and CB, respectively, at pH 7). It is interesting to note that the oxidizing power of h^+ in TiO_2 is equivalent to that of the hydroxyl radical. The usual mechanistic steps of TiO_2/UV process are (Matthews 1984; Izumi et al. 1980; Okamoto et al. 1985; Sakata, Kawai, and Hashimoto 1984; Pelizzetti et al. 1990; Serpone et al. 1986; Kraeutler and Bard 1978; Norton, Bernasck, and Bocarsly 1988; Turchi and Ollis 1990; Al-Ekabi and Serpone 1988):

$$TiO_2 + h\nu \rightarrow TiO_2 (e^-, h^+) \tag{24}$$

$$TiO_2 (h^+) + RX \rightarrow TiO_2 (RX^+)_{ads} \tag{25}$$

$$TiO_2 (h^+) + H_2O/OH^- \rightarrow TiO_2(OH)_{ads} + H^+ \tag{26}$$

$$TiO_2(e^-) + O_2 \rightarrow TiO_2(O_2^-\bullet)_{ads} \tag{27}$$

$$TiO_2(RX^+)_{ads} + TiO_2(OH)_{ads} \rightarrow TiO_2(R\text{-}OH)_{ads} + X^- \tag{28}$$

$$TiO_2(R-OH)_{ads} + OH \rightarrow CO_2 + H_2O \tag{29}$$

Both adsorbed H_2O/OH^- and molecular oxygen mediate the photocatalytic oxidation (PCO) process, as shown in steps (24)–(27). Steps (28) and (29) are too simplified, but they constitute a series of attacks by powerful OH radicals to finally disintegrate RX to CO_2 and so forth. Heteroatoms in the pollutant molecule are transformed into their corresponding anions, such as SO_4^{2-}, NO_3^-, and PO_4^{3-} as per the generalized reaction shown below:

$$C_xH_yN_zO_pCl_qP_rS_t \rightarrow CO_2 + H_2O + NO_3^- + PO_4^{3-} + SO_4^{3-} + Cl^-$$

In the simplest case of an intrinsic semiconductor, absorption of a photon ($h\upsilon$) of band-gap energy E_g or greater excites an electron (e_{cb}^-) to the conduction band, leaving an electronic vacancy (called a hole, h_{vb}^+) in the valence band. A potential exists between the interior bulk solid and the external surface of the solid under this condition, due to depletion of conduction-band electrons in the space-charge region. The resulting gradient causes the photo-excited hole and electron to migrate to the exterior surface. In the absence of an electrical circuit, at steady state, the electron and the hole arrive at the surface simultaneously. These electron-hole pairs at the surface can recombine, producing only luminescence, or can react with adsorbed species on the surface to produce redox reactions.

These separated charge carriers may also recombine, migrate to the surface while scanning several shallow traps. The traps may be identified as anion vacancies and/or Ti^{4+} for the electrons, and oxygen vacancies or other defect sites for the hole. Both charge carriers scan the surface, visiting several sites to reduce adsorbed electron acceptors (A_{ads}) and to oxidize adsorbed electron donors (D_{ads}). This occurs in competition with surface recombination of the surface-trapped electrons and holes (e_{st}^- and h_s^+) to produce light emission and/or photon emission. Oxygen on the particle surface acts as an electron acceptor, whereas OH^- groups and H_2O molecules are available as electron donors to yield the strongly oxidizing •OH radicals. Trapping of electrons and holes in pristine naked TiO_2 colloids of nanometer size takes place in less than 30 picoseconds. At concentrations of organic pollutant substrates normally found in the environment (a few tens of mg dm^{-3}), the •OH radicals are the primary oxidants. In the case of an aromatic substance, the corresponding •OH adducts (a cyclohexadienyl radical) ultimately break down into a variety of intermediate products on the way to total mineralization to carbon dioxide.

2.2   Reaction Pathways for Generation of Hydroxyl Radicals

The primary oxidant responsible for the photoassisted catalytic oxidation of organic compounds in aqueous solutions is the highly reactive hydroxyl radical. The mechanisms of hydroxyl radical formation have been discussed widely in the literature (Matthews 1984; Izumi et al. 1980; Okamoto et al. 1985; Sakata, Kawai, and Hashimoto 1984; Pelizzetti et al. 1990; Serpone et al. 1986; Kraeutler and Bard 1978; Norton, Bernasek, and Bocarsly 1988; Turchi and Ollis 1990; Al-Ekabi and Serpone 1988).

Two pathways have been proposed for the formation of hydroxyl radicals. Matthews (1984) proposed that the major pathway for hydroxyl radical formation is through the reaction of the valance-band hole with either water (H_2O) or hydroxide ions (OH^-) adsorbed on the catalyst surface in the presence of photon energy ($h\upsilon$). Okamoto and colleagues (1985) proposed an alternative pathway, suggesting that hydroxyl radical formation can also be promoted by the reaction of the conduction-band electron with adsorbed oxygen, $O_{2(ads)}$, to produce hydrogen peroxide (H_2O_2) from superoxide ions ($O_2^- •$), where $HO_2•$ is the perhydroxyl radical and HO_2^- is the perhydroxyl ion.

Proponents of the latter theory believe that oxygen plays an important role in photocatalytic reactions (Okamoto et al. 1985; Hsiao, Lee, and Ollis 1983). Surface-adsorbed oxygen not only reacts to produce hydroxyl radicals but also traps the conduction-band electrons, preventing electron-hole recombination. Hsiao and colleagues (1983) performed experiments using aqueous suspensions of TiO_2 to determine the effect of PCO on dichloromethane and dibromomethane in both the presence and absence of O_2. They found that the rate of disappearance of these compounds was significantly slower when O_2 was not present in solution and that the participation of O_2 is a kinetically important step. Similar results were reported for the oxidation of phenol using an aqueous suspension of TiO_2 (Okamoto et al. 1985). Several researchers have since reported that molecular oxygen is necessary for complete oxidation of organic compounds (Matthews 1984; Kraeutler and Bard 1978; Turchi and Ollis 1990; Al-Ekabi and Serpone 1988; Rao and Natarajan 1994).

It is possible, thermodynamically, that many organic compounds in solution can be oxidized by direct contact with the valance band holes. The question whether h^+ or $\cdot OH$ is the continuing source of curiosity in the photocatalysis field. However, this pathway is unlikely because the concentration of H_2O or OH^- is much higher than the concentration of organic compounds. Nevertheless, holes may become important when the coverage of organic compounds is very high. Izumi and colleagues (1980) conducted PCO experiments using aerated pure organic solvents (no water present) and found that only partial oxidation occurred with a very slow production of carbon dioxide.

In another study, Cunningham and Srijaranai (1988) conducted photoassisted catalytic oxidation experiments with aqueous TiO_2 slurries using isopropanol and deuterated isopropanol. They found that the substitution of deuterium for hydrogen on isopropanol had no effect on the initial reaction rate and that direct reaction with the valance-band hole was not significant. They also conducted experiments where deuterium oxide (D_2O) was used instead of water as the solvent and found that the initial reaction rate decreased by a factor of three. The observed decrease in the rate when D_2O was used as a solvent was due to the higher electron voltage required for generating the $\cdot OD$ radical as compared to the $\cdot OH$ radical. Their results showed that the rate-limiting reaction step is the formation of the hydroxyl radical ($\cdot OH$) or the deuterium radical ($\cdot OD$) (Cunningham and Srijaranai 1988).

Turchi and Ollis (1990) summarized studies that have used electron spin resonance (ESR) to identify and verify the existence of hydroxyl radicals in near-UV illuminated aqueous solutions of TiO_2. Dimethyl picryl hydrazide (DMPH) is commonly used to trap $\cdot OH$. The DMPH$\cdot$-OH spin-trap complex displays a four-finger symmetric resonance with 1:2:2:1 intensity ratio. Turchi and Ollis pointed out that the radical–spin trap complex is detected in the aqueous phase but stated that it is not clear whether the actual trapping occurs on the catalyst surface or in solution. The perhydroxyl radical ($HO_2\cdot$) and $O_2^-\cdot$ were also detected in these studies.

In summary, experimental evidence supports the proposition that the hydroxyl radical is the primary oxidant responsible for the mineralization of organic compounds and that its formation is principally due to the reaction of the valence-band hole with water or surface OH^- groups. Although it can be stated that hydroxyl radicals are produced on the catalyst surface, the reaction pathway of the radicals in mineralizing organic compounds is still unclear. Several detailed reaction pathways have been proposed for the photoassisted catalytic oxidation of organic compounds in the aqueous phase (Izumi et al. 1980; Okamoto et al. 1985; Sakata, Kawai, and Hashimoto 1984; Pelizzetti et al. 1990; Serpone et al. 1986; Kraeutler and Bard 1978; Ollis et al. 1984; Childs and Ollis 1981; Matthews 1987a, 1987b; Hashimoto, Kawai, and Sakata 1984; Herrmann and Pichat 1980; Pruden and Ollis 1983; Fox, Ogawa, and Pichat 1988). As pointed out by Turchi and Ollis (1990), there are many possible reaction pathways between the radical species and the organic molecules. The adsorbed hydroxyl radical may interact with the catalyst or react with an adjacent adsorbed or aqueous-phase organic molecule or other radical species. Alternatively, the hydroxyl radical may diffuse into the aqueous phase and react with an adsorbed or aqueous-phase organic molecule or other radical species. The pathway may be some combination of all of these possibilities. The actual reaction pathway for the degradation of organic compounds is still unknown.

2.3   Advanced Oxidation Processes Based on PCO for (Waste)water Treatment

Global water supplies, formerly of good quality, are now potentially hazardous to mankind and aquatic life in particular. Extensive industrial development and insincere use of methods for disposal and storage of polluted effluents have resulted in progressive deterioration of water quality in many places. A number of contaminant chemicals have been found to enter ground and surface waters through routine industrial, domestic, and agricultural operations.

Conventionally, varieties of chemicals in effluents are removed by physicochemical and biological methods. However, these methods are associated with some severe problems such as sludge disposal and regeneration of adsorbents and thus cannot offer a permanent solution to the pollution hazard. In particular, the biological methods are indifferent to some toxic and hazardous chemicals and do not ensure desired removal efficiency. Photocatalysis may be chosen as a pretreatment method to induce molecular transformations for enhanced biological degradation (step A in figure 5.18) or alternatively as a posttreatment method for ensuring standards compliance (step B, in figure 5.18).

Based on the water purification, self-cleaning, air-cleaning, and self-sterilizing properties of the TiO_2/UV system, a number of process applications have been investigated. Selected applications are listed in table 5.7.

As discussed above, photocatalytic oxidation involves the use of photoactive n-type semiconductor powders and near-UV light. When these semiconductors are illuminated in water, a redox environment is established, indicating that oxidation of organic compounds can take place. In most cases, the crystalline anatase form of TiO_2 has been used as the semiconductor because of its high activity and stability. Laboratory studies (Okamoto et al. 1985; Al-Ekabi and Serpone 1988; Ollis et al. 1984; Childs and Ollis 1981; Matthews 1987a, 1987b; Herrmann and Pichat 1980; Pruden and Ollis 1983) have demonstrated that a wide variety of organic compounds, such as chlorinated alkanes and alkenes, polychlorinated phenols, aromatics, aldehydes, and organic acids, can be oxidized using near-UV or solar-illuminated TiO_2 aqueous suspensions. Parachlorobenzene, as well as 2,4-dichlorophenoxyacetic acid, under-

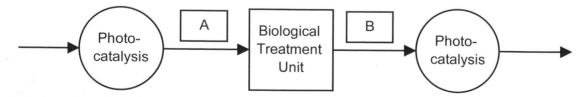

Figure 5.18 Role of photocatalysis in industrial wastewater treatment

Table 5.7 Selected applications of photocatalysis

Property	Category	Application
Self-cleaning	Materials for residential and office buildings	Exterior tiles, kitchen and bathroom interior furnishings, plastic surfaces, building stones and curtains, cloth for hospital
	Indoor/outdoor lamps	
	Materials for roads/subways	Tunnel wall & lamps, spray coatings for cars
Water purification	Drinking water	River water, groundwater, lakes, fish feeding tanks, drainage water, etc.
Air cleaning	Indoor air cleaners	Room/factory storeroom air cleaners
	Outdoor air purifiers	Photocatalysts for highway tunnel air cleaning, etc.
Self-sterilizing	Hospital and others	Tiles to cover floors, walls, public restrooms, bathrooms, etc.

goes rapid degradation with TiO_2/UV technique (Rao, Dube, and Natarajan 1993). Details with respect to each class of compounds are published (Rao and Natarajan 1994).

Surfactants can also be destroyed under solar or simulated sunlight in the presence of TiO_2. The authors applied this technique to treat water containing commercial soap/detergent formulations, which comprise mixtures of surfactants (Rao and Dube 1996). The technique may also be applied to remove various metal impurities. Thus, Ramachandriah and colleagues (1996) were able remove copper from industrial plating sludge and achieved about 80 percent removal as Cu/TiO_2 in two hours with a 400-W mercury vapor lamp and in six hours under natural sunlight.

Rao and Dube (1996) have also developed a compartmentalized photoelectrosynthetic cell constituting an illuminated TiO_2 photoanode and a platinum cathode and found H_2 evolution at the cathode with organic pollutants as sacrificial donors. Organic pollutants can act as sacrificial electron donors in photoelectrochemical cells, which may be used to generate hydrogen and to destroy water soluble organic pollutants simultaneously, both of which are useful reactions. The system appears to possess some practical interest as it combines energy- and environment-related reactions. The authors reported photodegradation of the herbicide 2,4-D, with a COD removal efficiency of 90 percent and above (Rao and Dube 1995).

In addition to some TiO_2 catalysts from the firm Aldrich, Degussa, and others, Indian commercial TiO_2 also shows significant photoactivity upon reduction in H_2 atmosphere at 450°C. Rao and Dube (1997) have reported the TiO_2-catalyzed photodegradation of Reactive Orange 84 and alizarin red S biological stain. While complete color removal took place within 30 minutes, about four hours is required to remove COD to the extent greater than 90 percent.

The photocatalytic degradation of Reactive Orange 84 in dye-house effluent using a single pass photoreactor has also been observed (Rao and Dube 1998). In this system, completely colorless treated solutions were obtained with less than a 5-ml/min flow rate. Simultaneously, some COD is also removed. The reactor was so designed as to retain catalyst particles in the reactor itself. Both color and COD are reduced from a dye-containing wastewater using the photocatalytic method (Dhodapkar et al. 2000). Through a detailed study of photodegradation of surfactants using MO_3/TiO_2 catalysts (M = Mo or W), the rate constant for their photocatalytic oxidation is found to be independent of the nature of surfactants (Rao and Dube 1996). The TiO_2 photocatalysis is indiscrete and does not significantly depend on the nature of organic reactant.

Photocatalysis has been used to induce biodegradability in 2-CP solutions (Rao et al. 2003). The 2-CP solutions are readily detoxified as the C–Cl bond is broken, and the treated solutions show increase in COD. Some new intermediates formed were found more biodegradable. The aspect of photocatalytic destruction of organic pollutants and various methods applicable to enhance the efficiency of photocatalytic process has been reviewed recently (Rao and Kaul 1999; Dhodapkar et al. 1999).

Table 5.8 lists many of the compounds that have been successfully destroyed using this process. Most were completely mineralized to such compounds as carbon dioxide, water, and hydrochloric acid (HCl). Several of the compounds listed in the table are important to the water industry. However, studies have shown that some nitrogen-containing compounds, such as atrazine, are difficult to mineralize with this process. For example, Pelizzetti and colleagues (1990) reported that atrazine was photocatalytically oxidized to cyanuric acid.

Currently, development of PCO for destroying organic contaminants is in the pilot-scale stage, and several pollution equipment companies and research laboratories are moving toward its commercialization. For example, one company recently received funding from the U.S. Environmental Protection Agency for a pilot-plant demonstration of its PCO system. In the process developed by this company, TiO_2 is coated onto a fiberglass mesh, which is wrapped around near-UV lamps and placed in a stainless-steel reactor. The contaminated water is then pumped through a jacket so that it is exposed to the illuminated TiO_2. The contaminants are oxidized (Al-Ekabi 1990). In another process, developed and pilot-tested by researchers from Sandia National Laboratories, a linear parabolic sun-tracking solar collector concentrates sunlight onto a UV-transmitting reactor containing TiO_2 powder suspended in contaminated water

Table 5.8 List of compounds that have been destroyed using PCO

chloroform	cyanide
2-chlorophenol	4-chlorophenol
1,2-dibromomethane	1,2-dichloroeth ane benzoic acid
biphthalates	carbon tetrachloride
chloromethane	nitrobenzene
methanol	tetrachloroethene
dichloroacetic acid	n-propanol
2-propanol	1,1,1-trichloroethane
2,4-dichlorophenol acetic acid	ethyl acetate
p-dichlorobenzene	o-dichlorobenzene
2-naphthol	sucrose
vinyl bromide	difluorodichloro-methane
hydrogen sulfide	

(Al-Ekabi 1990). The reactor is capable of mineralizing 5 ppm of trichloroethane (TCE) to 5 ppb in about four minutes at a rate of 42 liters per minute.

The PCO process is an emerging water and wastewater treatment technology. However, major engineering strides must be taken if this process is to become viable and cost-effective.

For water treatment, the primary goal of the photocatalytic oxidation process is to produce an oxidizing environment that will allow organic compounds to be destroyed or mineralized in a cost-effective manner. Meeting this goal depends on several process variables: (1) the catalyst type; (2) the incident light intensity and wavelength; (3) the reactor configuration; and (4) the physical and chemical properties of the water matrix. Providing a cost-effective process requires optimizing at least the first three variables, although understanding the impact of the physical and chemical properties of a given water matrix is necessary in the overall process operation. Optimization of those properties (such as pH and temperature) is usually not cost-effective for drinking-water treatment.

Several factors can influence the efficiency of electron-hole separation at the surface of a photoactive semiconductor and the subsequent oxidation/reduction reactions at the solid–liquid interface. Some of the important factors, which include crystal morphology, are surface character, substitution doping, and surface modification.

2.4 TiO_2 Materials as Photocatalysts

In terms of criteria for choosing appropriate semiconductor photocatalysts, the most important appears to be related to the location of the band edges at the surface. That is, the conduction and valence band edges have to be located at energies (potentials) such that reactions of hydroxyl radical formation using adsorbed H_2O and O_2 may be photodriven. Hence, the valence band has to be placed at potentials that are at least $+2.85$ V (with respect to the standard hydrogen electrode). Only then will the photo-generated holes have sufficient energy to oxidize water. Similarly, if O_2 were used as the electron acceptor, the conduction band would have to lie at a value negative of the standard potential for the reduction of O_2.

In addition to surface energetics, the stability of the semiconductor particle to photo-corrosion is a crucial factor. Finally, in terms of practical considerations, cost is also an issue. Titanium dioxide has been by far the most popular photocatalyst. Both anatase and rutile modifications have been used, although commercial samples (e.g., Degussa P-25) often contain a mixture of the two. The energy-band gaps of anatase (3.23 eV, 384 nm) and rutile (3.02 eV, 411 nm) combine with the valence-band positions to create a favorable situation for the photogeneration of highly energetic holes at the interface. However, anatase is superior to rutile for photocatalytic applications. First, the conduction-band location for anatase is more favorable for driving conjugate reactions involving electrons. Other variant reasons have been given. For example, the poorer photocatalytic activity of rutile has been attributed to its high e^-–h^+ recombination rate and its lower oxygen photoadsorption capacity. A more recent study concludes that very stable surface peroxo groups can be formed during photo-oxidation reactions at the anatase but not at the rutile surface. The implication is that the oxidation of organic compounds such as 4-chlorophenol proceeds through an indirect pathway involving these surface species. The decrease noted in the photocatalytic activity when titania (prepared by precipitation method) is annealed at temperatures higher than 600°C appears to have a similar mechanistic origin in the extent of hydroxylation of the oxide surface. Other semiconductors have also been employed, and a representative listing appeared in table 5.6. However, none of these (at least to date) appear to match the attributes of TiO_2.

The band gap values listed in table 5.6 are for "bulk" semiconductors or for colloidal particles of several hundred nanometers in size. Size quantization, however, can cause sizable shifts in these band-gap energies. For example, TiO_2 microcrystallites as small as two nanometers have been prepared in Nafion and clay interlayers, with a corresponding band-gap energy as high as 3.95 eV. Similar trends have been observed for Fe_2O_3, ZnS, and CdS. Very high activities have been observed for Fe_2O_3 nano-sized particles (relative to bulk Fe_2O_3 powder) for model photoreactions such as the photodecomposition of saturated carboxylic acids.

The vast majority of the semiconductor photocatalysts listed have rather high band gaps. Large-band-gap semiconductors in general tend to be more stable against photocorrosion. Oxidation of many pollutants, especially organic species, requires high potential, with the result that the valence band location at the semiconductor–electrolyte interface has to be rather positive as exemplified by TiO_2 and CdS.

The ability of a photocatalyst such as TiO_2 to perform oxidation/reduction reactions can be significantly influenced by its morphology. The two most common crystalline forms in which TiO_2 exists are anatase and rutile. Kraeutler and Bard (1978) found that the anatase form of TiO_2 was significantly more reactive than the rutile form in the photocatalytic decarboxylation of acetic acid–sodium acetate mixtures. Sclafani, Palmisano, and Schiavello (1990) evaluated the photoassisted catalytic oxidation rate of phenol in an aqueous solution using seven commercially available TiO_2 powders, six primarily anatase and one rutile. They found a wide range of phenol degradation activity for the anatase phase samples, whereas the rutile form was not active.

Sclafani, Palmisano, and Schiavello (1990) conducted further experiments to evaluate the impact of calcination temperature and time on phenol destruction. Two differently prepared TiO_2 powders were used. One method precipitated titanium trichloride ($TiCl_3$) in an aqueous solution of 25-wt% ammonia; the other method precipitated titanium tetrachloride ($TiCl_4$) in pure water. Both methods produced the anatase form of TiO_2. In the experiments, calcination temperatures ranged from 120°C to 800°C and calcination times from 3 to 336 hours.

For TiO_2 prepared by the first method (using $TiCl_3$), the phenol degradation activity was low and increased slightly with calcination temperature up to about 650°C. At this temperature, TiO_2 crystallinity was 70 percent anatase and 30 percent rutile. The calcination time did not appear to influence the catalyst activity. For TiO_2 prepared by the second method (using $TiCl_4$), the phenol degradation activity was significantly higher. A significant rutile phase (42 percent) was formed at 400°C and increased steadily to 100 percent at about 650°C. The highest oxidation rate for phenol was reported at 550°C, where the TiO_2 contained about 25 percent anatase and 75 percent rutile. The results also showed that the surface area of the TiO_2 decreased with increasing calcination temperature and that increasing calcination times decreased the anatase–rutile ratio.

Similar results were reported by Abrahams, Davidson, and Morrison (1985). In addition, they found that the particle size decreased with increasing calcination temperature. These results demonstrate exceptional variability in the photoactivity of TiO_2 due to morphological changes. The changes depend on the starting material and method of preparation.

Changing the crystal morphology of the semiconductor by substituting other transition metal ions into its crystal lattice can increase its photocatalytic activity. Substitution doping has been tested in solar energy applications for photo-dissociating water (Kraeutler and Bard 1978; Borgarello et al. 1981; Kiwi 1983; Pelizzetti and Visca 1983). For example, Borgarello and colleagues (1981) evaluated the impact of niobium (NbS^+) doping on both the anatase and rutile forms of TiO_2. For the anatase form, they found that increasing the NbS^+ doping level from 0.07 wt% to 0.4 wt% increased the activity by about 20 percent, whereas for the rutile form, the activity increased dramatically. Substituting the titanium (IV) ions (Ti^{4+}) in the TiO_2 with the higher-valence NbS^+ ions introduces extra valence electrons into the lattice and causes the NbS^+ ions to behave as electron donors. Their behavior extends the spectral response toward visible light wavelengths, changing the band gap of TiO_2 (Pelizzetti and Visca 1983) and allows the electrons to migrate more easily to the solid–liquid interface (Kiwi 1983). However, Malati and Wong (1984) showed that substituting chromium (III) ions (Cr^{3+}) for Ti^{4+} ions had the opposite effect. The Cr^{3+} ions act as electron acceptors and retard electron migration toward the solid–liquid surface. These investigators also showed that TiO_2 doped with Cr^{3+} extends the spectral response toward visible light further than NbS^+ does; however, lower quantum efficiencies were observed.

For a photoactive semiconductor particle immersed in an aqueous medium, the production of hydroxyl radicals at the exterior particle surface is influenced by the electron–hole separation at the surface. For a given incident light intensity and wavelength, minimizing the charge recombination at the catalyst surface will help maximize the production of hydroxyl radicals available for oxidation reactions, increase the quantum yield, and increase the efficiency of the overall process. In the presence or absence of oxygen, efficient charge separation at the surface of the particle can be carried out by depositing transition metals, such as platinum, and other metal oxides, such as ruthenium oxide (RuO_x), on the particle surface (Kraeutler and Bard 1978; Borgarello et al. 1981; Kiwi 1983; Pelizzetti and Visca 1983; Nakabayashi, Fujishima, and Honda 1983; Duonghong, Borgarello, and Gratzel 1981; Sakata and Kawai 1983). The conduction-band electrons are attracted to transition metals, whereas the valence-band holes are attracted to the metal oxides. The result is a high degree of electron–hole separation.

The use of band-gap semiconductors such as TiO_2, ZnO, ZrO_2, and CdS and their various modified forms as the gaseous and aqueous phase photocatalysts has been known. Further, it has long been known that certain materials such as noble metals (Pt, Pd, Au, Ag) and some metal oxides (RuO_2, WO_3 and SiO_2) facilitate electron transfer and prolong the length of time that electrons and holes remain segregated.

Sakata and Kawai (1983) evaluated the rate of hydrogen evolution for an ethanol–water solution containing TiO_2 loaded with several different transition metals at 3 wt%. The TiO_2 loaded with platinum, palladium, and rubidium showed significantly higher rates of hydrogen evolution than did TiO_2 alone. They also compared the hydrogen evolution rate from the photo-dissociation of water for TiO_2 alone, Pt-TiO_2 (weight ratio 10:100), and RuO_2-TiO_2-Pt (5:100:10). The rate of hydrogen evolution for RuO_2-TiO_2-Pt was about four times greater than for Pt-TiO_2, which was about 190 times greater than for TiO_2 alone. The quantum yields for bifunctional catalysts such as RuO_2-TiO_2-Pt have also been reported to be higher than for TiO_2 alone. Sakata and Kawai (1983) reported quantum yields of hydrogen production of 50 percent for water-alcohol mixtures containing RuO_2-TiO_2-Pt and illuminated with monochromatic light of 380 nm; for mixtures containing TiO_2 alone, quantum yields were only 3 percent. For a similar water–alcohol mixture under monochromatic light of 310 nm, Duonghong, Borgarello, and Gratzel (1981) reported quantum yields of hydrogen production of about 30 percent for a RuO_2-TiO_2-Pt catalyst.

Surface modification of TiO_2 with sulfate has been reported to be very useful in the photocatalytic oxidation of volatile organic compounds such as heptane, trichloroethylene, ethanol, acetaldehyde, and toluene (Muggli and Ding 2001). The coverage of all organics on SO_4^{2-}/TiO_2 was significantly higher than on P-25 TiO_2, and room temperature PCO is faster with P-25. The SO_4^{2-}/TiO_2 may be preferable to Degussa P-25 for PCO at elevated temperatures. On the other hand, Dhodapkar and colleagues (2003) have recently shown that carbonate present in wastewater could inhibit photodegradation strongly. The adsorption of organic compounds is significantly limited by preferential adsorption of CO_3^{2-} that also forms very stable Ti-CO_3^{1-} surface complexes.

2.5   Design of Photocatalysts and Photoreactors

Three types of mercury vapor lamps—low-, medium-, and high-pressure—are used as UV light sources in a typical photocatalytic experiment. However, medium-pressure mercury lamps (MPMLs) are more frequently used. The MPMLs commonly used for photocatalytic studies generate large amounts of thermal radiation at relatively higher temperatures, with electric-to-UV energy conversion efficiency of less than 15 percent. The heat must also be used for driving thermocatalytic reactions.

It is generally recognized that only a very thin layer on the photocatalyst surface can actually be excited to enter photocatalytic reactions. For most active photocatalysts, the physical thickness of this layer or skin does not exceed few microns. This is because UV radiation is completely absorbed within a skin only few microns thick on the exposed photocatalyst surface, in contrast to thermal radiation that can penetrate deep into the supported catalyst. The fact that most target species can also be adsorbed into the deep layers of the porous photocatalytic media encourages the use of multifunctional catalysts capable of utilizing both heat and light emitted by medium- and high-pressure UV lamps. Thus, a multipurpose catalyst can comprise a base material that acts as both a thermocatalyst and a support structure for the photocatalyst. Alternatively, a dual catalyst that can function as both thermocatalyst and photocatalyst simultaneously may be used.

Assuming band-gap energy of 3.1 eV for TiO_2, a threshold wavelength of about 400 nm is obtained. Thus, TiO_2 will absorb light having a wavelength equal to or lower than this value. While these wavelengths are much longer than the excitation wavelengths needed for the UV/H_2O_2 and UV/O_3 processes, the match of the TiO_2 absorption profile with the solar spectrum is unfortunately rather poor. However, solar photocatalysis would be a better candidate in terms of economics. Attempts to extend the TiO_2 spectral response into the red through dye sensitization has relied on initial excitation of the dye followed by subsequent electron transfer from the excited-state dye molecules into the TiO_2 conduction band. Such an approach should be especially well suited for photocatalytic treatment of dye-containing wastewater.

Once holes and electrons are photogenerated, they move about the crystal lattice freely. Thus, a major portion of the electrons and holes get a chance to recombine, and only a small portion reach the surface where they react with adsorbed species to form reactive radicals. An important factor in controlling the rate of $e^- - h^+$ recombination on the photocatalyst surface is the size of the photocatalyst particles. The smaller these particles are, the shorter the distance charge carriers must travel to reach the surface and the larger the exposed catalyst surface area is. Photocatalysts with a diameter of a few nanometers and BET surface areas of more than 100 m^2/g are commercially available (e.g., ST-01 and ST-31 grade titania produced by Ishihara Sangyo Kaisha, Ltd., of Japan).

The rate of recombination of holes and electrons is a function of the catalyst surface irradiance: the greater the surface irradiance, the greater the rate of recombination. It is generally recognized that hydroxyl radical generation is a rate-limiting step. The rate of surface reactions will then be equal to

$$r = k_{(C+d)}(h^+_{(VB)})$$

The rate of hole formation is $k_a q_i$, where q_i is catalyst surface irradiance (quanta/sec/cm^2). The rate of e–h recombination then is

$$k_e(h^+_{VB})(e^-_{cb}) = k_e(h^+_{vb})^2$$

When q_i is high, a large number of electrons and holes will be generated, and r will be equal to $k(q_i)^{1/2}$. At low values of q_i, when the surface concentration of holes is relatively small, the recombination term will be negligible and r will be $k_a q_i$. The surface irradiance value at which the reaction rate transitions from q_i to $(q_i)^{1/2}$ is called the Egerton-King threshold q_{EK} (equal to 2.5×10^{15} quanta/sec/cm^2 at wavelengths λ of 335, 365, and 404 nm). The q_{EK} for two commonly used UV light sources (low- and medium-pressure mercury lamps) is approximately equal to 1.95 mW/cm^2 (for λ = 254 nm) and 1.36 mW/cm^2 (for λ = 365 nm), respectively.

In order to limit the rate of recombination of electrons and holes and maximize the photoreactor performance, it is necessary to limit the catalyst surface irradiance to levels at or below the Egerton-King threshold. The rate of surface reactions r is proportional to $(q_i)^m$, where m varies between 0.5 and 1. To increase the rate of surface reactions for target pollutants, it may be necessary to allow q_i to exceed q_{EK} under certain conditions. Therefore, under practical situations, the requirement for an efficient utilization of the photogenerated charge carriers must be balanced against the need for optimum rate of the surface reactions involving primary and secondary reactants. This requires a careful photoreactor design that allows uniform irradiance over all photocatalyst surfaces at a level that is as close to q_{EK} as possible and an optimum rate of conversion of surface-borne target species to desirable final products.

Just like radiation and heat transfer, transport of the primary reactants to and final products from the catalyst surface affect the photo-processes' performance. The reactor engineering is closely coupled to the choice and configuration of the media and the type of light source used. A proper photoreactor design should provide for uniform irradiance on all catalytic surfaces as well as effective mass transport to and from active sites on the photocatalyst surface. In general, photoreactor designs fall into three categories, which are listed in table 5.9.

Mass transfer of species is generally improved by generating and enhancing turbulence in the flow. Many articles such as ribs, fins, pleats, beads, chips, flaps, strips, coils, baffles, baskets, wires, and so on have been conceived, used, and patented for generating mixing and turbulence. However, due care should be given when using

Table 5.9 Types of photoreactors

Type	Properties	Reactor Types
Category I (most photoreactors)	Good mass transfer but generally poor radiation field characteristics	Photocatalyst-coated monolith, photocatalyst-coated panel, and baffled annular photoreactor
Category II	Poor mass transfer but uniform catalyst surface irradiance	Annular photoreactor designs
Category III	Poor mass transfer and nonuniform catalyst surface irradiance	Externally lit annular photoreactor, catalyst coated on the inner wall

such articles for enhancing turbulence so that the radiation field characteristics are not altered. In view of this, it is advantageous to design the photoreactors in such a way that the bulk of the catalyst resides on the reactor wall. This limits the number and proximity of internals. In general, baffles receive more radiation than the walls; thus, most internals would be undesirable in the context of the uniform radiation field that is always desired.

At least six distinct types of catalytic (photocatalyst/support) media arrangements have been in use for related applications. The media, designated Type 0 to Type 5, are described in table 5.10.

Semiconductor materials may be available in three forms: powders, single crystals, and thin films. For practical applications, single crystals can be safely excluded as serious candidates for remediation processes in terms of both economic and technical perspectives. Semiconductor thin films are attractive for use in photovoltaic devices, but are inadequate for photocatalytic waste treatment applications unless special efforts are made to tailor their surface morphology. This leaves the option of reactors containing slurries and suspensions of semiconductor powders. These do offer the advantage of high surface dispersion that entails a higher particle–substrate encounter frequency. However, photocatalyst recovery after use becomes an issue of major practical concern.

Immobilized photocatalysts in the form of highly microporous particles that are attached to a solid support represent an effective compromise incorporating, to a degree, the positive features of powder-based slurries and suspensions. Thus TiO_2 has been immobilized on beads, hollow tubes, Vycor glass, woven fabric, silica gel, optical fibers, and sand. Porous thin films of TiO_2 have been synthesized thermally, by fusing ZnO or TiO_2 particles onto conducting glass or dispersing the powder on a polymeric binder.

Several immobilization approaches have evolved, most of them support-specific. A novel method for immobilization of the titania catalysts on a metallic support (SS 304) was developed by using the chemically inert and photo-stable fluroresin PTFE (Teflon) as a binder (Kuo 2000). The conditions with respect to baking temperature, baking time, and the ratio of resin/TiO_2 were optimized by applying response surface methodology (RSM). The resin-bonded titania photocatalyst showed twice as much photoactivity. Rao and Dube (1995) have been able to immobilize TiO_2 onto polyester fabric

Table 5.10 Types of (photo)catalytic media

Type	Properties	Remarks
Type 0	A suitable catalyst such as titania is used in colloidal form without any support or base materials. There exists a subcategory to Type 0 media; these employ, in addition to titania, an absorbent material that absorbs the target compounds. The contaminant-loaded adsorbent is then separated from the fluid and brought into contact with the aqueous slurry of a suitable photocatalyst. This is implicit that whenever adsorption target species is rate limiting. Preadsorption of contaminant followed by photocatalysis is a more suitable method. However, this is generally not the case. It is desirable to simplify the treatment process by eliminating additional adsorbent in favor of multifunctional catalytic media.	Type 0 catalytic media need to be separated and recovered after the photocatalytic reaction is over. If the activity persists, it may be reused.
Type 1	In this type of photocatalyst/support arrangement, the catalyst, such as TiO_2, is immobilized or bonded onto a ceramic, glassy (e.g., fiberglass mesh, woven glass tape, etc.) or metal oxide (e.g., silica gel), metallic (e.g., stainless steel), or synthetic polymeric support substrate (e.g., plastic, PE fabric, etc.)	The substrate has no function other than providing physical and structural support for the photocatalyst.
Type 2	In Type 2 media configuration, impregnated glassy mesh/matrix or porous ceramic monolith or beads, along with metallic and metal oxide substrates (in the form of plates, beads, etc.), are employed as the photocatalyst support to which titania is bonded utilizing a method known as the "sol-gel" technique. In this technique, titania colloid is generated by controlled hydrolysis of a titanium precursor salt. The titania from colloid is coated on supports by dip-thermal cycles are spin-coating technique.	The substrate has no function other than providing physical and structural support for the photocatalyst.
Type 3	Type 3 catalyst/support configuration is a variation of the Type 2 media that involves synthesis and use of metal oxide aerogels—most prominently, SiO_2 aerogels doped or co-gelled with other transition metal oxides such as titania—to produce photochemically active catalyst/support material. The preparation comprises two steps: A condensed metal oxide intermediate is formed. From this intermediate, aerogels are prepared having the desired density, clarity and UV transparency, moisture, and stability.	The competitive adsorption of organics at TiO_2 as well as unreactive SiO_2 sites may lead to incomplete mineralization of pollutants.
Type 4	In Type 4 photocatalyst/support media, a photocatalyst (e.g., doped and undoped modifications of TiO_2, CdS, etc.) is deposited by bonding or cementing onto fabric of a modified or unmodified natural or synthetic polymer material. Examples of polymers include (natural) wood, paper, kozo, gampi, kraft lignin, woven cotton, kenaf, linen, wool, and others.	There is an upswing in research in this area, as naturally available supports are cheap and devoid of catalyst–support interactions, unlike Type 3 catalytic media.

(continues)

Table 5.10 Continued

Type	Properties	Remarks
Type 5	Type 5 media include the broad field of moderate-temperature (150°C–350°C) thermal oxidation catalysts. A subclass of relevance to the present subject is supported-transition metal-oxide catalysts and cation-modified zeolites as dual-function sorbent/catalyst media. Such catalysts are useful for removing dilute VOCs from air at room temperature, and then act as a catalyst at higher temperature to both desorb and oxidize trapped VOCs. Due to their microporous crystalline structure, various forms of zeolites have been widely used as commercial adsorbents. Further, noble or base metal–supported titania or zirconia (1-wt% Pt, 5-wt% V, 5-wt% La, 89% TiO_2), with added promoters such as Mo, W, or V, are useful for catalytic oxidation of organic nitrogen-containing compounds.	An excellent choice if dual-function adsorbent/catalyst media are needed.

using polyvinyl alcohol as binder. Similarly, they experimented with Ti-TiO_2, prepared by thermal as well as flame oxidation of Ti sheet.

Sol-gel technology can also be used for fabricating porous films onto suitable supports. The physicochemical and photocatalytic properties in aqueous solutions of the supported TiO_2 films prepared by different sol-gel methods have been studied (Guillard et al. 2002). The properties are influenced by the nature of the stabilizing agents (e.g., acetic acid, acetylacetone, etc.), the nature of organo-titanium precursors (methoxo-, ethoxo-, isopropoxo-, or ethoxo-isoporopxo-), and solvent. The migration of cations native to supports into thin TiO_2 films drastically reduces the photoactivity of the supported catalysts. The migration is facilitated at higher temperatures and longer calcination times.

Unfortunately, immobilized photocatalyst reactor configurations resulted in lower photocatalytic activity relative to their slurry counterparts. This is traced to substrate mass transport problems crucial with immobilized catalysts. This area of research is now awaiting a breakthrough in the development of the supported form of TiO_2 with a hundredfold higher photoactivity for oxidation of organics unselectively.

2.6   Photocatalysis with Sunlight

Most of the solar photochemical processes are based on the utilization of high-energy, short-wavelength photons (300–400 nm) to promote photochemical reactions. In the solar heterogeneous photocatalytic detoxification process, the near-UV part of the solar spectrum (wavelengths shorter than 380 nm) is used to photoexcite a semiconductor catalyst. Although scientific research on solar detoxification has been conducted for at least three decades (Blake 1999), industrial/commercial applications, engineering systems, and engineering design methodologies have only been developed recently (Blanco and Malato 2000; Malato et al. 2002).

The hardware needed for solar photochemical applications is similar to that used for solar thermal applications. The original solar photoreactors were based on line-focus parabolic trough concentrators and nonconcentrating collectors (see figure 5.19).

The first engineering-scale solar photochemical facility for water detoxification in Europe was developed by CIEMAT (Blanco and Malato 1992) using 12-axis PTC. The sunlight is reflected by a parabola onto the reactor tube at the focus, through which contaminated water is circulated. The reactor receives a large amount of radiation energy per unit volume and an optical efficiency of greater than 50 percent is obtained (Blanco et al. 1991; Blanco and Malato 1992). Apart from a variety of organic contaminants, industrial wastewaters are treated using the PTC reactors. The main disadvantages are that (1) the photoreactors use only direct radiation and (2) the collectors are expensive and have low optical efficiencies, at least for TiO_2 applications.

Nonconcentrating solar reactors are in principle cheaper than PTCs, as their components are simple and easy to maintain and also do not have expensive tracking devices and so forth. There are three types of nonconcentrating reactors:

- Free-falling film, in which the process fluid falls slowly over a tilted plate with a catalyst attached to the surface, which faces the sun and is open to atmosphere (Goslich, Dillert, and Bahnemann 1997; Blanco et al. 1994)

- Pressurized flat-plate, which consists of two plates between which the fluid circulates using a separating wall (Well et al. 1997; Dillert et al. 1999; Bahnemann et al. 1999)

- Solar ponds—small, shallow, on-site pond reactors (Gimenez, Curco, and Marco 1997; Gimenez, Curco, and Queral 1999)

The fluid flow in nonconcentrating solar reactor is usually laminar. This presents problems with mass transfer and vaporization of reactants. In this sense, tubular reactors do have decisive advantage.

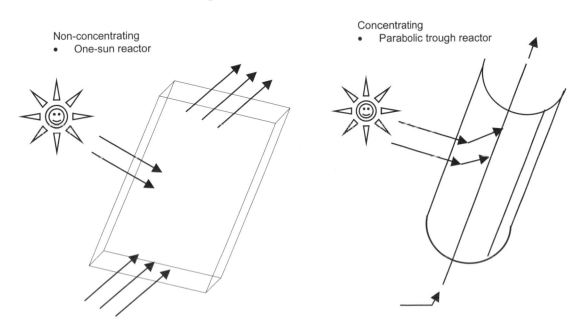

Figure 5.19 Design concepts for solar water detoxification photoreactors

Compound parabolic concentrator is a hybrid reactor possessing characteristics of both trough and one-sun reactors. They are static collectors with involute reflective surface around a cylindrical reactor tube. They are simple and cost-effective. The reflector design allows almost all the UV-irradiation to reach the reactor aperture where the tube reactor is placed.

Some chlorophenols (Malato et al. 1997; Dillert et al. 1999; Minero et al. 1993) have been subjected to photocatalytic degradation in a two-axis solar pilot-plant of PTC design at Almeria, Spain (260 liters, 20 mg/dm^3 chlorophenol, 0.2 mg/dm^3 TiO$_2$, and 32 m^2 collector surface area). The mineralization treatment rate was found to be 0.015, 0.012, and 0.03 g/m^2-hr TOC for PCP, DCP, and CP, respectively. The mean intensity of solar UV light was considered to be 30 WUV/m^2. When applied to real wastewater containing pesticides, about 20–30 mg/dm^3 TOC was reduced using 56 kJ/dm^3 solar UV energy (Blanco et al. 1994; Malato et al. 2000; Malato et al. 2002). Considering that the mean solar UV intensity at Almeria is 18.6 W$_{UV}$/m^2, a collector of approximately 1,000 m^2 is estimated to be necessary for treating 5,000 m^3 of wastewater per year.

2.7 Calculation of Energy Efficiency

The photodegradation of aqueous organic pollutants is an energy-intensive process; thus, definition of figures of merit based on the energy input seems logical. The appropriate figure of merit is the electrical energy per mass EE/M, defined as the electrical energy (kWh) required to remove one kilogram of the contaminant (Bolton and Cater 1994):

$$EE/M = (P \times t) / (100V \times (C_o - C_t)),$$

where:

P = power input of the UV lamp (kW)

t = treatment time (hr)

V = volume of the treated wastewater (m^3)

C_o = initial pollutant concentration (kg/m^3)

C_t = final pollutant concentration at time t (kg/m^3)

Calculation of the electrical energy requirements has proven to be very useful for AOPs employing artificial light sources (UV lamps), because a major fraction of the operating costs of AOPs is the electrical energy costs. However, in a solar light–driven treatment system, the electrical energy component is practically absent and the solar figure of merit has to be defined based on the collector area necessary to achieve a certain rate of degradation. In this case, the equivalent of the electrical energy input is the solar energy input.

If the incident solar radiation is standardized to 1 kW/m^2, the appropriate solar figure of merit similar to the EE/M definition is the collector area per mass CA/M, defined as the required collector area (m^2) to remove one kilogram of contaminant (Bolton, Ravel, and Cater 1996):

$$CA/M = (A \times E \times t) / (V \times (C_o - C_t))$$

where A is the irradiated surface area of the photoreactor (m^2), E is the average irradiation intensity reaching the photoreactor surface (W/m^2), and the other parameters are as defined above.

Solar figure of merit, estimated based on the mineralization kinetics of the synthetic dye-house effluents in kilograms of TOC removed during a one-hour treatment via the TiO_2/UV-A process, is shown in table 5.11.

It is clear that catalyst activity has profound impact on the figure of merit. For example, the Microanatas/UV-A system requires 42 m² less collector area per kg TOC removed in comparison to the standard P25TiO_2/UV-A system.

2.8   Titanium Dioxide Photocatalysis: Self-Cleaning of Building Materials

Cleaning of building materials such as tiles, facades, and glass panes causes practical problems of inaccessibility, as well as consumption of energy and chemicals. Using thin film coatings of photocatalyst materials on the surfaces to be cleaned is smarter than cleaning based on lowering the contact angle with soap and detergents. Thus, very thin films of photoactive TiO_2 shone with UV light have helped to lower the contact angle to less than 1. These coatings are superhydrophilic, and water lies flat on the surface in sheets instead of forming droplets. The superhydrophilicity is retained for approximately two days even after the UV light is removed.

In addition, UV-irradiated TiO_2 has the ability to oxidize and decompose many types of bacteria and organic and inorganic materials. This effect is used for cleaning tile walls and ridding them of bacteria (Fujishima et al. 2000). In a reasonably well-lit room, these coatings can decompose a hydrocarbon layer of approximately 1 m thickness every hour (Fujishima et al. 2000).

Some of the applications reported in the literature are TiO_2-containing paper, microfilms, self-cleaning TiO_2 subway tunnel lamps, and flow-through water purification. Further, self-cleaning window glass has been developed, as well as TiO_2-containing, self-cleaning paint. The TiO_2 photocatalysts can also be applied for killing bacteria, which can lead to the development of self-sterilizing surfaces (Kikuchi et al. 1997).

3   Conclusion

The feasibility has been demonstrated for application of direct or mediated electrochemical oxidation to solving problems of industrial pollution. Several components of industrial effluents cyanides, dyes, ammonium ions, sulfide, tannins, and other organics (TOC and COD) of real and synthetic wastewaters—have been proven to be

Table 5.11   Solar figures of merit for synthetic dye-house effluents

Process type	Treatment conditions	EE/M (kWh/kg TOC)	CA/M (m²/kg TOC)*
P25TiO_2/UV-A	1-g/L TiO_2; pH = 7	—	244.00
Pt-P25/UV-A	1-g/L TiO_2; pH = 10.5	—	233.92
Microanatas/UV-A	1-g/L TiO_2; pH = 10.5	—	201.88
PC 500/UV-A	1-g/L TiO_2; pH = 4.0	—	259.97

* Incident solar irradiance taken as 1 kW/m²
Source: Arslan et al. 2000

oxidized effectively by electrochemically generated active chlorine, formed by chloride oxidation in an undivided electrochemical batch reactor with parallel plate electrodes. Enhancement of mass transport rates in the reactor by mechanical or hydraulic agitation increased the depletion rates of many contaminants (e.g., dyes, cyanides, sulfides) but decreased that of other pollutants (such as ammonium ions). The reason for this behavior involves a complex interaction among electron transfer, mass transport, and homogeneous chemical kinetics occurring simultaneously in the electrochemical reactor, the rate constants for the homogeneous process affecting whether the chemical oxidation occurs within the anode reaction layer or in the bulk solution. Consequently, the selectivity of the reactor changes depending on the hydrodynamic conditions. These findings can be useful in establishing the operational conditions of the electrochemical reactor, once a pollutant considered a priority for depletion has been chosen.

The photocatalytic process using TiO_2 and UV light may be considered feasible for the treatment of wastewater containing hazardous contaminants for which biological waste treatment plants are impractical at medium or low pollutant concentrations. Some of the advantages of the TiO_2/UV method over other AOP processes include:

- It does not require the use of expensive oxidants (such as H_2O_2 or O_3); instead, it uses cheap and abundant oxygen.

- The oxide photocatalysts, especially TiO_2, are available at a low price and can be recycled on a technical scale due to its stability in aqueous media.

- The light required to activate the catalysts may be a natural UV component of sunlight, in addition to low- and medium-pressure mercury vapor lamps that generate UV-B and UV-A light. The use of the TiO_2/UV process with UV-A light opens the possibility of using concentrated solar light.

- TiO_2/UV-based oxidation is powerful and indiscriminate, leading to mineralization of a majority of organic pollutants.

The use of sunlight instead of artificial light for this process would dramatically lower the process costs and thus provide a major step toward industrial applications. The feasibility of the solar photocatalytic process for decolorization of textile dye wastewater has been demonstrated at NEERI, as well as at many other places in the world. Since the rate of the process is linearly dependent on UV–photon flux, the associated investment is also dependent on the photon collector surface. Thus, further research and development studies are needed to design appropriate photocatalysts and photoreactors so that optimum utilization of sunlight is achieved for decontamination of large volumes of industrial wastewater.

4   Nomenclature

A	irradiated surface area (m²) of the photoreactor
A_a	anode area (m²)
A_{ads}	adsorbed electron acceptor
A_c	cathode area (m²)
(C)	pollutant concentration (mg/dm³)

C_0	initial pollutant concentration (kg/m^3)
CB	conduction band
(Cl_2)	concentration of active chlorine (mg/dm^3)
COD	chemical oxygen demand (g/dm^3)
C_t	final pollutant concentration at time t (kg/m^3)
D	diffusivity (m^2/sec)
D_{ads}	adsorbed electron donor
E	average irradiation intensity reaching photoreactor surface (W/m^2)
e^-	photogenerated electron
EC	power consumption (kWh/m^3)
Eg	Band-gap energy
F	Faraday's constant: 96,485 C/eq
F_o	molar flow of a pollutant (mol/sec)
h^+	photogenerated hole
h^ν	photon
I	applied current (A)
I_L	limiting current (A)
j	current density (A/m^2)
j_L	limiting current density (A/m^2)
k	second-order chemical rate constant (m^3/mol-sec)
k_a	rate of hole formation
k_e	rate e–h recombination
$k_{m,a}$	mass transport rate coefficient at anode (m/sec)
$k_{m,c}$	mass transport rate coefficient at cathode (m/sec)
k_{obs}	apparent pseudo-first-order kinetic constant (min^{-1})
l	characteristic length parameter (m)
λ	wavelength
m	number of moles converted
n	charge number of reaction
P	power input of the UV lamp (kW)
Q	charge (C)
q_{EK}	Egerton-King threshold
q_i	surface irradiance ($1/quantas$-cm^2)
r_A	rate of disappearance of a pollutant (mol/dm^3-sec)
Re	Reynolds number: vl/ν
Sc	Schmidt number: ν/D
SEC	specific energy consumption (kWh/mol)
Sh	Sherwood number: $k_m l/D$
t	reaction time (min) or treatment time (hr)

U   cell voltage (V)

v   velocity (m/sec)

ν solution viscosity (m²/sec)

V   reactor volume (dm³) or volume of the treated wastewater (m³)

VB   valence band

V_{fb} flat-band potential

X   conversion

Φ_e current efficiency for an electrode reaction

ω number of resolutions of the stirrer

5   References and Recommended Resources

Abdo, M. S. E., S. Rasheed, and Al-Ameeri. 1987. Anodic oxidation of a direct dye in an electrochemical reactor. *J. Environ. Sci. Health* A22 (1): 27–45.

Abrahams, J., R. S. Davidson, and C. L. Morrison. 1985. Optimization of the properties of titanium dioxide. *J. Photochem. Photobiol.* 29:353.

Adams, R. N. 1969. *Electrochemistry at solid electrodes.* New York: Marcel Dekker.

Al-Ekabi, H. 1990. Catalytic oxidation of waste organics. *Chemical Engineering* 97 (9): 27.

Al-Ekabi, H., and N. Serpone. 1988. Kinetic studies in heterogeneous photocatalysis. Part 1, Photocatalytic degradation of chlorinated phenols in aerated aqueous solutions over TiO_2 supported on glass matrix. *J. Phys. Chem.* 92:5726.

Alvarez-Gallegos, A., and D. Pletcher. 1998. The removal of low-level organics via hydrogen peroxide formed in a reticulated vitreous carbon cathode cell. Part 1, The electrosynthesis of hydrogen peroxide in aqueous acidic solutions. *Electrochimica Acta* 44:853–61.

Arikado, T., C. Iwakura, H. Yoneyama, and H. Tamura. 1976. Anodic oxidation of potassium cyanide on the graphite electrode. *Electrochimica Acta* 21:1021–27.

Arslan, I., I. A. Balcioglu, T. Tuhkanen, and D. Bahnemann. 2000. H_2O_2/UV-C and Fe_2^+/H_2O_2/UV-C versus TiO_2/UV-A treatment for reactive dye wastewater. *J. Environ. Eng.* (October): 903–11.

Bahnemann, D., R. Dillert, J. Dzengal, R. Goslich, G. Sagawe, H.-W. Schumacher, and V. W. Benz. 1999. *J. Adv. Oxid. Tech.* 4:11.

Bejan, D., J. Lozar, G. Falgayrac, and A. Saval. 1999. Electrochemical assistance of catalytic oxidation in liquid phase using molecular oxygen: Oxidation of toluenes. *Catalysis Today* 48 (4): 363–69.

Blake, D. M. 1999. *Bibliography of work on the photocatalytic removal hazardous compounds from water and air.* Update no. 3. Springfield, Va.: National Technical Information Service, U.S. Department of Commerce.

Blanco, J., and S. Malato. 1992. In *Proceedings of the 6th Int. Symp. on Solar Thermal Concentrating Technology, Mojacar, Spain*, 1183. Madrid: CIEMAT.

———. 1994. In *Proceedings of the Solar Engineering 1994, ASME, USA*, ed. D. E. Klett, R. E. Hogan, and T. Tanaka, 103.

———. 2000. Solar detoxification, Natural Sciences, World Solar Programme, UNESCO, 1996–2005. Available at www.unesco.org/science/wsp.

Blanco, J., S. Malato, D. Bahnemann, D. Bockelmann, D. Weich Grebe, F. Carmona, and F. Martinez. 1994. In *Proceedings of the 7th International Symposium on Solar Thermal Concentrating Technology, Moscow, Russia, Institute for High Temperature of Russian Academy of Science (IVTAN), 1994*, ed. O. Popel, S. Fris, and E. Shchedrova, 468.

Blanco, J., S. Malato, M. Sanchez, A. Vidal, and B. Sanchez. 1991. In *Proceedings of the ISES Solar World Congress, Denver, USA*, ed. M. E. Arden, S. M. A. Burley, and M. Coleman, 2097. Oxford: Pergamon Press.

Bolton, J. R., and S. R. Cater. 1994. Homogeneous photodegradation of pollutants in contaminated water: An introduction. In *Surface and Aquatic Photochemistry*, ed. G. Heltz, R. G. Zepp, and D. Crosby, 467–90. Boca Raton, Fla.: Lewis.

Bolton, J. R., M. Ravel, and S. R. Cater. 1996. Homogeneous solar photodegradation of contaminants in water. In ASME Symp. Ser., 53–60. New York: ASME.

Bonfanti, F., S. Ferro, F. Lavezzo, M. Malacarne, G. Lodi, and A. De Battisti. 2000. Electrochemical incineration of glucose as a model organic substrate. *J. Electrochem. Soc.* 147:592–96.

Borgarello, E., J. Kiwi, E. Pelizzetti, M. Visca, and M. Gratzel. 1981. Sustained water cleavage by visible light. *J. Amer. Chem. Soc.* 103:6324.

Boxall, C., and G. H. Kelsall. 1992. Hypochlorite electrogeneration. Part 2, Thermodynamics and kinetic model of anode reaction layer. Electr. Engin. I.Chem.E. Symp. Series, I.Chem.E., Rugby, UK, 127, 59–70.

Burstein, G. T., C. J. Barnett, A. R. Kucernak, and K. R. Williams. 1997. Aspect of the anodic oxidation of methanol. *Catalysis Today* 38 (4): 425–37.

Carriere, J., J. P. Jones, and A. D. Broadbent. 1993. Decolourization of textile dye solutions. *Ozone Sci. Eng.* 15:189–200.

Casella, I. G., and M. Gatta. 2000. Anodic electrodeposition of copper oxide/hydroxide films by alkaline solutions containing cuprous cyanide ions. *Journal of Electroanalytical Chemistry* 494:12–20.

Cegarra, J., P. Puente, and J. Valldeperas. 1981. Tintura delle materie tessili. Paravia, Italy: Ed. Texilia.

Chiang L. C., Chang J. E., and Wen T. C. 1995. Indirect oxidation effect in electrochemical oxidation treatment of landfill leachate. *Water Res.* 29 (2): 671–78.

Childs, L. P., and D. F. Ollis. 1981. Photoassisted heterogeneous catalysis: Rate equations for oxidation of 2-methyl-2-butyl alcohol and isobutene. *J. Catal.* 67 (1): 35.

Chou S., Huang Y.-H., Lee S.-N., Huang G.-H., and Huang C. 1999. Treatment of high-strength hexamine containing wastewater by electro-Fenton method. *Water Res.*, 33 (3): 751–59.

Comninellis, Ch. 1992. Electrochemical treatment of wastewater containing phenol. *IChemE* 70, part B: 219–24.

Comninellis, Ch., and A. Nerini. 1995. Anodic oxidation of phenol in the presence of NaCl for wastewater treatment. *J. Appl. Electrochem.* 25:23–28.

Cunningham, J., and S. Srijaranai. 1988. Isotope effect for hydroxyl radical involvement in alcohol photo-oxidation sensitized by TiO_2 in aqueous suspension. *J. Photochem. Photobiol.* A43:329.

Czarnetzki, L. R., and L. J. Janssen. 1992. Formation of hypochlorite, chlorate and oxygen during NaCl electrolysis from alkaline solutions at an RuO_2/TiO_2 anode. *J. Appl. Electrochem.* 22 (2): 315–24.

Dhodapkar, R. S., N. N. Rao, S. P. Pande, and S. N. Kaul. 1999. Photocatalytic destruction of organic pollutants in water. Chap. 26 of *Advancements in Industrial Wastewater Treatment*, 453–66.

———. 2000. Photocatalytic route for reduction of color and chemical oxygen demand from dye containing wastewater. *Environment Conservation Journal* 1 (1): 13–20.

Do J. S. and Chen M. L. 1994. Decolourization of dye-containing solutions by electro-coagulation. *J. Appl. Electrochem.* 24:785–90.

Do J. S. and Yeh W. C. 1995. In-situ degradation of formaldehyde with electrogenerated hypochlorite. *J. Appl. Electrochem.* 25 (5): 483–89.

———. 1996. Paired electrooxidative degradation of phenol with in-situ electrogenerated hydrogen peroxide and hypochlorite. *J. Appl. Electrochem.* 26 (6): 673–78.

Dube, S., and N. N. Rao. 1996. Rate parameter independence on the organic reactant: A study of adsorption and photocatalytic oxidation of surfactants using MO_3/TiO_2 (M = Mo or W) catalysts. *J. Photochem. Photobiol.* A93:71.

Duonghong, D., E. Borgarello, and M. Gratzel. 1981. Dynamics of light-induced water cleavage in colloidal systems. *J. Amer. Chem. Soc.* 103 (16): 4685.

El-Ghaoui, E. A., and R. E. Jansson. 1982. Application of trickle tower to problems of pollution control. Part 2, The direct and indirect oxidation of cyanide. *J. Appl. Electrochem.* 12:69–73.

El-Ghaoui, E. A., R. E. Jansson, and C. Moreland. 1982. Application of trickle tower to problems of pollution control. Part 3, Heavy-metal cyanide solutions. *J. Appl. Electrochem.* 12:75–80.

Feng, J., L. L. Houk, D. C. Johnson, S. N. Lowery, and J. J. Carey. 1995. Electrocatalysis of anodic oxygen transfer reactions: The electrochemical incineration of benzoquinone. *J. Electrochem. Soc.* 142 (11): 3626–32.

Fleszar, B., and J. Ploszynska. 1985. An attempt to define benzene and phenol electrochemical oxidation mechanism. *Electrochimica Acta* 30 (1): 31–42.

Foller, P., and W. Tobias. 1982. Electrolytic process for the production of ozone. U.S. Patent no. 4,316,782.

Foti, G., D. Gandini, Ch. Comninellis, A. Perret, and W. Haenni. 1999. Oxidation of organics by intermediates of water discharge on IrO_2 and synthetic diamond anodes. *Electrochemical and Solid-State Letters* 2 (5): 228–30.

Fox, M. A., H. Ogawa, and P. Pichat. 1988. Regioselectivity in the semiconductor-mediated photooxidation of 1,4-pentane diol. *J. Org. Chem.* 54 (16): 3847.

Fujishima A. Tata, N. Rao, and D. A. Tryk. 2000. Titanium dioxide photocatalysis, *J. Photochem. Photobiol.* C 1:1.

Fukui Y. and Yuu S. 1984. Development of apparatus for electro-flotation. *Chem. Eng. Sci.* 39:939–45.

Gawronski, R. 1996. *Processes of water treatment* (in Polish). Warsaw: Oficyna Wydawnicz P.W.

Gijbers, H. F. M., and L. J. J. Janssen. 1989. Distribution of mass transfer over a 0.5-m-tall hydrogen-evolving electrode. *J. Appl. Electrochem.* 19:637.

Gimenez, J., D. Curco, and M. A. Queral. 1999. *Catalysis Today* 54:267.

Gimenez, J., D. Curco, and P. Marco. 1997. *Water Sci. & Technol.* 35 (4): 207.

Goncalves, M. S., A. M. F. Oliveira-Campos, E. M. M. S. Pinto, P. M. S. Plasencia, and M. J. R. P. Quciroz. 1999. Photochemical treatment of solutions of azo dyes containing TiO_2. *Chemosphere* 39:781–86.

Goslich, R., R. Dillert, and D. Bahnemann. 1997. *Water Sci. & Technol.* 35 (4): 137.

Guillard, C., B. Beaguraud, C. Dutriez, J. M. Herrmann, H. Jaffrezic, N. Jaffrezic-Renault, and M. Lacroix. 2002. Physicochemical properties and photocatalytic activities of TiO_2: Films prepared by sol-gel methods. *Appl. Catal.* 39:331.

Gurnham, C. F., ed. 1965. *Industrial waste control.* New York: Academic Press.

Hashimoto, K., T. Kawai, and T. Sakata. 1984. Photocatalytic reactions of hydrocarbons and fossil fuels with water. *J. Phys. Chem.* 88 (18):4083.

Heaton, J. 1994. *The chemical industry.* Blakie Academic & Professional.

Herrmann, J. M., and P. Pichat. 1980. Heterogeneous photocatalysis: Oxidation of halide ions by oxygen in ultraviolet irradiated aqueous suspension of titanium dioxide. *J. Chem. Soc.* 76 (5): 1138.

Hine, F., M. Yasuda, T. Iida, and Y. Ogata. 1986. On the oxidation of cyanide solutions with lead dioxide–coated anode. *Electrochimica Acta* 31:1389–95.

Ho S. P. Y., Wang C., and Wan C. 1989. Electrolytic decomposition of cyanide effluent with electrochemical reactor packed with stainless steel fiber. *Water Res.* 24:1317–21.

Hsiao, C.-Y., C. L. Lee, and D. F. Ollis. 1983. Heterogeneous photocatalysis: Degradation of dilute solutions of dichloromethane, chloroform, and carbon tetrachloride with illuminated TiO_2 photocatalyst. *J. Catal.* 82:418.

Hsiao, Y. L., and K. Nobe. 1993. Hydroxylation of chlorobenzene and phenol in a packed-bed flow reactor with electrogenerated Fenton's reagent. *J. Appl. Electrochem.* 23:943–45.

Hu C. C., Lee C. H., and Wen T. C. 1996. Oxygen evolution and hypochlorite production on Ru-Pt binary oxides. *J. Appl. Electrochem.* 26:72–82.

Hwang J.-Y., Wang Y.-Y., and Wan C.-C. 1987. Electrolytic oxidation of cuprocyanide electroplating waste waters under different pH conditions. *J. Appl. Electrochem.* 17:684–94.

Izumi, I. W., W. Dun, K. Wilbourn, F. Fan, and A. J. Bard. 1980. Heterogeneous photocatalytic oxidation of hydrocarbons on platinized TiO_2 powders. *J. Phys. Chem.* 84 (24): 3207.

Jain, G., S. Saryanarayan, P. Nawghare, S. N. Kaul, and L. Szpyrkowicz. 2001. *Intern. J. Environ. Studies* 58:313.

Kelsall, G. H., and I. Thompson. 1993. The redox chemistry of H_2S oxidation by the British gas Stretford process. Part 2, Electrochemical kinetics of HS⁻/S redox systems. *J. Appl. Electrochem.* 23 (4): 287–95.

Kelsall, G. H., S. Savage, and D. Brandt. 1991. Cyanide oxidation at nickel anodes. Part 2, Voltammetry and coulometry of Ni/CN-H_2O systems. *J. Electrochem. Soc.* 138:117–24.

Kikuchi, Y., K. Sunada, T. Iyoda, K. Hashimoto, and A. Fujishima. 1997. Photocatalytic bactericidal effect of TiO_2 thin films: Dynamic view of the active oxygen species responsible for the effect. *J. Photochem. Photobiol.* A106:51.

Kiriakidou, F., D. I. Kondarides, and X. E. Verykios. 1999. The effect of operational parameters and TiO_2-doping on the photocatalytic degradation of azo-dyes. *Catalysis Today* 54:119–30.

Kiwi, J. 1983. In *Energy resources through photochemistry and catalysis*, ed. M. Gratzel. New York: Academic Press.

Kowal, A., S. N. Port, and R. J. Nichols. 1997. Nickel hydroxide electrocatalysts for alcohol oxidation reactions: An evaluation by infrared spectroscopy and electrochemical methods. *Catalysis Today* 38 (4): 483–92.

Kraeutler, B., and A. J. Bard. 1978. Heterogeneous photocatalytic decomposition of saturated carboxylic acids on TiO_2 powder: Decarboxylation route alkanes. *J. Amer. Chem. Soc.* 100:2239, 5985.

Krasnobrodko, I. G. 1988. *Destruction treatment of wastewater from dyes* (in Russian). Leningrad: Ed. Chimia.

Kuo, W. G. 1992. Decolorizing dye wastewater with Fenton's reagent. *Water Res.* 26:881–86.

Kuo, W. S. 2000. Preparation and photocatalytic activity of metal-supported resin-bonded titania. *J. Environ. Sci.* A35 (3): 419.

Levenspiel, O. 1999. *Chemical reaction engineering.* New York: Wiley.

Lin M. L., Wang Y. Y., and Wan C. C. 1992. A comparative study of electrochemical reactor configurations for the decomposition of copper cyanide effluent. *J. Appl. Electrochem.* 22:1197–1200.

Lin S. H. and Chen M. L. 1997. Treatment of textile wastewater by chemical methods for reuse. *Water Res.* 868–76.

Lin S. H. and Lin C. M. 1993. Treatment of textile waste effluents by ozonation and chemical coagulation. *Water Res.* 12:1743–48.

Lin S. H. and Peng C. F. 1994. Treatment of textile wastewaters by electrochemical method. *Water Res.* 2:277–82.

Lin S. H. and Wu C. L. 1998. Electrochemical removal of nitrite and ammonia removal for aquaculture. *Water Res.* 32:1059–66.

Lin S. H. and Wu L. 1996. Electrochemical removal of nitrite and ammonia for aquaculture. *Water Res.* 30:715–21.

Lu G., Qu J., and Tang H. 1999. The electrochemical production of highly effective polyaluminum chloride. *Water Res.* 33:807–13.

Malati, M. A., and W. K. Wong. 1984. Doping TiO_2 for solar energy applications. *Surface Tech.* 22:305–22.

Malato, S., J. Blanco, C. Richter, and M. Vincent. 1996. *Solar Energy* 56 (5): 401.

Malato, S., J. Blanco, A. Vidal, and C. Richter. 2002. Photocatalysis with solar energy at a pilot-plant scale: An overview. *Appl. Catal.* B37:1–15.

Malato, S., J. Gimenez, C. Richter, D. Curco, and J. Blanco. 1997. Low-concentrating CPC collectors for photocatalytic water detoxification: Comparison with a medium concentrating solar collector. *Water Sci. & Technol.* 35 (4): 157.

Marinec, L., and F. B. Lectz. 1978. Electro-oxidation of ammonia in wastewater. *J. Appl. Electrochem.* 8:335–45.

Matthews, R. W. 1984. Hydroxylation reactions induced by near-ultraviolet photolysis of aqueous TiO_2 suspensions. *J. Chem. Soc.* 80 (2): 457.

———. 1987a. Photooxidation of organic impurities in water using thin films of titanium dioxide. *J. Phys. Chem.* 91 (2): 3328.

———. 1987b. Solar-electric water purification using photocatalytic oxidation with TiO_2 as a stationary phase. *Solar Energy* 38 (6): 405.

Merli, C., E. Petrucci, A. Da Pozzo, and M. Pernetti. 2003. Fenton-type treatment: State of the art. *Annali di Chimica* 93 (9–10): 761–70.

Muggli, D. S., and L. Ding. 2001. Photocatalytic oxidation using SO_4^{2-}/TiO_2. *Appl. Catal.* B32:181.

Murphy, O., G. D. Hitchens, L. Kaba, and C. E. Vcrostko. 1992. Direct electrochemical oxidation of organics for wastewater treatment. *Water Res.* 26:443–51.

Nakabayashi, S., A. Fujishima, and K. Honda. 1983. Experimental evidence for hydrogen evolution site in photocatalytic process on Pt/TiO_2. *Chem. Phys. Lett.* 102 (2): 464.

Nameri, N., A. R. Yeddu, H. Lounici, D. Belhocine, H. Grib, and B. Bariou. 1998. Defluoridation of septentrional Sahara water of North Africa by electrocoagulation process using bipolar aluminum electrodes. *Water Res.* 32:1604–12.

Naumczyk, J., L. Szpykowicz, M. De Faveri, and F. Zilio Grandi. 1996. Electrochemical treatment of tannery wastewater containing high-strength pollutants. *Trans. IChemE* 74, part B: 59–68.

Naumczyk, J., L. Szpykowicz, and F. Zilio Grandi. 1996. Electrochemical treatment of textile wastewater. *Water Sci. & Technol.* 11:17–24.

Norton, A. P., S. C. Bernasek, and A. B. Bocarsly. 1988. Mechanistic aspects of the photo-oxidation of water at the n-TiO$_2$/aqueous interface: Optically induced transients as a kinetic probe. *J. Phys. Chem.* 92 (21): 6009.

Okamoto, K., Y. Yamamoto, H. Tanaka, M. Tanaka, and A. Itaya. 1985. Heterogeneous photocatalytic decomposition of phenol over TiO$_2$ powder. *Bull. Chem. Soc. Japan* 58:2015.

Ollis, D. F., Hsiao C.-Y., L. Budiman, and Lee C. L. 1984. Heterogeneous photoassisted catalysis: Conversions of perchloroethylene, dichloroethane, chloroacetic acids and chlorobenzenes. *J. Catal.* 88 (1): 89.

Otsuka, K., and I. Yamanaka. 1998. Electrochemical cells as reactor for selective oxygenation of hydrocarbons at low temperature. *Catalysis Today* 41:311–25.

Oturan, M. A. 2000. An ecologically effective water treatment technique using electrochemically generated hydroxyl radicals for in-situ destruction of organic pollutants: Application to herbicide 2,4-D. *J. Appl. Electrochem.* 30:475–82.

Panizza, M., and G. Cerisola. 2001. Removal of organic pollutants from industrial wastewater by electrogenerated Fenton's reagent. *Water Res.* 35 (6): 3987–92.

Pelizzetti, E., V. Maurino, C. Minero, V. Carlin, E. Pramauro, O. Zerbinati, and M. L. Tasato. 1990. Photocatalytic degradation of atrazine and other s-triazine herbicides. *Environ. Sci. Technol.* 24 (100): 1559.

Pelizzetti, E., and M. Visca. 1983. Examples of photogeneration of hydrogen and oxygen from water. In *Energy resources through photochemistry and catalysis*, ed. M. Gratzel. New York: Academic Press.

Picket, D. J., and C. J. Wilson. 1982. Mass transfer in a parallel plate electrochemical cell: The effect of change of flow area and flow cross-section at the cell inlet. *Electrochimica Acta* 27:591–94.

Pletcher, D., and F. C. Walsh. 1993. *Industrial electrochemistry.* London: Blackie Academic & Professional.

Polcaro, A. M., and S. Palmas. 1997. Electrochemical oxidation of chlorophenols. *Ind. Eng. Chem. Res.* 36:1791–98.

Pruden, A. L., and D. F. Ollis. 1983. Photoassisted heterogeneous catalysis: The degradation of trichloroethene in water. *J. Catal.* 82 (2): 404.

Qiang, Z., J.-H. Chang, and C.-P. Huang. 2002. Electrochemical generation of hydrogen peroxide from dissolved oxygen in acidic solutions. *Water Res.* 36:85–94.

Rajalo, G., and T. Petrovskaya. 1996. Selective electrochemical oxidation of sulphides in tannery wastewater. *Environ. Technol.* 17:605–12.

Rajeshwar, K., J. G. Ibanez, and G. M. Swain. 1994. Electrochemistry and environment. *J. Appl. Electrochem.* 24:1077–91.

Ralph, T. R., M. L. Hitchman, J. P. Millington, and F. C. Walsh. 1996. Mass transport in an electrochemical laboratory filterpress reactor and its enhancement by turbulence promoters. *Electrochimica Acta* 41:591–603.

Ramachandriah, G., S. K. Thampy, P. K. Narayan, D. K. Chuahan, N. N. Rao, and V. K. Indushekhar. 1996. Separation and concentration of metals present in industrial effluent and sludge samples by using electrodialysis, coulometry and photocatalysis. *Separation Science & Technology* 31 (4): 523–32.

Rao, N. N. 1996. Environmental applications of photocatalysis: A new member of advanced oxidation processes. *TERI Information Digest on Energy* 6 (3).

Rao, N. N., and S. Dube. 1995. Application of Indian commercial TiO_2 powder for destruction of organic pollutants: Photocatalytic degradation of 2,4-dichlorophenoxyacetic acid (2,4-D) using suspended and supported TiO_2 catalysts. *Ind. J. Chem. Technol.* 2:241–48.

———. 1996a. Photocatalytic degradation of some commercial soap/detergent products using suspended TiO_2 catalysts. *J. Mol. Catal.* A104:L197–99.

———. 1996b. Photoelectrochemical generation of hydrogen using organic pollutants as sacrificial electron donors. *Int. J. Hydrogen Energy* 21 (2): 95–98.

———. 1997. TiO_2-catalyzed photodegradation of Reactive Orange 84 and alizarin red S biological stain. *Ind. J. Chem. Technol.* 4:1–6.

———. 1998. Photocatalytic degradation of Reactive Orange 84 (RO84) in dye-house effluent using single-pass reactor. *Studies in Surface Science and Catalysis* 113:1045–50.

Rao, N. N., S. Dube, and P. Natarajan. 1993. Photocatalytic degradation of some chlorohydrocarbons in aqueous suspensions of MO_3/TiO_2 (M = Mo or W). *Trace Metal Environment* 3:695–700.

Rao, N. N., A. Dubey, J. Rajesh, P. Khare, S. Mohanty, and S. N. Kaul. 2003. Photocatalytic degradation of 2-chlorophenol: A study of kinetics, intermediates and biodegradability. *J. Hazard. Mater.* 101 (2).

Rao, N. N., and S. N. Kaul. 1999. Methods to enhance the efficiency of photocatalytic process. *Bull. Catal. Soc. India* 9 (1):81–91.

———. 2000. Photocatalytic activity of H_2 reduced Indian commercial TiO_2: Photodegradation of 2,4-dichlorophenoxyacetic acid. *Bull. Catal. Soc. India* 10 (1–2): 95–112.

Rao, N. N., and P. Natarajan. 1994. Particulate models in heterogeneous photocatalysis. *Current Science* 66 (10): 742.

Rao, N. N., K. M. Somasekhar, S. N. Kaul, and L. Szpyrkowicz. 2001. Electrochemical oxidation of tannery wastewater. *J. Chem. Technol. Biotechnol.* 76:1124–31.

Reutergardh, L. B., and M. Iangphasuk. 1997. Photocatalytic decolourization of reactive azo dye: A comparison between TiO_2 and CdS photocatalysis. *Chemosphere* 35:585–96.

Sakata, T., and T. Kawai. 1983. Photosynthesis and photocatalysis with semiconductor powders. In *Energy resources through photochemistry and catalysis*, ed. M. Gratzel. New York: Academic Press.

Sakata, T., T. Kawai, and K. Hashimoto. 1984. Heterogeneous photocatalytic reactions of organic acids and water: New reaction paths besides the photo-Kolbe reaction. *J. Phys. Chem.* 88:2344.

Sarasa, J., M. P. Roche, M. P. Ormad, E. Gimeno, A. Puig, and J. L. Ovelleiro. 1998. Treatment of wastewater resulting from dyes manufacturing with ozone and chemical coagulation. *Water Res.* 9:2721–27.

Saunders, F. M., P. J. Gould, and R. C. Southerland. 1983. The effect of solute competition on ozonolysis of industrial dyes. *Water Res.* 17:1407–19.

Schwager, F., P. M. Robertson, and N. Ibl. 1980. The use of eddy promoters for the enhancement of mass transport in electrolytic cells. *Electrochimica Acta* 25:1655–65.

Sclafani, A., L. Palmisano, and M. Schiavello. 1990. Influence of the preparation methods of TiO_2 on the photocatalytic degradation of phenol in aqueous dispersion. *J. Phys. Chem.* 94 (2): 829.

Scott, K. 1991. *Electrochemical reaction engineering.* London: Academic Press.

Serpone, N., E. Borgarello, R. Harris, P. Cahill, M. Borgarello, and E. Pelizzetti. 1986. Photocatalysis over TiO_2 supported on a glass substrate. *Solar Energy Materials* 14:121.

Simond, O., V. Schaller, and Ch. Comninellis. 1997. Theoretical model for the anodic oxidation of organics on metal electrodes. *Electrochimica Acta* 42 (13–14): 2009–12.

Singer, P. C., and W. B. Zilli. 1975. Ozonation of ammonia in wastewater. *Water Res.* 9 (2): 127–34.

Stork, A., and B. Hutin. 1981. Mass transfer and pressure drop performance of turbulence promoters in electrochemical cells. *Electrochimica Acta* 26:127–37.

Sudoh, M., H. Kataguchi, and K. Koide. 1985. Polarization characteristics of packed-bed electrode reactor for electroreduction of oxygen to hydrogen peroxide. *J. Chem. Eng. Japan* 18:364–71.

Sudoh, M., T. Kodera, T. Sakai, J. Q. Zhang, and K. Koide. 1986. Oxidative degradation of aqueous phenol effluent with electrogenerated Fenton's reagent. *J. Chem. Eng. Japan* 19 (6): 513–17.

Szpyrkowicz, L. 2002. Electrocoagulation of textile wastewater bearing disperse dyes. *Annali di Chimica* 92:1025–34.

Szpyrkowicz, L., R. Cherbanski, E. Molga, and G. H. Kelsall. 2001. Effect of mixing on mediated electrochemical oxidation of disperse dyes in an undivided electrochemical reactor. In proceedings of ICheaP-5 (Italian Conference on Chemical and Process Engineering), Florence, 73–79.

Szpyrkowicz, L., C. Juzzolino, S. Daniele, and M. De Faveri. 2001. Electrochemical destruction of thiourea dioxide in an undivided parallel plate electrodes batch reactor. *Catalysis Today* 66:519–27.

Szpyrkowicz, L., C. Juzzolino, and S. N. Kaul. 2001. A comparative study on oxidation of disperse dyes by electrochemical process, ozone, hypochlorite and Fenton reagent. *Water Res.* 35:2129–36.

Szpyrkowicz, L., C. Juzzolino, S. N. Kaul, and S. Daniele. 2000. Electrochemical oxidation of dyening baths bearing disperse dyes. *Ind. Eng. Chem. Res.* 39:3241–48.

Szpyrkowicz, L., S. N. Kaul, E. Molga, and M. De Faveri. 2000. Comparison of the performance of a reactor equipped with a Ti/Pt and an SS anode for simultaneous cyanide removal and copper recovery. *Electrochimica Acta* 46:381–87.

Szpyrkowicz, L., G. H. Kelsall, S. N. Kaul, and M. De Faveri. 2001. Performance of electrochemical reactor for treatment of tannery wastewater. *Chem. Eng. Sci.* 56 (3): 1579–86.

Szpyrkowicz, L., J. Naumczyk, and F. Zilio Grandi. 1994. Application of electrochemical processes for tannery wastewater treatment. *Toxicological and Environmental Chemistry* 44:189–202.

———. 1995. Electrochemical treatment of tannery wastewater using Ti/Pt and Ti/Pt/Ir electrodes. *Water Res.* 29 (2): 517–24.

Szpyrkowicz, L., F. Ricci, and S. Daniele. 2003. Removal of cyanides by electrooxidation. 93 (9–10): 833–40.

Szpyrkowicz, L., and F. Zilio Grandi. 1996. Performance of a full-scale treatment plant for textile dyeing wastewater. *Toxicological and Environmental Chemistry* 56:23–34.

Szpyrkowicz, L., F. Zilio Grandi, S. N. Kaul, and A. M. Porcaro. 2000. Copper electrodeposition and oxidation of complex cyanide from wastewater in an electrochemical reactor with a Ti/Pt anode. *Ind. Eng. Chem. Res.* 39 (7): 2132–39.

Szpyrkowicz, L., F. Zilio Grandi, S. N. Kaul, and S. Rigoni-Stern. 1998. Electrochemical treatment of copper cyanide wastewaters using stainless steel electrodes. *Water Sci. & Technol.* 38 (10): 261–68.

Tamura, T., T. Arikado, H. Yoneyama, and Y. Matsuda. 1974. Anodic oxidation of potassium cyanide on platinum electrode. *Electrochimica Acta* 19:273–77.

Tissot, P., and M. Fragniere. 1994. Anodic oxidation of cyanide on a reticulated three-dimensional electrode. *J. Appl. Electrochem.* 24:509–12.

Tomcsanyi, L., A. De Battisti, G. Hirschberg, K. Varga, and J. Liszi. 1999. The study of the electrooxidation of chloride at RuO_2/TiO_2 electrode using CV and radio-tracer techniques and evaluating by electrochemical kinetic simulation methods. *Electrochimica Acta* 44 (14): 2463–72.

Trasatti, S. 1987. Progress in the understanding of the mechanism of chlorination evolution at oxide electrodes. *Electrochimica Acta* 32 (3): 369–82.

Trinidad, P., and F. C. Walsh. 1996. Hydrodynamic behaviour of the FM01-LC reactor. *Electrochimica Acta* 41:493–502.

Turchi, C. S., and D. F. Ollis. 1990. Photocatalytic degradation of organic water contaminants: Mechanisms involving hydroxyl radical attack. *J. Catal.* 122 (1): 178.

Tzedakis, T., A. Savall, and M. J. Clifton. 1989. The electrochemical regeneration of Fenton's reagent in the hydroxylation of aromatic substrates: Batch and continuous processes. *J. Appl. Electrochem.* 19:911–21.

Van Benschoten and J. K. Edzwald. 1990. Chemical aspect of coagulation using aluminum salts. Part 1, Hydrolitic reactions of alum and polyaluminum chloride. *Water Res.* 24:1519–26.

Wang, Y. 2000. Solar photocatalytic degradation of eight commercial dyes in TiO_2 suspension. *Water Res.* 34:990–94.

Well, M., R. H. G. Dillert, D. Bahnemann, V. W. Benz, and M. A. Mueller. 1997. *J. Solar Energy Eng.* 119:114.

Wels, B., and D. C. Johnson. 1990. Electrocatalysis of anodic oxygen transfer reactions: Oxidation of cyanide at electrodeposited copper oxide electrodes in alkaline media. *J. Electrochem. Soc.* 137:2785–91.

Westerhoff, P., G. Aiken, G. Amy, and J. Debroux. 1999. Relationships between the structure of natural organic matter and its reactivity towards organic ozone and hydroxyl radicals. *Water Res.* 33 (10): 2265–76.

Westerterp, K. R. 1984. *Chemical reactor design and operation.* 2d ed. Manchester, England: J. Wiley & Sons. 503–506.

Wragg, A. A., and A. A. Leontaritis. 1997. Local mass transfer and current distribution in baffled and unbaffled parallel plate electrochemical reactors. *Chemical Engineering Journal* 66:1–10.

Note

1. Limiting current density $j_L = nFk_m(C)$, where:

    j_L = limiting current density (A/m^2)

    n = number of electrons involved in the reaction

    F = Faraday constant: 96,485 C/eq

    k_m = mass transport coefficient (m/sec)

    (C) = concentration of a compound of interest (mol/m^3)

Chapter 6
Treatment Technologies for Suspended Matter in Wastewater

Nicholas P. Cheremisinoff

I   Thickeners and Clarifiers

Sedimentation involves the removal of suspended solid particles from a liquid stream by gravitational settling. This unit operation is divided into *thickening*, that is, increasing the concentration of the feed stream, and *clarification*, removal of solids from a relatively dilute stream.

A *thickener* is a sedimentation machine that operates according to the principle of gravity settling. Compared to other types of liquid/solid separation devices, a thickener's advantages are:

- simplicity of design and economy of operation
- its capacity to handle extremely large flow volumes
- versatility, as it can operate equally well as a concentrator or as a clarifier

In a batch-operating mode, a thickener normally consists of a standard vessel filled with a suspension. After settling, the clear liquid is decanted and the sediment removed periodically. The operation of a continuous thickener is also relatively simple.

Figure 6.1 illustrates a cross-sectional view of a standard thickener. A drive mechanism powers a rotating rake mechanism. Feed enters the apparatus through a feed

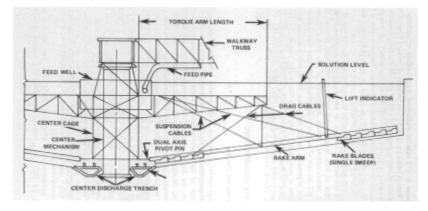

Figure 6.1 Cross-sectional view of a thickener

well designed to dissipate the velocity and stabilize the density currents of the incoming stream. Separation occurs when the heavy particles settle to the bottom of the tank. Some processes add flocculants to the feed stream to enhance particle agglomeration and promote faster or more effective settling. The clarified liquid overflows the tank and is sent to the next stage of a process. The solids are withdrawn from an underflow cone by gravity discharge or pumping.

Thickeners can be operated in a countercurrent fashion. Applications are aimed at the recovery of soluble material from settleable solids by means of continuous countercurrent decantation (CCD). The basic scheme involves streams of liquid and thickened sludge moving countercurrently through a series of thickeners. The thickened stream of solids is depleted of soluble constituents as the solution becomes enriched. In each successive stage, a concentrated slurry is mixed with a solution containing fewer solubles than the liquid in the slurry and then is fed to the thickener. As the solids settle, they are removed and sent to the next stage. The overflow liquid, which is richer in the soluble constituent, is sent to the preceding unit. Solids are charged to the system in the first-stage thickener, from which the final concentrated solution is withdrawn. Wash water or virgin solution is added to the last stage, and washed solids are removed in the underflow of this thickener.

The flow scheme for a three-stage CCD system is illustrated in figure 6.2. The feed stream, F, is mixed with overflow O_2 (from thickener 2) before entering stage 1. The overflow of concentrated solution, O_1, is withdrawn from the first stage. The underflow from the first stage, U_1, is mixed with third-stage overflow, O_3, and fed to the second stage. Similarly, the second-stage underflow, U_2, is mixed with wash water and fed to thickener 3. The washed solids are removed from the third stage as the final underflow, U_3.

Continuous clarifiers handle a variety of process wastes, domestic sewage, and other dilute suspensions. They resemble thickeners in that they are sedimentation tanks or basins whose sludge removal is controlled by a mechanical sludge-raking mechanism. They differ from thickeners in that the amount of solids and weight of thickened sludge are considerably lower. Figures 6.3 and 6.4 show examples of cylindrical clarifiers. In this type of sedimentation machine, the feed enters up through the hollow central column or shaft, referred to as a *siphon feed system*. The feed flows into the central feed well through slots or ports located near the top of the hollow shaft.

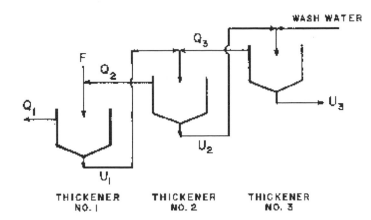

Figure 6.2 Flow scheme for three-stage CCD

Figure 6.3 Cylindrical clarifier

Siphon feed arrangements greatly reduce the feed stream velocity as it enters the basin proper. This tends to minimize undesirable crosscurrents in the settling region of the vessel.

Most cylindrical units are equipped with peripheral weirs; however, some designs include radial weirs to reduce the exit velocity and minimize weir loadings. Units are usually equipped with adjustable rotating overflow pipes. Although there are a fixed number of U.S. suppliers for these unique machines, the designs are universal, and nearly identical configurations and comparable operating parameters can be found throughout the world.

Gravity Thickening

The gravity thickening process involves the concentration of sludge to a more dense sludge in special circular tanks designed for this purpose. Its use is largely restricted

Figure 6.4 Clarifier in operation

to the watery excess sludge from the activated sludge process and in large plants of this type where the sludge is sent directly to digesters instead of to the primary tanks. It may also be used to concentrate sludge to primary tanks or to a mixture of primary and excess activated sludge prior to high-rate digestion. The thickening tank is equipped with slowly moving vertical paddles built like a picket fence. Sludge is usually pumped continuously from the settling tank to the thickener, which has a low overflow rate so that the excess water spills over and the sludge solids concentrate in the bottom. A blanket of sludge is maintained by controlled removal, which may be continuous at a low rate. A sludge with a solids content of 10 percent or more can be produced by this method. This means that with an original sludge of 2 percent, about four-fifths of the water has been removed and one of the objectives in sludge treatment has been attained.

Flotation Thickening

Flotation thickening units are becoming increasingly popular at sewage treatment plants, especially for handling waste-activated sludges. With activated sludge, these units have the advantage, relative to gravity thickening tanks, of offering higher solids concentrations and lower initial cost for the equipment.

Dissolved Air-Pressure Flotation

The objective of flotation thickening is to attach a minute air bubble to suspended solids and cause the solids to separate from the water in an upward direction. This works because the solid particles have a specific gravity lower than water when the bubble is attached. Dissolved-air flotation depends on the formation of small-diameter bubbles resulting from air released from solution after being pressurized to 40 to 60 psi. Since the solubility of air increases with pressure, substantial quantities of air can be dissolved.

Two general approaches to pressurization are used: (1) air charging and pressurization of recycled clarified effluent or some other flow used for dilution, with subsequent addition to the feed sludge; and (2) air charging and pressurization of the combined dilution liquid and feed sludge. Air in excess of the decreased solubility, resulting from the release of the pressurized flow into a chamber at near-atmospheric pressures, comes out of solution to form the minute air bubbles. Sludge solids are floated by the air bubbles, which attach themselves to and are enmeshed in the floc particles. The degree of adhesion depends on the surface properties of the solids. When released into the separation area of the thickening tank, the buoyed solids rise under hindered conditions analogous to those in gravity settling and can be called hindered separation or flotation. The upward-moving particles form a sludge blanket on the surface of the flotation thickener.

The primary variables for flotation thickening are:

*   pressure
*   recycle ratio
*   feed solids concentration
*   detention period
*   air-to-solids ratio
*   type and quality of sludge
*   solids and hydraulic loading rates
*   use of chemical aids

Similar to gravity sedimentation, the type and quality of sludge to be floated affects the unit performance. Flotation thickening is most applicable to activated sludges, but higher float concentrations can be achieved by combining primary with activated sludge. Equal or greater concentrations may be achieved by combining sludges in gravity thickening units.

2   Lamella Clarifier

Cross-flow lamellar clarification is a technology used in industrial environments to remove oils and solids from residual water. It takes advantage of the natural tendency of oils to float and the decantation principle for suspended solids that are denser than water. The originality of this process is the combination of natural flotation and clarification techniques in one system. A *strip decanter* performs as well as conventional clarifiers, but is more compact and occupies a smaller area.

The process is mainly used for dealing with oils and grease present in residual liquids emanating from industrial activities in the petrochemical, chemical, mechanical, metallurgical, and food-processing sectors. Depending on the specific design criteria (available space, quantity and quality of water to be treated, structural and hydraulic constraints, budget limits, and need for mobility), the equipment can be housed in a circular or rectangular reservoir or can consist of mechanical components attached to a below-grade concrete tank. If the oils are partially or completely emulsified, the cross-flow lamellar clarifier can be equipped with a coalescer, which uses a physical process to trigger separation of the oil and water phases. The coalescer is filled with various elements (rings, plastics, honeycombs, and other appropriate materials) that maximize the potential contact surface. The accretion of microglobules to these elements leads to phase separation.

System performance depends on the specific nature of the effluent to be treated and varies according to the type of industry. Depending on the particular situation, removal of free oils and greases, and of suspended solids, varies from 90 to 99 percent. With no chemical amendment (i.e., demulsifying agent), 20–40 percent removal of emulsified oils and greases can be achieved; the addition of an agent enables the process to achieve 50–99 percent removal, depending on the application. This technology is designed to handle effluents containing a maximum of 10,000 mg/l of grease and 3,000 mg/l of solids. Hydraulic load is not a limiting factor.

The cross-flow lamellar clarifier consists of the following basic units:

- primary screening chamber
- separator plate cell
- sludge silo
- oil and grease storage chamber

First, water is pretreated in the primary screening chamber to remove part of the floating oil and grease and to allow sedimentation of large solid particles (greater than 500 mm in diameter). Then the effluent feeds through the plate cell, where separation of phases is accomplished: oils are deflected upward by the plates to form a film on the surface of the water, sludges settle to the bottom, and the purified water flows horizontally to the reservoir outlet. Before leaving the system, water passes through a calibrated opening that controls the unit's hydraulic load. Sludges are recovered in a conical extraction silo that aids their compaction and provides easier handling. Dryness rates of 1–5 percent are achieved, depending on sludge type. The sludge is kept apart from water to be treated, so as not to draw it back into the process flow. Floating oils are recovered from the water surface and channeled into a storage reservoir located beneath the sludge silo.

Equipment installation and startup take less than a week. Running the system requires no energy input except for the effluent pump. Maintenance is limited to monthly testing and cleaning of the separator plates once every six months. If needed, various products (polymers, demulsifiers, and coagulants) can be added to the process to improve its performance. Running the system requires no special safety measures. If required, the equipment can be designed to provide a safe environment for treating effluents containing volatile compounds that are at risk of explosion.

Various parameters are taken into account in designing a cross-flow lamellar clarifier. These include:

- hydraulic load
- suspended matter load
- hydrocarbon load
- desired performance
- space available for clarifier's installation

The investment required to set up a treatment unit complete with coalescer varies from $750 to $2,500 per cubic meter of water to be treated, depending on unit size and available space. Operating costs are low.

3   Chemistry of Clarification

The term *subsidence* means settling. While a degree of clarification can be accomplished by subsidence, most industrial processes require better quality water than can be obtained from settling only. Most of the suspended matter in water would settle, given enough time, but in most cases the amount of time required would not be practical. Settling characteristics depend upon the:

- weight of the particle
- shape of the particle
- size of the particle
- viscosity and/or frictional resistance of the water, which is a function of temperature

The settling rates of various-size particles at 50°F (10°C) is illustrated in table 6.1. A great deal of the suspended matter found in wastewaters falls into the colloidal suspension range, and hence we cannot rely on gravitational force alone to separate out the pollutants.

The term *coagulation* refers to the first step in complete clarification. It is the neutralization of the electrostatic charges on colloidal particles. Because most of the smaller suspended solids in surface waters carry a negative electrostatic charge, the natural repulsion of these similar charges causes the particles to remain dispersed almost indefinitely. To allow these small suspended solids to agglomerate, the negative electrostatic charges must be neutralized. This is accomplished by using inorganic

Table 6.1 Some settling rates for different particles (assumed spherical) and sizes

Particle Diameter (mm)	Particle Type	Time to Settle One Foot
10.0	Gravel	0.3 seconds
1.0	Coarse sand	3 seconds
0.1	Fine sand	38 seconds
0.01	Silt	33 minutes
0.001	Bacteria	35 hours
0.0001	Clay particles	230 days
0.00001	Colloidal particles	65 years

coagulants (which are water-soluble inorganic compounds), organic cationic polymers, or polyelectrolytes. The most common and widely used inorganic coagulants are:

- Aluminum sulfate, $Al_2(SO_4)_3$, also known as alum
- Ferric sulfate, $Fe_2(SO_4)_3$
- Ferric chloride, $FeCl_3$
- Sodium aluminate, $Na_2Al_2O_4$

Inorganic salts of metals work by two mechanisms in water clarification: The positive charge of the metals serves to neutralize the negative charges on the turbidity particles. The metal salts also form insoluble metal hydroxides, which are gelatinous and tend to agglomerate the neutralized particles. The most common coagulation reactions are as follows:

$$Al_2(SO_4)_3 + 3Ca(HCO_3)_2 \leftrightarrow 2Al(OH)_3 + 3CaSO_4 + 6CO_2$$

$$Al_2(SO_4)_3 + 3Na_2CO_3 + 3H_2O \leftrightarrow 2Al(OH)_3 + 3Na_2SO_4 + 3CO_2$$

$$Al_2(SO_4)_3 + 6NaOH \leftrightarrow 2Al(OH)_3 + 3Na_2SO_4$$

$$Al_2(SO_4)_3 + (NH_4)_2SO_4 + 3Ca(HCO_3)_2 \leftrightarrow 2Al(OH)_3 + (NH_4)_2SO_4 + 3CaSO_4 + 6CO_2$$

$$Al_2(SO_4)_3 + K_2SO_4 + 3Ca(HCO_3)_2 \leftrightarrow 2Al(OH)_3 + K_2SO_4 + 3CaSO_4 + 6CO_2$$

$$Na_2Al_2O_4 + Ca(HCO_3)_2 + H_2O \leftrightarrow 2Al(OH)_3 + CaCO_3 + Na_2CO_2$$

$$Fe(SO_4)_3 + 3Ca(OH)_2 \leftrightarrow 2Fe(OH)_3 + 3CaSO_4$$

$$4Fe(OH)_2 + O_2 + 2H_2O \leftrightarrow 4Fe(OH)_3$$

$$Fe_2(SO_4)_3 + 3Ca(HCO_3) \leftrightarrow 2Fe(OH)_3 + 3CaSO_4 + 6CO_2$$

The effectiveness of inorganic coagulants is dependent upon water chemistry—in particular, pH and alkalinity. Their addition usually alters that chemistry. Table 6.2

Table 6.2 Coagulant, acid, and sulfate 1 ppm equivalents

Formula or chemical (1 ppm)	Alkalinity reduction (ppm)	SO$_4$ increase (ppm)	Na$_2$SO$_4$ increase (ppm)	CO$_2$ increase (ppm)	Total solids increase (ppm)
$Al_2(SO_4)_3 + 18H_2O$	0.45	0.45	0.64	0.40	0.16
$Al_2(SO_4)_3 + (NH_4)2SO_4 + 24H_2O$	0.33	0.44	0.63	0.29	0.27
$Al_2(SO_4)_3 + K_2SO_4 + 24H_2O$	0.32	0.43	0.60	0.28	0.30
$FeSO_4 + 7H_2O$	0.36	0.36	0.61	0.31	0.13
$FeSO_4 + 7H_2O + (SCl_2)$	0.54	0.36	0.51	0.48	0.18
$Fe_2(SO_4)_3$	0.76	0.76	1.07	0.64	0.27
H_2SO_4, 96%	1.00	1.00	1.42	0.88	0.36
H_2SO_4, 93.2% (66° Be)	0.96	0.95	1.36	0.84	0.34
H_2SO_4, 77.7% (66° Be)	0.79	0.79	1.13	0.70	0.28
$NaSO_4$	—	0.64	0.95	—	1.00
$Na_2Al_2O_4$	Increases 0.54	—	—	Reduces 0.47	0.90

illustrates the effect of the addition of 1 ppm of the various inorganic coagulants on alkalinity and solids concentration.

The use of metal salts for coagulation may increase the quantity of dissolved solids. One must consider the downstream impact of these dissolved solids. In addition, the impact of the carryover of suspended aluminum and iron compounds and their related effect on downstream processes must be considered.

Aluminum salts are most effective as coagulants when the pH range is between 5.5 and 8.0. Because they react with the alkalinity in the water, it may be necessary to add additional alkalinity (called *buffering*) in the form of lime or soda ash. Table 6.3 can be used as a guide.

Iron salts, on the other hand, are most effective as coagulants at higher pH ranges (between 8 and 10). Iron salts also depress alkalinity and pH levels; therefore, additional alkalinity must be added. Sodium aluminate increases the alkalinity of water, so care must be taken not to exceed pH and alkalinity guidelines. As is evident from the reactions discussed above, a working knowledge of the alkalinity relationships of water is mandatory. By using inorganic coagulants, we can wind up producing a voluminous, low-solids-content sludge that is difficult to dewater and dries very slowly. The properties and estimated quantities of the sludge to be generated need to be carefully determined, in part from pilot-scale and bench testing prior to the design and construction of a plant.

Polymers are often described as long chains with molecular weights from around 1,000 to 5,000,000. Along the chain or backbone of the polymer are numerous charged sites. In primary coagulants, these sites are positively charged and are available for adsorption onto the negatively charged particles in the water. To accomplish optimum polymer dispersion and polymer/particle contact, initial mixing intensity is critical. The mixing must be rapid and thorough. Polymers used for charge neutralization must not be overdiluted or overmixed. The farther upstream in the system these polymers can be added, the better their performance. Because most polymers are viscous, they must be properly diluted before they are added to the influent water. Special mixers such as static mixers, mixing tees, and specially designed chemical dilution and feed systems are all aids in polymer dilution. Static or motionless mixers in particular are popular for this application.

Once the negative charges of the suspended solids are neutralized, flocculation begins. Flocculation can be thought of as the second step of the coagulation process.

Table 6.3 Recommended alkali and lime 1 ppm equivalents

Chemical (1 ppm)	Formula	Alkalinity increase (ppm)	Free CO_2 reduction (ppm)	Hardness as $CaCO_3$ increases
Sodium bicarbonate	$NaHCO_3$	0.60	—	—
Soda ash (56% Na_2, 99.16% Na_2CO_3)	Na_2CO_3	0.94	0.41	—
Caustic soda (76% Na_2O, 98.06% Na_2CO_3)	$NaOH$	1.22	1.09	—
Quicklime (90% CaO)	CaO	1.61	1.41	1.61
Hydrated lime (93% $Ca(OH)_2$)	$Ca(OH)_2$	1.26	1.10	1.36

Charge reduction increases the occurrence of particle–particle collisions, promoting particle agglomeration. Portions of the polymer molecules that are not absorbed protrude for some distance into the solution and are available to react with adjacent particles, promoting flocculation. Bridging of neutralized particles can also occur when two or more turbidity particles with a polymer chain attached come together. It is important to remember that during this step, when particles are colliding and forming larger aggregates, mixing energy should be great enough to cause particle collisions but not so great as to break up these aggregates as they are formed. In some cases, flocculation aids are employed to promote faster and better flocculation. These flocculation aids are normally high-molecular-weight anionic polymers. Flocculation aids are normally necessary for primary coagulants and water sources that form very small particles upon coagulation. A good example of this is water that is low in turbidity but high in color (colloidal suspension).

Color is perhaps the most difficult impurity to remove from waters. In surface waters, color is associated with dissolved or colloidal suspensions of decayed vegetation and other colloidal suspensions. The composition of this material is largely tannins and lignins, the components that hold together the cellulose cells in vegetation. In addition to their undesirable appearance in drinking water, these organics can cause serious problems in downstream water purification processes. For example, expensive demineralizer resins can be irreversibly fouled by these materials. In addition, some of these organics have chelated trace metals, such as iron and manganese, within their structure, which can cause serious deposition problems in a cooling system.

There are many ways of optimizing color removal in a clarifier. The three most common methods are:

- Prechlorination (before the clarifier) significantly improves the removal of organics, as well as reducing the coagulant demand.
- The proper selection of polymers for coagulation has a significant impact on organic removal.
- Color removal is affected by pH. Generally, organics are less soluble at low pHs.

4   Equipment Options

Although we have discussed the major hardware, it is still worthwhile reviewing it in relation to the major classes of clarifier processes. The major categories of this process are:

- conventional
- upflow
- solids-contact
- sludge-blanket

Conventional clarification is the simplest form of the process. It relies on the use of a large tank or horizontal basin for sedimentation of flocculated solids. The basin normally contains separate chambers for rapid mixing and settling, the first two steps critical in achieving good clarification. An initial period of turbulent mixing is needed for contact between the coagulant and suspended solids. This is followed by a period of gentle stirring, which helps to increase particle collisions and floc size. Retention times are typically between 3 and 5 minutes, with 15 to 30 minutes for flocculation and 4 to 6 hours for settling. Coagulants are added to the wastewater in the rapid mix chamber or sometimes immediately upstream. The water passes through the mix chambers and enters the settling basin. The water passes out to the circumference, while the flocculated particles settle to the bottom. Accumulated sludge is scraped into a sludge collection basin for removal and disposal (or sometimes for postprocessing). The clean water flows over a weir and is held in a tank, which is referred to as a *clearwell*.

It is often advantageous to employ a zone of high solids contact to achieve a better-quality effluent. This is accomplished in an upflow clarifier, so called because the water flows upward through the clarifier as the solids settle to the bottom. Most upflow clarifiers are either solids-contact or sludge-blanket-type clarifiers, which differ somewhat in theory of operation. The coagulant is added either in the rapid mix zone or somewhere upstream of the clarifier.

In the solids-contact clarifier, raw water is drawn into the primary mixing zone, where initial coagulation and flocculation take place. The secondary mixing zone is used to produce a large number of particle collisions so that smaller particles are entrained in the larger floc. Water passes out of the inverted cone into the settling zone, where solids settle to the bottom, and clarified water flows over the weir. Solids are drawn back into the primary mixing zone, causing recirculation of the large floc. The concentration of solids in the mixing zones is controlled by occasional or continuous blowdown of sludge.

The sludge-blanket clarifier goes one step further, by passing the water up from the bottom of the clarifier through a blanket of suspended solids that acts as a filter. The inverted cone within the clarifier produces an increasing cross-sectional area from the bottom of the clarifier to the top. Thus, the upward velocity of the water decreases as it approaches the top. At some point, the upward velocity of the water exactly balances the downward velocity of a solid particle and the particle becomes suspended, with heavier particles suspended closer to the bottom. As the water containing flocculated solids passes up through this blanket, the particles are absorbed onto the larger floc, which increases the floc size and drops it down to a lower level. It eventually falls to the bottom of the clarifier to be recirculated or drawn off.

Although these processes seem relatively simple, especially in relation to many chemical manufacturing operations or unit processes, there are a number of operational problems that can make the life of an operator miserable. Excessive floc carryover is a very common problem. This is most often associated with hydraulic overload or unexpected flow surge conditions. You can tackle this problem by relying on equalize flow (metering the flow of the clarifier), which will help to dampen out surges. Unfortunately, hydraulic overload conditions are not the only causes of excessive floc carryover. Other reasons may include thermal currents, short-circuiting effects, low-density floc, and chemical feed problems.

Another common operator problem is simply no floc in the centerwell. This can result from underfeeding of chemicals or a loss of the sludge bed recirculation. You will have to investigate and apply trial-and-error field tests to resolve some of these problems. When new equipment is installed, it is wise to spend time during a shakedown and startup period to explore the operational limitations of the process and train operators on how to handle these types of problems.

5   Rectangular Sedimentation Tanks

The process concept for sedimentation tanks has hardly changed over the past 80 years. Dimensioning these vessels according to existing guidelines guarantees safe operation. With ever-tightening legislation, however, the question of expansion or upgrading of existing sewage treatment plants arises. Expansion is an expensive solution—and nearly impossible if the available space is scarce—so that new construction may be the only option. The basis of upgrading consists in changed process concepts to those which are able to exploit the unused potential of existing tanks. An essential prerequisite for upgrading plants is having sufficient settling volume for activated sludge. Besides the clarified-water discharge, the feeding method for the sludge/water mix and the skimmer system have an essential influence on the separation efficiency in tanks. The inlet height is approximately one-third to two-thirds of the tank depth, and the skimming direction in the counterflow. This concept is based on the empirical knowledge of normally minor turbulence in the tank. However, introducing process concepts with a bottom-near inlet and concurrent skimming can minimize such turbulence to such an extent that the sludge load and thus the separation efficiency can be increased.

It is important to recognize that the process-engineering installations in rectangular sedimentation tanks have a great influence on the performance of this final treatment stage. Of particular importance is the design of the inlet section, as turbulences are generated there by mixing with the wastewater inflow, which may significantly affect the sedimentation process. The density of flows has a strong influence on the separation efficiency. The density flow sinks to the tank bottom during inflow and passes to the tank end. The rising density flow initiates backflow of the clarified water on the surface. To increase the separation efficiency of tanks, the density flow should be minimized or the engineering process modified in a way that the density flow will be integrated with the sedimentation process.

Inlet dimensions are important. The density flow can be substantially influenced by the inlet structure by minimizing the potential and kinetic energy of the wastewater stream with a suitable feed design. The inlet should have near-bottom feed openings to have as small an intermixing zone as possible between the activated sludge/water mix and the tank content. The velocity gradient in the inlet section should be small to avoid floc disturbance by shearing forces. The inlet section sludge scraping also influences the separation efficiency, most notably by altering the degree of thickening in the sedimentation tank. For minimization of the turbulences in the secondary clarification tank and improved separation efficiency and for sludge thickening, practice has shown that scraping the sludge in the direction of the density flow works best. The scraping velocity should be low in order to prevent resuspension of the activated sludge flocs. To increase the degree of thickening and to minimize the volume flow of the return sludge, a minimum sludge residence time in the tank must be provided. Although a sludge hopper for thickening the activated sludge is not necessary, a sludge hopper at the tank end tends to increase the surface load of the tank.

6   Air Flotation Systems

Air flotation is one of the oldest methods for the removal of solids, oil and grease, and fibrous materials from wastewater. Suspended solids and oil and grease removals exceeding 99 percent can be attained with these processes. Air flotation is simply the production of microscopic air bubbles that enhance the natural tendency of some materials to float by carrying wastewater contaminants to the surface of the tank for removal by mechanical skimming. Many commercially available units are packaged rectangular steel tank flotation systems, shipped completely assembled and ready for simple piping and wiring on site. Models typically range from 10 to more than 1,000 square feet of effective flotation area for raw wastewater flows of more than 1,000 gallons per minute. Complete systems often include chemical treatment processes.

A dissolved air flotation (DAF) system can produce clean water in wash operations, where reduction of oil and grease down to 2 mg/l is achievable in certain applications. There is a broad and varied market for DAF. Any sites with excessive oil and grease (especially emulsified oils), plus high levels of suspended solids and metals, are good candidates for DAF systems. In addition to municipal and heavy industry applications, DAF has found a home with commercial vehicle washing, industrial laundries, and food processing. Vehicle wash applications need not be confined to automobiles and trucks but can extend to buses, tank cars, and many other types of vehicles. Another example is an industrial laundry, where there is the need for good waste treatment systems. A DAF system can be used in a variety of food processing applications, including vegetable oils, animal and seafood fats, red meat butchering, poultry processing, and kitchen and equipment washing. Oil drilling on- and offshore is under pressure from the U.S. Environmental Protection Agency. The dirty water that surfaces when drilling wells cannot be dumped without removal of the oils it contains. DAF technology has proven very effective in this industry. Two waste stream problems found in shipyards and aboard ships are oily bilge water and solids containing copper and other heavy metals used in marine paints. DAF is a proven method in these applications as well. In metal-finishing operations, DAF will remove cadmium, chromium, lead, zinc, and other toxic heavy metals from a waste stream. Finally, when combined with other treatment processes, DAF can be applied upstream of large water treatment systems to handle contaminated water before it mingles with the rest of the waste stream. Figures 6.5 and 6.6 show an installation at a textile manufacturing plant.

When the primary target is oil removal, it is important to distinguish between the forms of oil. There are two forms of oil that we find in wastewater. *Free oil* is oil that will separate naturally and float to the surface. *Emulsified oil* is oil that is held in suspension by a chemical substance (detergents, surfactants) or electrical energy. When making an evaluation, free oil will normally separate by gravity and float to the surface in approximately 30 minutes. Emulsified oil is held in a molecular structure called a *micelle* and will not separate on its own. Hence, there is the need for a more sophisticated method of treating suspensions containing emulsified oils.

A good way to see how DAF technology works is to fill a glass full of water from the tap and observe the tiny, almost microscopic bubbles in the water. DAF uses the same principle to introduce tiny bubbles into water. The bubbles in the glass form because the water inside the pipes, which is at high pressure, dissolves enough air that, when the pressure drops as the water falls into the glass, the water becomes supersaturated

Figure 6.5 Photo of DAF in operation

with air. As a result, the excess air precipitates out in the form of tiny bubbles. These bubbles are much smaller than we produce by other means of dispersing air in water.

The flotation process was developed in the mining and coal processing industries as a way of separating suspended solids from a medium such as water. As noted above, the flotation process has found uses in other fields, including wastewater treatment. The process introduces fine air bubbles into the mixture, so that the air bubbles attach to the particles and lift them to the surface.

Dissolved-air flotation uses a particular way of introducing the air bubbles into the flotation tank. A DAF machine dissolves air into the untreated water by passing the water through a pressurizing pump, introducing air, and holding the air–water mixture at high pressure long enough for the water to become saturated with air at the high pressure. Typical pressures are 20–75 psig. After saturating the water with air at high pressure, the water passes through a pressure-relief nozzle, after which air precipitates as tiny bubbles in the lower-pressure water.

This process for creating air bubbles has two advantages over other processes. DAF typically produces bubbles in the 40–70 m range, whereas in normal foam fractionation, a bubble of 500 m is considered small. The smaller bubbles have much more surface area for their volume than do the larger bubbles. A particular volume of air distributed as 50-m bubbles has 10 times the surface area that it does when distributed as 500-m bubbles. Looking at this another way, you need 10 times the airflow

Figure 6.6 Photo of DAF in operation

with 500-m bubbles that you need with 50-m bubbles in order to achieve the same air/water interfacial area.

The second, and probably more important, advantage of producing bubbles by precipitation is that the process provides a more positive attachment between air bubbles and the particles or globules that are to be removed. Particles and globules in the water act as nucleation sites for the precipitation process; the precipitating air seeks out these sites to begin bubble formation. This is better than relying on chance encounters between waste particles and large bubbles introduced by some other means.

A typical DAF process is not simply a physical separation technique. One must consider the entire treatment process, which is based on chemical coagulation, clarification, and rapid sand filtration. This process train is widely accepted and is very applicable to the treatment of colored and turbid surface waters for municipal and industrial applications. Normally the clarification stage employs DAF. The suspended solid matter in the chemically treated water is separated by introducing a recycle stream containing small bubbles that floats this material to the surface of the tank. This is achieved by recycling a portion of the clarified flow back to the DAF unit. The recycle flow is pumped to higher pressure and is then mixed with compressed air. The flow passes through a tank, where the air dissolves to saturation at the higher pressure. When the pressure is released at the clarifier, the dissolved air precipitates as a cloud of microbubbles, which attach to the particulate matter causing it to rise to the surface.

The DAF process is particularly well suited for the removal of floc formed in the treatment of low-alkalinity, low-turbidity colored water. This type of floc tends to be very fragile and voluminous, making traditional gravity sedimentation inefficient. Flotation processes do not require large, heavy floc in order to achieve efficient solids removal. This results in lower chemical dosages and reduced time required for flocculation. The compressive forces applied to the sludge by the buoyant bubble/floc agglomerates result in greatly reduced volumes of wastewater from the clarification process. This enhances the efficiency of chemical use and reduces the volume of residuals to be treated. An equally important benefit of the technology is the efficiency of the clarification process. Since the performance of the filtration process is directly affected by the amount of solids in the clarified water, the high degree of solids removal achieved in the DAF results in an overall increase in system performance. In order to meet stringent standards for turbidity removal in potable water applications, this high performance is essential.

To recap, DAF is the process of removing suspended solids, oils, and other contaminants via the use of bubble flotation. Air is dissolved into the water, then mixed with the waste stream and released from solution while in intimate contact with the contaminants. Air bubbles form, mix with the wastewater influent, and are injected into the DAF separation chamber. The dissolved air then comes out of solution, producing literally millions of microscopic bubbles. These bubbles attach themselves to the particulate matter and float them to the surface, where they are mechanically skimmed and removed from the tank. Most systems are versatile enough to remove not only finely divided suspended solids but also fats, oils, and grease (FOG). Typical wastes handled include various suspended solids, food/animal production/processing wastes, industrial wastes, hydrocarbon oils/emulsions, and many others. Clarification rates of 97 percent or more are achievable.

The conventional DAF saturation design relies on a recycle pump combined with a saturation vessel and air compressor to dissolve air into the water. This type of system is effective, but it has the drawbacks of being labor intensive and expensive and can destabilize its point of equilibrium, creating "burps" due to incorrect, loss, or creeping of the EQ setpoint in the saturation vessel. Such designs are slow to recover and can upset the flotation process. Air transfer efficiency is roughly 9 percent, with 80 percent entrainment. This operational methodology can result in an increase in chemical use, labor costs, downtime, effluent loadings, and production schedules due to the EQ loss. To overcome these shortcomings, some equipment suppliers have devised operational and control schemes that are best categorized as pollution prevention techniques.

Chemical pretreatment is often used to improve the performance of contaminant removal. The use of chemical flocculants is based on system efficiency, the specific DAF application, and cost. Commonly used chemicals include trivalent metallic salts of iron or aluminum, such as $FeCl_2$, $FeSO_4$, or $AlSO_4$. Organic and inorganic polymers (cationic or anionic) are generally used to enhance the DAF process.

The most commonly used inorganic polymers are the polyacrylamides. Chemical flocculant concentrations employed normally range from 100 to 500 mg/l. The wastewater pH may require adjustment between 4.5 and 5.5 for the ferric compounds or between 5.5 and 6.5 for the aluminum compounds, using an acid such as H_2SO_4 or a base such as NaOH. In many applications, the DAF effluent requires additional pH adjustment, normally with NaOH to assure that the effluent pH is within the limits specified by the POTW. The pH range of the effluent from a DAF is typically between 6 and 9.

The mechanism by which flocculants work and enhance DAF is as follows. Attachment of most bubbles to solid particles can be effected through surface energies, while others are trapped by solids or by hydrous oxide flocs as the floc spreads out of the water column. Colloidal solids are normally too small to allow the formation of sufficient air–particle bonding; they must first be coagulated by a chemical such as the aluminum or iron compounds mentioned above. The solids are essentially absorbed by the hydrous metal oxide floc generated by these compounds. Often, a coagulant aid is needed in combination with the flocculant to agglomerate the hydrous oxide floc, increase particle size, and improve the rate of flotation. Mechanical/chemical emulsions can also be broken through the application of pH and polymer reactions.

The material recovered from the surface of the DAF is referred to as the *float*. The float often contains 2–10 percent solids. These solids need to be dewatered before ultimately finding a home for them.

In general, for many applications, air flotation is the system of choice. Microbubbles are produced by inducing air into a vortex as the floc is formed. This controlled induction of air allows the microbubbles to permanently attach to the floc, resulting in the clarifier. In fully integrated systems, the clarifier has a built-in sludge-holding section.

7   Separation Using Coalescers

A coalescer achieves separation of an oily phase from water on the basis of density differences between the two fluids. These systems obviously work best with nonemulsified oils. Applications historically have been in the oil and gas industry, and hence the most famous oil/water separator is the API (American Petroleum Institute) separator.

Modern-day designs are more sophisticated than the early, simple separators of a few decades ago that were introduced by the petroleum industry. Commercial systems comprise cylindrical vessels, rectangular vessels, and aboveground and underground installations.

There are variations of coalescer designs, each achieving different degrees of separation depending upon the application and properties of the influent. One fairly popular design used primarily in the oil and gas industry is a so-called deoiler cyclone vessel and cyclone system. These systems take advantage of the hydroclone (or cyclone) principle of separation. More correctly stated, the cyclonic action results in an increase in the oil droplet size, enabling a more efficient separation of the phases. Deoiling cylcones have steep droplet-cut size curves. The typical performance curve for a high-efficiency cyclone separator (or hydroclone) shows that a small increase in the droplet size from 5 to 10 m typically increases the separation from 15 to more than 90 percent. The inlet chamber of a conventional deoiling hydroclone is usually the largest chamber in the coalescer vessel and has a residence time up to about 20 seconds. Some commercial units employ specially designed low-intensity precoalescing internals and inlet vanes that take advantage of this residence time by optimizing the flow distribution. This enhances the coalescence of droplets and enables a precoalescing stage.

In general, the technology largely relies on many years of experience and empiricism. Before investing in a system, it is wise to run batch tests and perhaps pilot tests using vendor facilities. Units may range in size, complexity of internals, and configuration (e.g., rectangular, slant-rib designs that borrow concepts from the classical Lamella separator, or classical cylindrical designs). There are varying degrees of claims for removal efficiency. Cost can vary greatly with coalescers, depending not only on throughput requirements and add-on controls and monitors but with construction as well.

Because of strict environmental regulations tied to underground and aboveground tanks, double-wall vessel construction is often needed. Double-wall vessels are normally constructed in several ways. One common construction is achieved by wrapping a secondary steel wall completely around the primary vessel. Each double-wall vessel is constructed using the same basic fabrication techniques as on single-wall vessels. The area between the vessel walls, known as the *interstice*, can be monitored by a leak detection system installed in the monitor tube, located on the vessel head. Other variations of vessel construction exist and careful consideration to the advantages, disadvantages, and impacts on cost must be considered. Remember to select a separator based in part on vessel construction quality that meets API, Underwriter Laboratories (UL), or Steel Tank Institute (STI) ACT-100 or STI-P3 specifications. Of course, the supplier must also conform to ISO 9000 quality assurance standards.

8   Recommended Resources

Baeyens, J., Y. Mochtar, S. Liers, and H. De Wit. 1995. Plugflow dissolved air flotation. *Water Environment Research* 67 (7): 1027–35.

Fuerstenau, M. C., ed. 1976. *Flotation: A. M. Gaudin memorial volume.* 2 vols. New York: AIME.

Gaudin, A. M. 1957. *Flotation.* 2d ed. New York: McGraw-Hill.

Grainger-Allen, T. J. N. 1970. Bubble generation in froth flotation machines. *Trans. IMM* 79:C15–22.

Haarhoff, J., and S. Steinbach. 1996. A model for the prediction of the air composition in pressure saturators. *Water Res.* 30 (12): 3074–82.

Hedherg, T., J. Dahlqvist, D. Karlsson, and L.-O. Soerman. 1998. Development of an air removal system for dissolved air flotation. *Water Sci. & Technol.* 37 (9): 81.

Klassen, V. I., and V. A. Mokrousov. 1963. *An introduction to the theory of flotation.* Trans. J. Leja and G. W. Poling. London: Butterworths.

Leja, J. 1982. *Surface chemistry of froth flotation.* New York: Plenum Press.

Liers, S., J. Baeyens, and J. Mochtar. 1996. Modeling dissolved air flotation. *Water Environment Research* 68 (6): 1061.

Rykaart, E. M., and J. Haarhoff. 1995. Behaviour of air injection nozzles in dissolved air flotation. *Water Sci. & Technol.* 31 (3–4): 25–35.

Vrablik, E. R. 1959. Fundamental principles of dissolved air flotation of industrial wastes. *Industrial Waste Conference Proceedings.* 14th Industrial Waste Conference, 1959, Purdue University, Ann Arbor, pp. 743–79.

Walter, J., and U. Wiesmann. 1995. Comparison of dispersed and dissolved air flotation for the separation of particles from emulsions and suspensions. *Das Gas-
   und Wasserfach* 136 (2): 53.

Zabel, T. 1985. The advantages of dissolved-air flotation for water treatment. *J. Am.
   Water Works Assoc.* 77 (5): 42–46.

Zlokarnic, M. 1998. Separation of activated sludge from purified waste water by induced air flotation. *Water Res.* 32 (4): 1095–1102.

Chapter 7
Financial Tools for Environmental Technologies

Nicholas P. Cheremisinoff

Pollution management, whether through the use of control technologies or pollution prevention, requires management to make a decision on the commitment of financial resources for capital outlay and ongoing costs. While pollution prevention (P2) generally eliminates ongoing costs associated with wastes from control technologies that treat or control emission streams, the nature of technologies requires a capital outlay and operations and maintenance (O&M) costs to be carried for P2 technologies as well.

In evaluating technology options, the costs associated with all aspects of pollution management need to be made transparent and accounted for. Often corporations underestimate the true costs associated with waste and pollution management and hence select technologies that may meet regulatory standards but in fact are not cost-effective. In conducting enterprise-specific audits among 20 different companies both in the United States and abroad, the author has noted that the majority of companies tend to underestimate their actual costs for environmental management by at least 20 percent and sometimes as much as 40 percent. Common reasons for this are often associated with the fact that companies tend to bury cost issues associated with environmental compliance within overhead, do not recognize hidden costs within their organization, or heavily discount so-called soft costs.

In prior publications, the author has noted four tiers of environmental costs: usual, hidden, less tangible, and future liability. Oftentimes enterprises tend to lose track of indirect and hidden costs as part of their overhead accounting systems. And most often, because corporations are generally focused on short-term financial benchmarks and milestones, future liabilities associated with waste and pollution management are so heavily discounted that initially more costly technology investments are eliminated as viable options. Future liability costs frequently create nightmares even for the largest of corporations. One only has to look to the large number of multimillion-dollar toxic torts faced by companies such as Lockheed-Martin and Koppers Industries because less costly control and P2 technologies were chosen without giving consideration to future impacts to the environment and public risk associated with their operations.

This chapter outlines the general environmental cost accounting principles that should be considered when evaluating the economics of technology options.

I   Total-Cost Accounting

Total-cost accounting, also known as *life-cycle costing,* is a method aimed at analyzing the costs and benefits associated with a piece of equipment, a technology, or a practice over the entire life of the proposed application. The concept was first applied in procuring weapons systems. Experience showed that the upfront purchase price of such systems was a poor measure of the total cost. Instead, costs such as those associated with maintainability, reliability, disposal, and salvage value, as well as employee training and education, had to be given equal weight in making financial decisions. By the same token, justifying pollution prevention or pollution control requires that all benefits and costs be clearly defined in the most concrete terms possible and projected over the life of each option.

Present Worth or Present Value

Present worth, also known as *present value,* refers to the fact that time is money. The preference between a dollar now and a dollar one year from now is driven by the fact that a dollar in hand can earn interest. Present value can be expressed by a simple formula:

$$P = F/(1 + i)^n$$

where P is present worth or present value, F is future value, *i* is the interest or discount rate, and *n* is the number of periods. For example, if we are to be given $1,000 in one year, at 6 percent interest compounded annually this amount would have a computed present value of:

$$P = \$1,000/(1 + 0.06)^1 = \$943.40$$

Because this money can "work" at 6 percent interest, there is no difference between $943.40 now and $1,000 in one year because they both have the same value now. If the money is to be received in three years, the present value would be:

$$P = \$1,000/(1 + 0.06)^3 = \$839.62$$

Present-value calculations allow both costs and benefits that are expended or earned in the future to be expressed as a single lump sum at their current or present value.

In considering either multiple payments or cash into and out of a company, the present values are additive. For example, at 6 percent interest, the present value of receiving both $1,000 in one year and $1,000 in three years would be $943.40 + $839.62, or $1,783.06. Similarly, if one were to receive $1,000 in one year and then pay $1,000 in three years, the present value would be $943.40 − $839.62, or $103.78.

Financial Analysis Factors

It is common practice to compare investment options based on the present-value expression. We may also apply one or all of the following four factors when comparing investment options:

- Payback period
- Internal rate of return
- Benefit-to-cost ratio
- Present value of net benefits

The payback period of an investment is essentially a measure of how long it takes to break even on the cost of that investment. In other words, how many weeks, months, or years does it take to earn the investment capital that was laid out for a project or a piece of equipment?

Those projects with the fastest returns are highly attractive. The technique for determining payback period again lies within present value; however, instead of solving the present-value equation for the present value, the cost and benefit cash flows are kept separate over time.

The project's anticipated benefit and cost are tabulated for each year of the project's lifetime. Then these values are converted to present values by using the present-value equation, with the company's discount rate plugged in as the discount factor. Finally, the cumulative total of the benefits (at present value) and the cumulative total of the costs (at present value) are compared on a year-by-year basis. At the point in time when the cumulative present value of the benefits starts to exceed the cumulative present value of the costs, the project has reached the payback period. Ranking projects then becomes a matter of selecting those projects with the shortest payback period. The minimum payback time standard is a good financial comparison, but it should not be the deciding basis for project selection.

Although this approach is straightforward, there are dangers in selecting technologies based upon a minimum payback time. For example, because the benefit stream generally extends far into the future, discounting makes its payoff period very long. Another danger is that the highest costs and benefits associated with most environmental projects are generally due to catastrophic failure, also a far-future event. Because the payback period analysis stops when the benefits and costs are equal, the projects with the quickest positive cash flow will dominate. Hence, for a technology with a high discount rate, the long-term costs and benefits may be so far into the future that they do not even enter into the analysis. In essence, the importance of life-cycle costing is lost in using the minimum payback time alone, because it only considers costs and benefits to the point where they balance, instead of considering them over the entire life of the project.

A more enlightening term for a technology is the *internal rate of return* (IRR) or *return on investment* (ROI). The IRR is defined as the interest rate that would result in a return on the invested capital equivalent to the project's return. For example, if we had a project with an IRR of 30 percent, that is financially equivalent to investing resources in the right stock and having its price go up by this percentage.

This method is based in the net present value of benefits and costs; however, it does not use a predetermined discount rate. Instead, the present-value equation is solved for the discount rate i. The discount rate that satisfies the zero benefit is the rate of return on the investment, and technology selection is based on the highest rate. From a simple calculation standpoint, the present value expression is solved for i after setting the net present value equal to zero and plugging in the future value, obtained by subtracting the future costs from the future benefits over the lifetime of the project. This approach is frequently used in business; however, the net benefits and costs must be determined for each time period and brought back to present value separately. Computationally, this could mean dealing with a large number of simultaneous equations, which can complicate the analysis.

The benefit to cost (B/C) ratio is a benchmark that is determined by taking the total present value of all of the financial benefits of a project and dividing it by the total present value of all the costs. If the ratio is greater than unity, then the benefits outweigh the costs, and we may conclude that the project is economically worthwhile.

The present values of the benefits and costs are kept separate and expressed in one of two ways. First, there is the pure B/C ratio, which implies that if the ratio is greater than unity, the benefits outweigh the costs and the project is viable. Second, there is the net B/C ratio, which is the net benefit (benefits minus costs) divided by the costs. In this latter case, the decision criterion is that the benefits must outweigh the costs, which means that the net ratio must be greater than zero (if the benefits exactly equaled the costs, the net B/C ratio would be zero). In both cases, the highest B/C ratios are considered the best projects.

The B/C ratio can be misleading. If the present value of a project's benefits were $10,000 and its costs were $6,000, the B/C ratio would be:

$$B/C = \$10,000/\$6,000 = 1.67$$

But what if, upon further reassessment of the project, we find that some of the costs are not "true" costs, but instead simply offsets to benefits? In this case, the ratio could be changed considerably. For argument's sake, let's say that $4,500 of the $6,000 total cost is for waste disposal, and that $7,000 of the $10,000 in benefits is due to waste minimization; one could then use them to offset each other. Mathematically then, both the numerator and denominator of the ratio could be reduced by $4,500, with the following effect:

$$B/C = (\$10,000 - \$4,500) / (\$6,000 - \$4,500) = 3.7$$

Without changing the project, the recalculated B/C ratio would make the project seem to be considerably more attractive.

The *present value of net benefits* (PVNB) shows the worth of a technology or project in terms of a present-value sum. The PVNB is determined by calculating the present value of all benefits, doing the same for all costs, and then subtracting the two totals. The result is an amount of money that would represent the tangible value of undertaking the project. This comparison evaluates all benefits and costs at their present values. If the net benefit (the benefits minus costs) is greater than zero, the project is worth undertaking; if the net is less than zero, the project should be abandoned on a financial basis. This technique is firmly grounded in microeconomic theory and is ideal for total-cost analysis and P2 financial analysis.

Even though it requires a preselected discount rate, which can greatly discount long-term benefits, the PVNB assures that all benefits and costs over the entire life of the project are included in the analysis. Once an enterprise knows the present value of all options with positive net values, the actual ranking of projects using this method is straightforward: those with the highest PVNBs are funded first.

There are no hard and fast rules as to which factors one may apply in performing life-cycle costing or total-cost analysis; however, conceptually, the PVNB method is preferred. There are nevertheless many small-scale projects where the benefits are so well defined and obvious that a comparative financial factor as simple as an IRR or the payback period will suffice.

2   Baseline Costs

To determine the cost of a project, a baseline is needed for comparative purposes. A baseline defines for management the option of maintaining the status quo. Changes in material consumption, utility demands, staff time, and so forth for options being considered can be measured as either more or less expensive than the baseline.

In determining baseline costs, four tiers of costs should be considered:

> *Tier 0:* Usual or normal costs, such as direct labor, raw materials, energy, equipment, etc.
>
> *Tier 1:* Hidden costs, such as monitoring expenses, reporting and record keeping, permit requirements, environmental impact statements, legal fees, etc.
>
> *Tier 2:* Future liability costs, such as remedial actions, personal injury under Occupational Safety and Health Administration (OSHA) regulations, property damage, etc.
>
> *Tier 3:* Less-tangible costs, such as consumer response and confidence, employee relations, corporate image, etc.

Tier 0 and Tier 1 costs are direct and indirect costs, respectively. They generally include engineering, materials, labor, construction, contingencies, and so on, as well as waste collection and transportation services, raw-material consumption (increase or decrease), and production costs.

Tier 2 and Tier 3 represent intangible costs. They are more difficult to define and include potential corrective actions under the Resource Conservation and Recovery Act (RCRA); possible site remediation at third-party sites under Superfund; liabilities that could arise from third-party lawsuits for personal injury or property damages; and benefits of improved safety and work environments. Although these intangible costs cannot be accurately predicted, they can be very important and should not be ignored. A present-value analysis that contains such uncertain factors generally requires a little ingenuity in assessing the full merits of a technology.

To establish the baseline, the current cost of doing business must first be determined. Once the present costs are known, all potential alternatives are then related to this baseline cost. There are three basic steps to establishing a baseline cost:

Step 1. Add up all the relevant input and output materials for the process, and compute their appropriate dollar value (note that we are focusing only on mass, not on energy or other issues).

Step 2. Check to make sure that the material balance makes sense. Specifically, is the volume of cleaning solvent purchased accounted for in the losses, product, inventory, and/or the waste? Account for such losses as evaporative. Once accomplished, determining the baseline costs becomes a matter of pricing each input and output, and then multiplying their volumes by the appropriate unit.

Step 3. Examine the expected business outlook and the most likely changes, such as business expansions, new accounts, rising prices, cutbacks, and so forth.

While Tier 2 costs include potential liabilities such as remedial actions, personal injuries covered under OSHA, property damage, and so on, Tier 3 costs are even harder to predict. For example, a typical Tier 3 cost could be the cost of sales lost due to adverse public reaction to a pollution incident, such as a leaking underground storage tank, a PCB spill, or a fire and explosion incident. Tier 2 and Tier 3 variables include the types of incidents that could occur, the severity of each incident, the ability of the firm to control or respond to the emergency, the public's reaction to the incident, and the company's ability to address the public's concerns.

In many cases, there is a probability that can be connected with a particular event. This can be incorporated into a calculation of *expected value*. The expected value of an event is the probability of an event occurring, multiplied by the cost or benefit of the event. Once all expected values are determined, they are totaled and brought back to present value, as is done with any other benefit or expense. Hence, the expected value measures the central tendency, or the average value that an outcome would have.

For example, there are a number of games at county fairs that involve betting on numbers or colors, much like roulette. The expected value of such a game can be computed as:

(Benefit of success) × (Probability of success) − (Cost of failure) × (Probability of failure)

If the required bet of a particular game is $1, the prize is worth $5, and there are 10 possible selections, the expected value would be

$$(\$5) \times (0.1) - (\$1) \times (0.9) = -\$0.40$$

On the average, the player will lose—meaning the game operator will win—40 cents on every dollar wagered.

For Tier 2 and 3 expenses, the analysis is the same. For example, there is a great deal of data available from OSHA studies regarding employee injury in the workplace. In justifying a material-substitution project, if the probability of injury and a cost can be found, the benefit of the project can be computed.

The concept of expected value is not complicated, though the calculations can be cumbersome. For example, even though each individual's chance of injury may be small, the number of employees, their individual opportunity costs, the various probabilities for each task, and so forth, could mean a large number of calculations. However, if one considers the effect of the sum of these small costs—or the large potential costs of environmental lawsuits or site remediation under either RCRA or the Comprehensive Environmental Response, Compensation, and Liability Act (CERCLA)—the expected value computations can be quite important in the financial analysis.

3   Revenues, Expenses, and Cash Flow

Because it is the goal of any business to make a profit, the costs-and-benefits cash flows for each option can be related to the basic profit equation:

$$\text{Revenues} - \text{Expenses} = \text{Profits}$$

The most important aspect of this is that profits can be increased by either an increase in revenues or a decrease in expenses. Pollution prevention technologies often lower expenditures and increase profit.

Revenue is money coming into the firm from the sale of goods or services, from rental fees, from interest income, and so on. The profit equation shows that an increase in revenue leads to a direct increase in profit, and vice versa, if all other revenues and expenses are held constant. Note that we are going to assume that all other expenses and revenues are held constant in the discussions below.

Revenue impacts must be closely examined. For example, firms often can cut wastewater treatment costs if water use (and, in turn, the resulting wastewater flow) is limited to nonpeak times at the wastewater treatment facility. However, this limitation on water use could hamper production. Consequently, even though the company's actions to regulate water use could reduce wastewater charges, revenue might also be decreased unless alternative methods can be found to maintain total production. Conversely, a change in a production procedure could increase revenue. For example, moving from liquid to dry paint stripping can not only reduce water consumption but also affect production output. Because cleanup time from dry paint-stripping operations (such as bead blasting) is generally much shorter than from using a hazardous, liquid-based stripper, it could mean not only the elimination of the liquid waste stream but also less employee time spent in the cleanup operation. In this case, production is enhanced and revenues are increased. Another potential revenue effect is the generation of marketable by-products from a recycling practice.

Expenses are monies that leave the firm to cover the costs of operations, maintenance, insurance, and so on. There are several major cost categories that technologies can have an effect on, including:

*   Insurance expenses
*   Depreciation expenses
*   Interest expenses
*   Labor expenses
*   Training expenses
*   Auditing and demo expenses
*   Floor-space expenses

Insurance expenses. Depending upon the project, insurance expenses could either increase or decrease. For example, OSHA sets limits on workers' exposure to a number of chlorinated solvents. If one technology option eliminates a hazardous chlorinated solvent from production operations, the enterprise could realize a savings in employee health coverage, liability insurance, and similar costs. Likewise, if a company uses a nonflammable solvent instead of a flammable one, it could decrease its fire insurance premium. Conversely, projects can increase insurance expenses. For example, if a company adds a heat-recovery still to a process operation, its fire insurance premiums could increase.

Depreciation expenses. If a technology involves the purchase of capital equipment with a limited life (such as storage tanks, recycle or recovery equipment, new solvent-bath systems, new fabric dyeing baths, etc.), the entire cost is not charged against the current year. Instead, depreciation expense calculations spread the equipment's procurement costs (delivery charges, installation, startup expenses, etc.) over a period of time by taking a percentage of the cost each year over the life of the equipment.

For example, if the expected life of a piece of equipment is 10 years, each year the enterprise would charge an accounting expense of 10 percent of the procurement cost of the equipment. This is known as the method of *straight-line depreciation*. Although there are other methods available, all investment projects under consideration at any given time should use a single depreciation method to accurately compare alternative projects' expense and revenue effects. Because straight-line depreciation is easy to compute, it is the preferred method. Note that even though a firm must use a different depreciation system for tax purposes (e.g., the Accelerated Cost Recovery System, or ACRS), it is acceptable to use other methods for bookkeeping and analysis. In any event, any capital equipment must be expensed through depreciation.

Interest expenses. Investment in technologies implies that one of two things must occur: either a company must pay for the project out of its own cash or it must finance the cost by borrowing money from a bank, issuing bonds, or some other means. When a firm pays for a project out of its own cash reserves, the action is sometimes called an *opportunity cost*. If the enterprise must borrow the cash, there is an interest charge associated with using someone else's money. It is important to recognize that interest is a true expense and must be treated, like insurance expenses, as an offset to the project's benefits. The magnitude of the expense will vary with bank lending rates, the interest rate offered on the corporate notes issued, and other factors. In any case, there will be an expense.

The reason enterprises account for equipment purchases as a cost is this: if cash is used for the purpose of pollution control or prevention, it is unavailable to use for other opportunities or investments. Revenues that *could* have been generated by the cash (for example, interest from a certificate of deposit at a bank) are treated as an expense and thus reduce the value of the project.

Although the reasoning seems sound, opportunity costs are not really expenses. Though it is true that the cash will be unavailable for other investments, opportunity cost should be thought of as a comparison criterion and not an expense. The opportunity forgone by using the cash is considered when the project competes for funds and is expressed by one of the financial analysis factors discussed earlier (net value of present worth, payback period, etc.). It is this competition for company funds that encompasses opportunity cost, so opportunity cost should not be accounted directly against the project's benefits.

Many enterprises apply a minimum rate of return, or *hurdle rate*, to express the opportunity-cost competition between investments. For example, if a firm can draw 10 percent interest on cash in the bank, then 10 percent would be a valid choice for the hurdle rate as it represents the company's cash opportunity cost. In analyzing investment options under a return-on-investment criterion, then, not only would the highest returns be selected but also any project that pays the firm a return of less than the 10 percent hurdle rate would not be considered.

Labor expenses. In the majority of situations, a project will cause a company's labor requirements to change. This change could be a positive effect that increases available productive time or there could be a decrease in employees' production time.

When computing labor expenses, the Tier 1 costs could be significant. For example, if a material-substitution project eliminates a hazardous input material that had been used to clean up hazardous waste, there could be a significant decrease in the labor required to complete and track manifests, a decrease in the costs of labeling, a decrease in transportation costs, elimination of handling and storing hazardous waste drums, and so on. Hence, both direct (Tier 0) expenses (for example, five hours per week of preventive maintenance on the P2 equipment) and secondary (Tier 1) expenses can have an effect on manpower costs.

Labor expense calculations can range from simple to comprehensive. The most direct and basic approach is to multiply the wage rate by the hours of labor. More-comprehensive calculations include the associated costs of payroll taxes, administration, and benefits. Many companies routinely track these costs and establish an internal "burdened" labor rate to use in financial analysis.

Training expenses. New technologies require additional operator training. In computing the total training costs, consider as an expense both the direct costs and the staff time spent in training. In addition, any other costs for refresher training, or for training for new employees, that is above the level currently needed must be included in the analysis. Computing direct costs is simply a matter of adding the costs such as tuition, travel, and per diem for the employees. Similarly, to compute the labor costs, simply multiply the employees' wage rates by the number of hours spent away from the job in training.

Auditing and demo expenses. Labor and other expenses associated with defining a P2 project are often overlooked. Although these tend to be small for low-investment projects, some projects may require extensive auditing; pilot or plant trials can incur significant up-front costs from production down-times, personnel, monitoring equipment, and laboratory measurements, as well as engineering design time and consultant-time charges. Some companies may prefer to absorb these costs as part of their R&D budget.

Floor-space expenses. As with any opportunity costs, the floor-space costs must be based on the value of alternative uses. For example, multiple rinse tanks have long been used to reduce water use in electroplating operations. If a single-dip rinse tank of 50 square feet is replaced with a cascade-rinse system of 65 square feet, then the floor-space expense is the financial worth of the extra 15 square feet; it must be included as an expense in the financial analysis for the project.

Unfortunately, computing floor-space opportunity cost is not always straightforward, as it is in the case of training costs. In instances where little square footage is required, there may be no other use for the floor space, which implies a zero cost. In other cases, where the area is currently being used only for storage of extra parts, bench stock, or feed materials, the costs may involve determining the value of having a drum of chemical or an extra part closer to the operator. Alternatively, as the square footage required increases, calculating floor-space costs becomes more straightforward. For example, if a new building is needed to house new equipment, it's easy to compute a cost. Similarly, if installing the equipment at the production site displaces enough storage room to require additional sheds be built, the cost is again easy to compute. As a default, the cost of floor space can be estimated from information available from realtors. The average square-foot cost for a new or used warehouse (or administrative or production space) that would be charged to procure the space on the local market is the average market worth of a square foot of floor space. Unless there is a specific alternative proposal for the floor space, this market analysis should work as a proxy.

Cash flow. Although cash flow does not have a direct effect on a company's revenues or expenses, the concept must be considered. If the project involves procurement costs, they often must be paid upon delivery of the equipment—yet cash recovery could take many months or even years.

Three things about a project can affect a firm's available cash. First, cash is used at the time of purchase. Second, it takes time to realize financial returns from the project, through either enhanced revenues or decreased expenses. Finally, depreciation expense is calculated at a much slower rate than the cash was spent. As a result of the investment, a company could find itself cash-poor. Conversely, P2 efforts can have a very positive effect on cash flow. For example, eliminating a hazardous waste via an input-material substitution could result in an increase in cash available, because the enterprise would not have to pay for hazardous-waste disposal every three months or so. Hence, even though cash flow does not directly affect revenues and expenses, it may be necessary to consider when analyzing projects.

4   Interest and Discount Rates

In determining the value of a technology investment, the discount rate used becomes very important. If the potential project benefits are accrued far into the future, or if a larger discount rate is used, the effect on the present value (and hence the apparent value of the project) could be dramatic.

Figure 7.1 illustrates the relationship between the percentages of future worth regained over time at varying interest rates. On average, enterprises prefer a return on investment, or hurdle rate, in the range of 10 to 15 percent. At 10 percent, more than half of a future benefit stream can be lost due to the time-value of money within the first 10 years. This factor works against the acceptability of projects that provide benefits far in the future. Hence, to justify project investments with long-term-benefit cash flows, it is often necessary to move to Tier 2 or Tier 3 criteria.

5   Income Taxes

Though most companies use only revenue and expense figures when comparing investment projects, income-tax effects can enter into each calculation if either revenues or expenses change from the baseline values. More expenses mean lower profits and less taxes, and higher revenues lead to increased profits and more taxes. If an enterprise needs to know the effect of income taxes on profit, the computations are simple and can be done during or after the analysis.

As with expenses and revenues, the enterprise does not need to compute the total tax liability for each option. Instead, it only needs to look at the options' effects on revenues and expenses and at the difference in tax liability resulting from deviations from the baseline. The profit equation reflects gross or pretax profits. Income tax is based on the gross profit figure from this equation and cannot be computed until the enterprise knows what effect the options will have on its revenues and expenses.

Taxes act to soften the impact of P2 projects on net profit, due to changes in revenue or expenses as follows. An example of this is shown in table 7.1. For the purposes of illustration, the income tax rate is taken as constant, at 40 percent of gross profit. If revenues increase by $100 with no other changes, pretax profits also increase $100. Because income taxes take $40 of this increase, however, the effect on net profit is to soften the $100 revenue increase to a $60 net profit increase. Similarly, if expenses were to increase $100, pretax income would decrease $100, but the tax liability would be $40 less. In this latter case, the $100 pretax loss would be softened to a $60 net-profit decrease.

Table 7.1 Effect of changes in revenues and expenses on pretax and net profits

Revenue increase	
Initial condition:	
Beginning pretax profit	$100
Tax liability	$40
Net profit without pollution prevention project	$60
Post–pollution prevention:	
Revenue increase subsequent to project	$100
New pretax profit	$200
New tax liability	$80
New net profit	$120
Increase in net profit due to + $100 in revenues	+ $60
Expense increase	
Initial condition:	
Beginning pretax profit	$100
Tax liability	$40
Net profit without pollution prevention project	$60
Post–pollution prevention:	
Expense increase subsequent to project	$100
New pretax profit	$00
New tax liability	$00
New net profit	$00
Decrease in net profit due to + $100 in expenses	− $60

Source: D. G. Stephen, *A Primer for Financial Analysis of Pollution Prevention Projects*, EPA Cooperative Agreement no. CR-815932, April 1993

The profit impact of an increase or decrease in revenues or expenses is limited by 1 minus the tax rate t, that is, $(1 - t)$. If the tax rate is different from 40 percent, it can be inserted into $(1 - t)$ and used to calculate the impact. For example, for a 33 percent tax rate, a $100 increase in revenue would increase profit by $(1 - 0.33)$, or $67.

Tax credits are a special case allowed by the Internal Revenue Service at various times. For example, during the energy crunch of the 1970s, certain capital expenses that reduced energy consumption (such as solar energy projects) were given special treatment as tax credits. Unlike personal tax deductions, tax credits are deducted directly from the tax obligation of a firm. As a result, in this special tax-credit case, capital expenses that would otherwise lower pretax income can be subtracted directly from the tax liability and increase profit.

6 Total-Cost Assessment

Total-cost accounting, also referred to as *full-cost environmental accounting*, is applied in management accounting to represent the allocation of all direct and indirect costs to specific products, to the lives of products, or to operations. *Total-cost assessment* has come to represent the process of integrating environmental costs into the capital-budgeting analysis. It is generally defined as a long-term, comprehensive analysis of the entire range of costs and savings associated with the investment by the enterprise making the investment.

Life-cycle cost assessment represents a methodical process of evaluating the life-cycle costs of a product, product line, process, system, or facility—starting with raw-material acquisition and going all the way to disposal—by identifying the environmental consequences and assigning monetary value. We shall expand on this important subject shortly.

When evaluating environmental technologies, one must not simply focus on the need to meet compliance. This approach limits solutions to end-of-pipe selections that in the end have no economic advantages and simply add to the costs for manufacturing. Instead, one should seek to identify more potential costs and savings from a range of technology options. The way to accomplish this is by expanding the cost and savings inventories in the analysis. Tables 7.2 and 7.3 provide lists of capital and operating costs that environmental managers can use to determine the financial costs and savings associated with a particular project opportunity.

The challenge in applying an expanded cost/savings inventory for investment analysis is that some of the cost data associated with a particular piece of equipment or process may be difficult to obtain. Estimating many costs can be a challenge, because they may be grouped with other cost items in existing overhead accounts—for example, waste-disposal costs for existing processes are often placed in a facility overhead account. An expanded cost inventory would call for these costs to be directly allocated to the process that produces them. Consequently, it's likely that information for all the cost categories will not be identified during analyses. Analysts can use the list of categories provided in tables 7.2 and 7.3 to incrementally expand their existing financial analyses whenever possible.

One approach to uncovering more of the true economic benefits of projects is to expand the evaluation of costs and savings over a longer time period, usually five or

Table 7.2 Partial inventory of potential capital-cost items

Purchased equipment	Site preparation (labor, supervision, materials)	Shakedown and startup (labor, supervision, materials)

Equipment
Delivery
Sales tax and VAT
Insurance
Price for initial spare parts
	Site studies (EIS, other)
	Demolition and cleaning
	Old equipment/garbage
	  disposal
	Grading/landscaping
	Equipment rental
	Tie-ins to existing utilities and
	  infrastructure
		In-house
		Contractor/vendor/consultant
		  fees
		Trials/manufacturing variance
		Training

Materials	Construction/Installation (labor, supervision, materials)	Regulatory and permitting (labor, supervision, materials)

Piping
Electrical
Instrumentation
Structural
Insulation
Other
	In-house
	Contractor/vendor/consultant
	  fees
	Equipment rentals
		In-house
		Contractor/vendor/consultant
		  fees
		Permit fees

Utility systems and connections	Planning and engineering (labor, supervision, materials)	Working capital

General plumbing
Electricity
Steam
Water (e.g., cooling, process)
Fuel (e.g., gas, oil)
Plant air
Inert gas supplies
Refrigeration
Sewerage
	In-house planning and
	  engineering (e.g., design,
	  shop drawings, cost
	  estimating, etc.)
	Contractor/vendor/consultant
	  fees
	Procurement
		Raw materials
		Other materials and supplies
		Product inventory
		Protective equipment

Contingency	Back-end
Future compliance costs Remediation	Closure and decommissioning Inventory disposal Site survey

Table 7.3 *Partial inventory of potential operating costs*

Direct materials	Waste management (labor, supervision, materials)	Insurance, future liability, fines and penalties, cost of legal proceedings (e.g., transaction costs), personal injury
Raw materials (e.g., wasted raw-material costs/savings)	Pretreatment	Property damage
Solvents	On-site handling	Natural resource damage
Catalysts	Storage	Superfund
Transport	Treatment	
Storage	Hauling	
	Insurance	
	Disposal	

Direct labor	Utilities	Revenues
Operating (e.g., worker productivity changes)	Electricity	Sale of product (e.g., from changes in manufacturing throughput, market share, corporate image)
Supervision	Steam	
Manufacturing clerical	Water (e.g., cooling, process)	
Inspection/QA/QC	Fuel	Marketable by-products
	Plant air	Sale of recyclables
	Inert gas	
	Refrigeration	
	Sewerage	

Regulatory compliance (labor, supervision, materials)

Permitting	Labeling and packaging	Financial assurance
Training (e.g., Right-to-Know training, Hazmat, etc.)	Manifesting	Value of marketable pollution permits (e.g., SO_x)
Monitoring and inspections	Recordkeeping	Avoided future regulation (e.g., CAAA)
Notifications	Reporting	
Testing	General fees and taxes	
	Closure and postclosure care	

more years. This is because many of the costs and savings can take years to materialize and because the savings from projects often occur every year for an extended period of time—for example, some projects realize recurrent savings as a result of less waste, requiring less management and disposal every year. Conventional project analysis, however, often confines costs and savings to a three- to five-year time frame, a time period shorter than the useful life of the equipment being evaluated. Using this traditional time frame in project evaluation will exclude some of the areas of savings generated by projects.

Although expanding cost inventories and time horizons can greatly enhance the ability to accurately portray the economic consequences of a single project, the financial-performance indicators—payback period, net present value, and internal rate of return—are needed to allow comparisons to be made between competing projects alternatives. To review, the payback-period analysis focuses on determining the length of time it will take before the costs of a new project are recouped. A useful formula used to calculate payback period is:

Payback period (in years) = startup costs / (annual benefits − annual costs)

For example, if startup costs are $800 and the annual benefits and costs are $600 and $400, respectively:

Payback period = $800 / ($600/yr − $400/yr) = 4 years

Those investments that recoup their costs before a predefined "threshold" period of time are determined to be projects worthy of funding. Readers will remember from earlier discussions that the payback-period analysis does not discount costs and savings over future years. Furthermore, costs and savings are not considered if they occur in years later than the threshold time in which a project must pay back to be justified.

A more thorough analysis is based upon net present value (NPV). The NPV method is particularly useful when comparing projects against alternatives that result in higher annual waste-management and disposal costs. The increased costs of status-quo operations (or of the investment options that do not reduce wastes) will tend to lower their NPV. The method also easily accommodates the expanded cost inventory when analyzing all costs and benefits.

An additional financial analysis term to consider is the internal rate of return, or IRR (also known as return on investment, or ROI). The purpose of IRR calculations is to determine the interest rate at which NPV is equal to zero. If the rate exceeds the hurdle rate (the minimum acceptable rate of return on a project), the investment is deemed worthwhile. The following formula may be used:

$$\text{Initial Cost} + \text{Cash Flow in Year } 1/(1 + r)^1 + \text{Cash Flow in Year } 2/(1 + r)^2 + \text{Cash Flow in Year } 3/(1 + r)^3 + \ldots + \text{Cash Flow in Year } n/(1 + r)^n = 0,$$

where r is the discount rate (let's assume that it's the IRR) and n is the number of years of the investment. A trial-and-error procedure can be applied to solve the above equation for r.

Table 7.4 helps to put the theory presented thus far into practice. Table 7.4 constitutes a *Pollution Prevention Project analysis worksheet* (P3AW). Organizing such a worksheet in a spreadsheet program can help analyze the costs and benefits associated with current operations, potential projects, and alternative project opportunities.

The use of worksheets such as table 7.4 helps to demonstrate ways of capturing more cost categories by better allocating costs to specific activities, expanding the cost areas included in the analysis, and expanding the time horizon over which the project is analyzed. We have abbreviated the potential costs and revenues in table 7.4 for ease of use.

The P3AW enables its user to calculate two measures of financial performance: a simple payback analysis, and a net present value calculation (which incorporates the time-value of money). Both of these calculations can help an enterprise compare competing project options or a proposed project against the status quo. IRR calculations are not included on the worksheet, but they may be readily added by the reader.

It is important to note that when completing the worksheet, some data might not be available to complete all the sections. However, completing even only a few of the

Table 7.4 Pollution Prevention Project (P3) analysis worksheet

			Startup	1	2	3	4	5	6	7	
CASH OUTFLOW	1	**Capital Costs**									
		Equipment									
		Utility connections									
		Construction									
		Engineering									
		Training									
		Other									
		Subtotal									
	2	**Operating costs**									
		Materials									
		Labor									
		Utilities									
		Waste management									
		Compliance									
		Liability									
		Other									
		Subtotal									
CASH INFLOW	3	**Revenues**									
		Sale of products									
		Sale of by-products									
		Sale of recyclables									
		Other									
		Subtotal									
	4	**Payback**	years	*Equals Sec. 1 divided by (Sec. 2–Sec. 3)*							
	5	**Cash flow (CF)**									
			Cash flow is calculated by subtracting cash outflows from cash inflows during each year of the investment (i.e., Sec. 3 minus Sec. 2 minus Sec. 1 subtotals)								
	6	**PV factors**									
			For present values (PVs) for different investment durations, refer to the text for discussions								
	7	**CF × PV**									
	8	**NPV**	*Equals the sum of all values from Sec. 7*								

Note: Modified from USEPA, *Federal Facility Pollution Prevention Project Analysis: A Primer for Life Cycle and Total Cost Assessment Concepts*, EPA 300-B-95–008, July 1995.

sections of the worksheet with data that would not have normally been collected will significantly enhance the accuracy of evaluating project opportunities.

Before attempting to organize information and data needed for the P3AW, enterprises should define the objective of the analysis. In addition, the analysis ultimately will be used by decision makers, that is, top management, who will decide whether or not a proposed project will be funded. Like any other sales pitch, presentation means a lot. Therefore, knowing the audience, the decision-making criteria for company projects, and the format to best present the analysis are important items to consider. With some planning, you can make certain that the scope of the analysis is appropriate and that the completed analysis will be presented in a readily understood and accepted manner. When using P3AWs to compare project alternatives, or to compare a potential project to current operations, it is best to use a separate worksheet for each option under consideration. The following guidelines will help the reader use the P3AW (refer to table 7.4).

Sections 1–3. First, identify the economic consequences associated with the activity under review. Specific items (such as categories of cash outflows) noted in the P3AW may not necessarily represent a complete list of costs incurred for the facility under review. As such, tables 7.2 and 7.3 will help identify additional capital- and operating-cost categories. If the focus is on a payback analysis, completing information for only the initial year is acceptable, provided that the data are available to describe annual costs and annual savings. If the focus is on analyzing the financial performance using an NPV calculation, then you will need to obtain estimates of future costs and benefits. It is not necessary to make adjustments for inflation if the calculations are addressed through a nominal discount factor. This option is described below. Note also that, to allow comparisons with other project options, there are two measures of economic performance included on the P3AW (payback analysis and net present-value analysis).

Section 4. This section focuses on calculating the number of years that it will take to recoup the initial capital expenditure. This value is obtained by dividing the initial investment needed to establish the project by the net annual benefits (which are obtained by subtracting the annual cash outflows from the expected annual cash inflows). If only the payback analysis is important, then the remaining sections of the P3AW may be skipped.

Section 5. For each year included in the evaluation, calculate the annual net cash flow by subtracting the capital expenditures subtotal (in section 1) and the annual cash outflows (subtotals from sections 3, 4, and 5) from the annual cash inflows (section 2).

Section 6. To calculate the NPV, you need to determine the value of future cash flows, starting from today. To accomplish this, you can use present-value (PV) factors to discount future cash flows. The discount will be specific to your local region, but for a comparative basis, we can use the discount rates used by the U.S. government. Table 7.5 provides PV factors for nominal discount rates (based on 1995 figures).

Section 7. Multiply the net cash flows (section 5) by the PV factors (section 6) to determine the present value today of the cash flow in each year.

Section 8. Sum all the annual discounted cash flows to determine the NPV of the project. If the value is positive, the project is cost-beneficial. If more than one investment is being studied, then the project with the greatest NPV is the most attractive.

Table 7.5 Present-value factors for nominal discount rates

Year	7.3%	7.6%	7.7%	7.9%	8.1%
1	0.93197	0.92937	0.92851	0.92678	0.92507
2	0.86856	0.86372	0.86212	0.85893	0.85575
3	0.80947	0.80272	0.80048	0.79604	0.79163
4		0.74602	0.74325	0.73776	0.73231
5		0.69333	0.69012	0.68374	0.67744
6			0.64078	0.63368	0.62668
7			0.59496	0.58729	0.57972
8				0.54429	0.53628
9				0.50444	0.49610
10				0.46750	0.45893
11					0.42454
12					0.39273
13					0.36330
14					0.33608
15					0.31090
16					0.28760
17					0.26605
18					0.24611
19					0.22767
20					0.21061
21					0.19483
22					0.18023
23					0.16673
24					0.15424
25					t0.14268
26					0.13199
27					0.12210
28					0.11295
29					0.10449
30					0.09666

Once you've completed the analysis, you may prepare a report explaining the results. The report should highlight the economic benefits of the proposed P2 projects. Also discuss the noneconomic benefits, as these may tip the scales in favor of the P2 opportunity if the financial analysis is marginal or too close to call.

7 Life-Cycle Analysis

Decision-makers require analytical tools that accurately and comprehensively account for the environmental consequences and benefits of competing projects. These environmentally based project-review tools must be flexible, be easy to use, and require limited staff and funding so that they can be easily incorporated into the review process.

Life-cycle assessment (LCA) provides a means to evaluate environmental consequences and impacts. LCA identifies and evaluates "cradle-to-grave" natural resource requirements and environmental releases associated with processes, products, packaging, and services. LCA concepts also can be particularly useful in ensuring that identified opportunities are not causing unwanted secondary impacts by shifting burdens to other places within the life cycle of a product or process.

LCA concepts can be useful in acquiring a broader appreciation of the true environmental impacts of current manufacturing practices and of proposed opportunities. It has taken a good two decades for many environmental professionals to become more aware that the consumption of manufactured goods and services can have adverse impacts on the supplies of natural resources as well as the quality of the environment. These effects occur at virtually all stages of the life cycle of a product, starting with raw-materials harvesting, continuing through materials manufacturing and product fabrication, and concluding with product consumption and disposal. LCA is essentially a tool that enables us to evaluate the environmental consequences of a product or activity across its entire life.

LCA is made up of the following elements:

• *Goal definition and scoping* is a screening process that involves defining and describing the product, process, or activity; establishing the context in which the assessment is to be made; and identifying the life-cycle stages to be reviewed for the assessment.

• *Inventory analysis* involves identifying and quantifying energy, water, and materials usage and the environmental releases (e.g., air emissions, solid waste, wastewater discharges) during each life-cycle stage.

• *Impact assessment* is used to assess the human and ecological effects of material consumption and environmental releases identified during the inventory analysis.

• *Improvement assessment* involves evaluating and implementing opportunities to reduce the environmental burdens as well as the energy and material consumption associated with a product or process.

LCA can be used in process analysis, materials selection, product evaluation, product comparison, and even in policymaking. LCA can be used by acquisitions staff, new-product design staff, and staff involved in investment evaluations.

What makes this type of assessment unique is its focus on the entire life cycle, rather than a single manufacturing step or environmental emission. The theory behind this approach is that operations occurring within a facility can also cause impacts outside the facility's gates that need to be considered when evaluating project alternatives. Examining these upstream and downstream impacts can identify benefits or drawbacks to a particular opportunity that otherwise may have been overlooked. For example, examining whether to invest in plastic bottle cartons for a beverage bottling facility or to use wooden crates for staging and storing incoming bottles should include a comparison of all major impacts, both inside the facility (e.g., disposing of the wooden crates) and outside the gate (e.g., additional wastewater discharges from off-site washing of the reusable plastic cartons).

Gaining a complete understanding of a proposed project's environmental effects requires identifying and analyzing inputs and releases from every life-cycle stage. However, securing and analyzing this data can be a frustrating and, perhaps, endless task. Process engineers in plants often are faced with immediate priorities and may not have the time or resources to examine each life-cycle stage or to collect all pertinent data. Despite this shortcoming, it is worthwhile to discuss the steps required to begin applying LCA concepts and principles to project analysis and to give examples that demonstrate steps within selected life-cycle stages.

Before beginning to apply LCA concepts to projects under review, it is important to first determine the purpose and the scope of the study. In determining the purpose, facility managers should consider the type of information needed from the environmental review (for example, does the study require quantitative data or will qualitative information satisfy the requirements?). Once the purpose has been defined, the boundaries or the scope of the study should then be determined. What stages of the life cycle are to be examined? Are data available to study the inputs and outputs for each stage of the life cycle to be reviewed? Are the available data of an acceptable type and quality to meet the objectives of the study? Are adequate staff and resources available to conduct a detailed study?

This definition and scoping activity links the purpose and scope of the assessment with available resources and time and allows reviewers to outline what will and will not be included in the study. In some cases, the assessment may be conducted for all stages of the life cycle (i.e., raw-materials acquisition, manufacturing, use/reuse/maintenance, and recycling/waste management). In many instances, the analysis may begin at the point where equipment and/or materials enter the facility. Other times, primary emphasis may be placed on a single life-cycle stage, such as identifying and quantifying waste and emissions data. In all cases, managers should ensure that the boundaries of the LCA address the purpose for which the assessment is conducted and the realities of resource constraints. Whenever possible, managers should include in the analysis all life-cycle stages in which significant environmental impacts are likely to occur.

Conducting an LCA that includes all life-cycle stages will provide decision makers with the most complete understanding of environmental consequences. However, if resources are limited and an in-depth, quantitative analysis is not practical, a simplified approach may be taken. This alternative approach makes use of a simple *life-cycle checklist* to identify and highlight certain environmental implications associated with competing projects. A checklist that uses qualitative data instead of quantitative inputs can be very useful when available information is limited or when used as a first step in conducting a more thorough LCA. In addition, the life-cycle checklist should include questions regarding the environmental effects of current operations and/or potential projects that cover materials and resources consumed and wastes/emissions generated.

Conducting a more-thorough review of the environmental consequences of projects will require more and, most likely, dedicated resources. A more in-depth analysis would be aimed at identifying and evaluating the resource and material inputs, and the environmental releases, associated with each life-cycle stage. This is a resource-intensive operation and much more comprehensive than applying a life-cycle checklist. As first steps, we recommend that you define and scope out the analysis to fit available resources while including all significant areas of environmental impact.

8   Future Costs and Benefits

Developing reliable estimates of future costs and benefits can be a difficult task. Quantitatively estimating future costs for items such as property cleanup and environmental compliance for a facility's decommissioning and post-closure can be extremely difficult. A generalized approach to this problem is to group future costs into one of two categories: *recurring costs* and *contingent costs*.

Recurring costs include items that are currently incurring costs and are anticipated to continue incurring costs into the foreseeable future, based upon regulatory requirements. These include permits, monitoring, and compliance with regulatory requirements. The first step in estimating the future costs of these items is to determine how much the facility is currently paying. Then, estimate how much the cost can reasonably be expected to escalate in the future. For example, if monitoring costs are currently $10,000 and are expected to rise with inflation, a conservative estimate would be a 4 percent annual increase. Consequently, the monitoring costs for the following year could be estimated at $10,400, assuming that monitoring requirements do not become more stringent. Note that when using the P3AW (table 7.4), it is not necessary to escalate these values, because nominal discount rates from table 7.5 can be incorporated into the worksheet. In other words, table 7.4 already takes inflation into account when calculating present values.

Contingent costs include catastrophic future liabilities, such as remediation and cleanup costs. Although current activities can lead to these future costs, quantitative estimates of these liabilities are difficult to obtain. Often, the only way to include these future liabilities in the budgeting process is to qualitatively describe estimated liabilities without attempting to define them using dollar amounts. If a P2 option is being considered, make sure to include a comparison that highlights the areas in which future liability would be reduced by implementing the P2 option. For example, this approach could be used to describe the future benefit of switching from lead-based to water-based paint. Most likely, the best option may be to fully describe the potential liability if the change is not made and, if possible, document the remediation cost that could result if a liability event occurred today.

Index

About the Editor

NICHOLAS P. CHEREMISINOFF has provided environmental and pollution prevention consulting to industry, policy makers and regulators, lending institutions, and international donor agencies for nearly three decades. His international experiences include assignments in central and eastern europe, Russia, the Balkans, Latin America, and the Middle East. He has consulted on numerous overseas assignments for the U.S. Agency for International Development, the World Bank, the U.S. Export and Import Bank, the European Union, the U.S. Trade and Development Agency, and others. In the United States, Dr. Cheremisinoff has focused on responsible care issues as an expert witness on environmental litigations and toxic torts. Dr. Cheremisinoff received his B.S., M.Sc., and Ph.D. degrees in chemical engineering from Clarkson College of Technology.